BIRKHÄUSER

Systems & Control: Foundations & Applications

Harold J. Kushner

Numerical Methods for Controlled Stochastic Delay Systems

Birkhäuser
Boston • Basel • Berlin

Harold J. Kushner
Division of Applied Mathematics
Brown University
Providence, RI 02912
USA

ISBN: 978-0-8176-4534-2 e-ISBN: 978-0-8176-4621-9
DOI: 10.1007/978-0-8176-4621-9

Library of Congress Control Number: 2008923928

Mathematics Subject Classification (2000): 34K28, 34K35, 34K50, 60F17, 60H20, 65C20, 65Q05, 90C39, 93E20, 93E25

Printed on acid-free paper.

9 8 7 6 5 4 3 2 1

www.birkhauser.com

To the Memory of My Parents

Harriet and Hyman Kushner

Contents

Preface

This book deals with numerical methods for control and optimal control problems for nonlinear continuous-time stochastic systems with delays. It is an extension to the model with delays of the Markov chain approximation methods of [58]. For the nondelay problem, these methods are a widely used and powerful class of numerical approximations of optimal costs or other functionals of controlled or uncontrolled stochastic processes in continuous time and have significant applications to deterministic problems. A comprehensive development is in [58].[1]

There are numerous sources of delays in the modeling of realistic physical and biological systems. Many examples arise in communications and queueing, due to the finite speed of signal transmission, the nonnegligible time required to traverse long communications distances, or the time required to go through a queue [90]. Other examples arise because of mechanical transportation delays as, for example in hydraulic control systems, delays due to noninstantaneous human responses or chemical reactions, or delays due to visco-elastic effects in materials. The books [44, 45] contains many concrete examples in mechanics, physics and control, as well as in biology and medecine. These examples are for the most part uncontrolled and deterministic. But many of them would be more realistic if noise were added. Many examples, together with a great deal of information on deterministic delay systems are in [77]. The excellent reference [46] contains a thorough development of the problems of optimal control of deterministic and stochastic delay systems up to its original date of publication (1992), with many examples from biology, mechanics, and elsewhere, as well as a discussion of approximation in policy space algorithms for approximating the optimal cost and control. Other examples can be found in [17, 39, 68, 77, 78]. Examples arise in biological systems due to the time

[1] This book is concerned with optimization and control problems, and with the computation of the expected values of system functionals of interest. Methods for the pathwise numerical solution of the delay equation itself for deterministic and stochastic models are discussed in [2, 6, 32, 47, 48, 69, 81].

delay in the body's adaptive loops, the finite speed of blood flow, or the time required for enzyme or other chemical reactions to occur (see, e.g., [4, Chapter 2]). Models of ecological interactions have been a main source of dynamical models with delays, and applications to financial mathematics are beginning to appear [11]. Very little information is available concerning solutions when the models are nonlinear and stochastic, and numerical methods should be a main source of such information. The reference [75] is concerned with discrete-time approximations to determinstic control problems governed by differential inclusions.

There is a huge literature on control problems for delay systems for the linear model (deterministic or stochastic) with a quadratic cost criterion, and many good computational methods have been developed. Some of the approximations have been done in the spectral domain, based on finite-order rational approximations to the transfer function of the system. Others work with the state-space formulation, where the key issue is the finite-dimensional approximation of the Ricatti equation, which is often done via an approximation to the semigroup of the system. A selection of the available results can be found in [3, 15, 22, 26, 28, 35, 36, 37, 38, 40, 63, 66, 72, 95] and in their references. Although these techniques and algorithms have been very useful for the linear problem, it is not clear (as for the problem without delays), how to adapt them to the nonlinear models that are of concern to us. For this reason, we confine attention to analogs of the approaches that have been found to be very useful for the general no-delay problem, namely the Markov chain approximation method.

The models of the systems of concern in the book are diffusion and reflected diffusion processes, and the results can be extended to cover jump-diffusions. The control might be "ordinary" in the sense that it is a bounded measurable function, or it might be impulsive, or what is known as a "singular" control. All of the usual cost functionals are covered; the discounted cost, stopping on reaching a boundary, optimal stopping, ergodic, etc. Any or all of the path, control, boundary reflection process, or driving Wiener process, might appear in delayed from. Examples where the boundary reflection process might be delayed occur in communications/queueing models, where there is a communications delay. (See Section 1.2 for an example.) If a buffer overflows (corresponding to a lost packet), a signal is sent to the source, which receives it after a delay, and then adjusts its rate of transmission accordingly. The buffer overflow is a component of the boundary reflection process. Models with delays of such boundary reflection terms have not been treated previously.

For the nondelay problem, the approach of the Markov chain approximation method starts by approximating the original controlled process by a controlled Markov chain on a finite state space. The approximation parameter is denoted by h and it might be vector-valued. The original cost functional is also approximated so that it is suitable for the chain. The approximating chain must satisfy a simple condition called "local consistency." This is quite unrestrictive and means simply that from a local point of view and for small

h, the conditional mean and covariance of the changes in state of the chain are proportional to the local mean drift and covariance of the original process, modulo small errors. Many straightforward ways of getting the approximating chains are discussed in [58], where it is seen that the approach is very flexible. The approximation yields a control problem that is close to the original, which gives the method intuitive content that can be exploited for the construction of effective algorithms. After getting the approximating chain, one solves the Bellman equation for the optimal cost (or simply the equation for the value function of interest if there is no control), and proves that the solution converges to the desired optimal cost or value function as h goes to zero. One tries to choose the approximation so that the associated control or optimal control problem can be solved with a reasonable amount of computation and that the approximation errors are acceptable.

The proofs of convergence of the Markov chain approximation method as $h \to 0$ are purely probabilistic. We always work with the processes. No tools from PDE theory or classical numerical analysis are used. The idea behind the proof can be described as follows. For the optimal control problem, starting with the approximating chain with its optimal control, one gets a suitable continuous-time interpolation, and shows that in the sense of weak or distributional convergence, there is a convergent subsequence whose limit is an optimally controlled process of the original diffusion type, and with the original cost function and boundary data. The mathematical basis is the theory of weak convergence of probability measures, and this powerful theory provides a unifying approach for all of the problems of interest. The development in this book depends heavily on the results and methods in [58]. We try to be as self-contained as possible, and do review all of the essential ideas, but it would be beneficial to be familiar with the basic ideas in that source before reading this book.

The probabilistic nature of the methods of process approximation and of the mathematical proofs of convergence allows us to use our physical intuition concerning the original problem in all phases of the development. This gives us great flexibility in the details of the approximation and in the construction of algorithms. These advantages will carry over to the problem with delays. In fact, the probabilistic approach to the approximation and convergence is particularly important when there are delays, since virtually nothing is known about the analytical properties of the associated (infinite-dimensional) Bellman equations for nonlinear problems.

When doing numerical work on general nonlinear systems, it is most convenient if the system is bounded. Many types of systems are *a priori* bounded, owing to the physical constraints on the state variables. For example, systems arising in communications or in approximations to queueing models might be bounded due to the boundedness of the buffers and the possible rates of transmission. Other systems are intrinsically bounded due to saturation effects. Models of many communications and queueing systems involve boundaries on the state space that are reflecting, where the reflection directions are determined by the internal routing of the data in the system [56].

There are two standard ways of bounding a state space if it is not already bounded due to the physical constraints imposed on the model. One might stop the process, with an associated stopping cost, if it attempts to leave a prespecified region. Or one might confine it to a given region via a reflecting boundary (the latter method is common in ergodic cost problems). Both approaches are dealt with. If the boundary is added for numerical purposes, then one might have to experiment with it to assure that it is large enough so that it does not materially affect the quantities of main interest. For simplicity, we confine attention to the diffusion model, with the noise variance not being controlled. The methods can be extended to cover jump diffusions and controlled variance and jumps; One adapts the methods that are used in [58] for such problems analogously to the way that the methods for the covered problems are adapted.

For models without delays, the system state takes values in a subset of some finite-dimensional Euclidean space, and the control is a functional of the current state. For models with delays, the state space must take the path of the delayed quantities (over the delay intervals) into account, and this makes the problem infinite-dimensional. So a major issue in adapting the Markov chain approximation method to models with delays concerns suitable "finite" approximations to the "memory segments" so that a reasonable numerical method can be devised, and much attention is given to this problem.

The methods of approximation that are developed are natural and seem to be quite promising. They deal with issues of approximation that are fundamental. However to date there has been little numerical experience, and considerable further work is required. Yet, judging from the experience with no-delay problems, the methods that are developed are very likely to be the foundation of useful algorithms. There are many additional difficulties to be overcome before effective numerical methods for nonlinear stochastic delay equations become a reality. One is rarely interested in an optimal control. Since the model of interest is often not known precisely, or the implementation of an optimal control might be difficult, what is desired is an understanding of the structure of the control, and how it can be approximated. For the no-delay case in low dimensions this is facilitated by being able to visualize the control via graphical methods. This would be a considerable challenge when there are delays.

Numerical optimization methods are often used as a means of exploring the possible tradeoffs among competing criteria. One solves the optimization problem repeatedly, varying the weights of the various components of interest, to see how a decrease in the value of one component affects the values of the other components, under conditions of optimality, as in [59]. Such information can be invaluable to the system designer, even if optimality is not sought for its own sake.

Next the contents of the various chapters of the book will be described.

Outline of the book. Suppose that the effect of the control action is delayed by an amount $\bar{\theta}$. This can cause serious instabilities. To effectively control in such a case, in determining the current control action one must take into account the control actions that were made in the recent past but whose effects have not yet been seen by the controller, those up to $\bar{\theta}$ units of time back from the present time. Chapter 1 contains some simple examples that dramatically illustrate this point. It also describes the class of examples for which there is a state transformation that reduces the problem to one in finitely many dimensions. The narrowness of this class makes numerical methods all the more important.

Chapter 2 is a summary of the main results that will be needed from the theory of weak convergence of a sequence of random processes, and of the so-called martingale problem for characterizing the limit of a weakly convergent sequence. The theory of weak convergence is an extension to a sequence of random processes of the theory of convergence in probability of a sequence of random variables, and is a fundamental tool for approximation and limit theorems. The primary processes of concern in the proofs of convergence are continuous-time interpolations of the approximating chains, and we will need to show that they have limits that are (in fact, optimal) controlled diffusions. Weak convergence theory, together with the methods of the so-called martingale problem for characterizing the limit procesess as the the desired diffusions, provides the essential tools. With their use, the proofs of convergence are purely probabilistic. For the no-delay case this probabilistic approach to the proofs of convergence of numerical algorithms is the most powerful and flexible. For the delay case, there does not seem to be any alternative since, as noted above, the Bellman equation is infinite-dimensional and virtually nothing is known about it.

Chapter 3 describes the controlled dynamical system models that will be of main interest. The subject of delay equations is vast, whether deterministic, stochastic, or controlled or not; for example, see [27, 39, 44, 45, 51, 57, 68, 73, 74, 77, 78]. The behavior can be quire bizarre, as seen in the examples in [6, 74]. The numerical approximations that are of interest require that the path take values in some compact subset G of a Euclidean space, and this motivates the models. The process can be either stopped on first (if ever) reaching the boundary of G, or else be prevented from leaving G by a boundary reflection process, both being standard models in applications. The stochastic differential equations with path and control delays are reviewed for the cases where the process is either reflected from a boundary or not. Relaxed controls, which are very helpful when dealing with approximation and convergence in control problems, and the Girsanov transformation, an approach to constructing control systems from uncontrolled systems, are discussed. The Girsanov transformation method will be crucial in dealing with the ergodic cost problem in Chapter 5. When the control and possibly the reflection process and/or the driving Wiener process appear in a delayed form, the most direct approaches to the numerical approximation could require an impossibly large memory. One promising way of alleviating this is discussed in Chapter 9 and a dynamical model that is particularly useful for that approach is introduced in Chapter 3.

The existence of an optimal control is also shown. The proof of this fact is important because it is a template for the proofs of convergence of the system and numerical approximations in subsequent chapters. Proofs of the existence of solutions to uncontrolled stochastic delay equations of the diffusion type (without reflecting boundaries) and some of their properties can be found in [39, 68, 73, 74]. For the singular control problem, the definition of the model and the existence of an optimal control are dealt with via a very useful "time transformation" method, which is necessary owing to the possibly wild nature of the associated paths and controls.

Numerical methods involve working with approximations of the original problem whether there are delays or not. The design and success of a numerical approximation is dependent upon the sensitivity of the original model to perturbations in its structure since the numerical algorithm itself is an approximation to the original model. This issue of sensitivity is a particularly acute problem when there are delays owing to the great sensitivity of many such models to parameter variations. See, for example the examples in [74]. One must always be aware of this issue of sensitivity in constructing a numerical approximation. Nevertheless it is important to simplify the original dynamical model as much as possible without sacrificing the essential aspects of the results. Fortunately, for many problems of interest, approximations that are useful for numerical purposes can be obtained.

The key difference between the problem with and without delays is that the state space for the problem with delays involves the "memory segments" of the components whose delayed values appear in the dynamics. The first step in the construction of a numerical approximation involves approximating the original dynamical system. In our case, this entails approximating the delays and dynamics so that the resulting model is simpler, and ultimately finite-dimensional. Chapter 4 is is devoted to a set of model simplifications that have considerable promise when the path or path and/or control are delayed. A variety of approximations are presented, eventually leading to finite-dimensional forms that will be used as the basis of numerical algorithms in Chapters 7–9. To help validate the approximations, simulations that compare the paths of the original and approximated system are presented, and it is seen that the approximations can be quite good.

Delay equations might have rapidly time-varying terms, even rapidly varying delays. This complicates the numerical problem. But, under suitable conditions, there are limit and approximation theorems that allow us to replace the system by a simpler "averaged" one and some such results are presented at the end of Chapter 4.

Chapter 5 is concerned with the average cost per unit time (ergodic cost) problem for nondegenerate reflected diffusion models, where only the path is delayed. The aim is to prepare ourselves for the needs of the numerical algorithms for this case. Hence the issues of model complexity and simplification that were of concern in Chapter 4 are also of concern here. There are only a few results on the ergodic theory for general delay equations. Some, dealing with the problems of existence and convergence of the distributions to invariant measures are [12, 18, 83, 86]. Since they are not quite adequate for the needs of the numerical and approximation problems for the systems of interest, the necessary results are developed, using methods based on the Girsanov transformation and the Doeblin condition, and to the extent possible following the procedures laid out in [56, Chapter 4]. Of particular interest is the demonstration that the various model approximations developed in Chapter 4 can also be used for the ergodic cost problem.

The Markov chain approximation method for the model with no delays is outlined in Chapter 6. We review of the key parts of [58] that will be needed in the sequel. All of the usual process models and cost functions can be handled. For efficiency, the development and analysis of the numerical algorithms in the following chapters is organized to take advantage of the results in [58], wherever possible, and it would be helpful if the reader has some familiarity with that reference. The notation will be slightly different from that in the references [31, 50, 58], since we wish to adapt or simplify it for the particular purposes of this book. The basic and unrestrictive local consistency condition, methods of approximation, continuous time interpolations, and the discounted, singular, impulsive control, and ergodic cost function are covered. The numerical algorithms are based on the finite-state Markov chain approximation. But the convergence proofs are based on continuous-time interpolations of the approximating chains. These interpolations are used for the convergence proofs only and not for the numerical algorithms.

Owing to the local consistency condition, the dynamical system that is represented by a continuous-time interpolation of the chain "resembles" the original controlled diffusion process. Thus we would expect that the optimal cost or the values of the functionals of interest would be close to those for the diffusion. This is quantified by the convergence theorems. There are two (asymptotically equivalent) methods of getting the approximating chains that are of interest, called the "explicit" and "implicit" methods. They differ in the way that the time variable is treated, and each can be obtained from the other. The first method was the basic approach for the nondelay problem. The second method will play a useful role in reducing the memory requirements when there are delays.

The adaptation of the methods of the Markov chain approximation method to the models with delays begins in Chapter 7 and is continued in Chapter 8. It is shown in Chapter 7 that any method of constructing the approximating chain for the no-delay problem can be readily adapted to the delay problem, with the transition probabilities taking the delays into account. The only change in the local consistency condition is the use of the "memory segment" arguments in the drift and diffusion functions.

The algorithms in Chapters 7–9 are well motivated and seem to be quite reasonable. But since the subject is in its infancy, what is presented should be taken as a first step, and will hopefully motivate further work. When constructing a numerical approximation algorithm, there are two main issues that must be kept in mind. The algorithm must be numerically feasible and it must be such that there is a proof of convergence as the approximating parameter goes to zero. These issues inform the structure of the development.

We start the development in Chapter 7 by working with numerical approximations to the original model. Then we turn our attention to the various approximations to the original model that were developed in Chapter 4, with an eye to the feasibility of their numerical approximations, taking the two main issues cited above into consideration. It will be seen that variations of the implicit approximation method of Chapter 6 can be advantageous in dealing with the memory problem. The continuous-time interpolations that are used for the convergence proofs are somewhat more complicated that those for the no-delay case, owing to the need to represent the "memory segment" argument in a way that is convenient for use in the proofs of convergence.

The development is continued in Chapter 8, where classes of numerical approximations that we call the periodic and periodic-Erlang are given. The chapter also contains the proofs of convergence for the algorithms in both chapters. Where possible, the proofs follow the general lines that were used for the no-delay case in [58]. As noted above, one interpolates the chain to a continuous-time process in a suitable manner, shows that the Bellman equation for the interpolation is the same as for the chain, and then that the interpolated processes converge to an optimal diffusion as the approximating parameter goes to zero.

The methods of Chapters 7 and 8 are promising if only the path is delayed or if the control is delayed but the control-value space has only a few points. The memory requirements can become onerous if the reflection process and/or the Wiener process also appear in delayed form, or if the control-value space has more than a few points. Chapter 9 takes an alternative approach that reduces the memory requirements for general nonlinear stochastic problems where the control and reflection terms, as well as the path variables, are delayed. The approach was suggested by the work in [94] for linear deterministic system with a quadratic cost function. Effectively, the delay equation is replaced by a type of stochastic wave equation with no delays, and its numerical solution yields the optimal costs and controls for the original model. The representation is equivalent to the original problem in that any solution to one yields a solution to the other. The details of the appropriate Markov chain approximation are given and the convergence theorem is proved. Theoretically, with the use of appropriate numerical approximations, the dimension of the required memory vector is much reduced, although there is little practical numerical experience as yet.

Because of the large sizes of the state spaces that arise in the numerical approximations, it would seem that the topic is well suited for one of the various approaches that are known as approximate dynamic programming, or even linear programming with suitably sampled constraints. See., e.g., the references in [30, 87, 82]. The success of such approaches usually depends on detailed insight into the "physics" of the problem, so that the approximation can be tailored apprpriately. At this time, it is not at all clear how to use such approaches for our problem, but one must always seek approaches that simplify the problem while yielding meaningful results.

Numbering and cross referencing. Cross reference numbering *within* a chapter does not include the chapter number. For example, within Chapter 5, Equation 4 of Section 3 of Chapter 5 is called Equation (3.4), and Subsection 6 of Section 3 of Chapter 5 is called Subsection 3.6, with the analogous usage for Theorem, Figure, and Assumption. Cross references *between* chapters do include the chapter number. For example, in Chapter 5, a reference to Equation 4 of Section 3 of Chapter 2 is called Equation (2.3.4), and Subsection 6 of Section 3 of Chapter 2 is called Subsection 2.3.6, with the analogous usage for Theorem, Figure, and Assumption.

A glossary of the more frequently used symbols appears at the end of the book.

The author gratefully acknowledges the long-term support of the Army Research Office and the National Science Foundation on numerical methods in stochastic control.

Providence, Rhode Island, USA

Harold J. Kushner
February 2008

1

Examples and Introduction

1.0 Outline of the Chapter

Consider a system where the effects of the control are delayed. Letting the control depend only on the current state of the system will lead to oscillations in the path, and possible instability, unless the delay is small. The cause is often overcontrol, due to the fact that the current choice for the control ignores control actions that have been made, but whose effects have not yet been seen due to the delay. The purpose of this chapter is simply to illustrate the improvement in behavior that can be achieved when the current control properly takes past control actions into account. The examples are simple and there is no concern for optimality in any sense, but they illustrate the point. The issue of finite-dimensional representations of a dynamical system with delays is intriguing. This is possible under certain conditions and is discussed briefly. The conditions are narrow, but the results are interesting since they do cover some applications of interest and can provide benchmarks to evaluate numerical methods.

1.1 An Introductory Example: Controlling the Temperature of a Fluid Flow

This example is intended to be illustrative of the importance of a control that depends on the past. It is a classical and perhaps the most familiar example of the effects of delay. One wishes to keep the temperature $x(t)$ of water at the end of a pipe at a desired value T. At the source, there are tanks of cool and hot water, and valves that connect these to the pipe. The control at time t is a signal $u(t)$ sent from the measurement point at the end of the pipe to the valve controllers. The signal is received instantaneously at the valves, which act instantaneously, but it takes a time $\bar{\theta}$ for any correction to be seen at the measurement point. A positive value of the control signal increases the

H.J. Kushner, *Numerical Methods for Controlled Stochastic Delay Systems*,

doi: 10.1007/978-0-8176-4621-9_1,

fraction of the flow that is hot water, and a negative value the fraction that is cool water. The system dynamics are $dx(t) = u(t - \bar{\theta})dt + \sigma dw(t)$, where $w(\cdot)$ is a Wiener process that serves as a disturbance in the system that might be due to turbulence in the pipe or to uneven cooling. Let us start with a noiseless model. Then the mixed flow moves through the pipe as a wave with a sharp wavefront.

A naive or inexperienced controller simply measures the discrepancy $T - x(t)$ between the desired and the measured temperatures and sends the corresponding signal $u(t) = c_0[T - x(t)], c_0 > 0$. The familiar result is an oscillation in the measured temperature. This is illustrated in Figure 1.1, where $\bar{\theta} = 10, c_0 = .15, T = 5$, and $x(0) = 15$. Both the measured temperature and the control are plotted. The behavior is typical, unless the delays are short. The oscillations will decrease as the gain c_0 decreases. For $c_0 = .05$ there will be minimal oscillation but the desired level 5 is not well-approximated until approximately time 80.

The oscillations occur because at each time the control signal simply depends on the current error. It does not account for the fact that other control signals that depended on the measured error were recently sent and whose effects have not yet been seen. So we have overcontroled. This is the key issue that distinguishes the problem with delays from that without delays. By making the gain smaller, the effect of ignoring the past is decreased since the rate of change of the temperature is smaller.

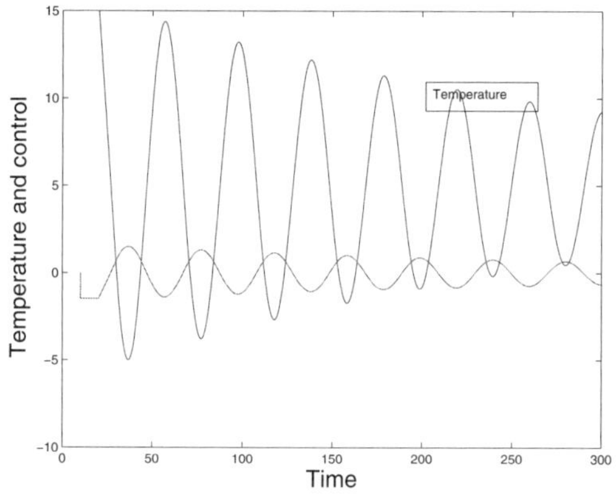

Figure 1.1. Deterministic model. Control not delay-dependent.

If the wave front was not sharp, but was approximatable by a wedge, then a possible model is $dx(t) = dt \int_{-\bar{\theta}}^{0} u(x(t+\theta))d\mu_c(\theta) + \sigma dw(t)$, where $\mu_c(\cdot)$ is a measure whose effect is to spread out the wave front.

A more experienced controller would account for his past actions in choosing the current value of the control. This is done in the case plotted in Figure 1.2, where $u(t)$ has the form $u(t) = c_1(T - x(t) - S(t)), c_1 = .15$, and

$S(t) = \int_{t-\bar{\theta}}^{t} u(s)ds$. There was no attempt to optimize the control. The aim is simply to illustrate the advantage of accounting for recent control actions. See [92] where a similar control is used.

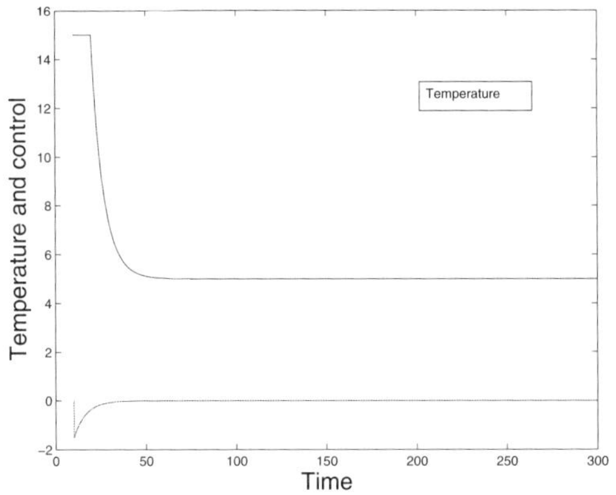

Figure 1.2. Deterministic model. Control delay-dependent.

The comparisons between the effects of a naive control and one that takes the past into account are even more dramatic when the control is required to be bang-bang, taking values $\bar{u}$ or $-\bar{u}$ for some $\bar{u} > 0$. Figure 1.3 is the case of Figure 1.1, but with $u(t) = c_0 \operatorname{sign}(T - x(t)), c_0 = .15$. As c_0 decreases, the amplitude of the oscillations decreases, but never goes to zero, and it takes it takes increasingly longer for the path to settle down to its asymptotic oscillatory form. Compare this to the situation in Figure 1.4, which is the case of Figure 1.2, with control $u(t) = c_1 \operatorname{sign}(T - x(t) - S(t)), c_1 = .15$.

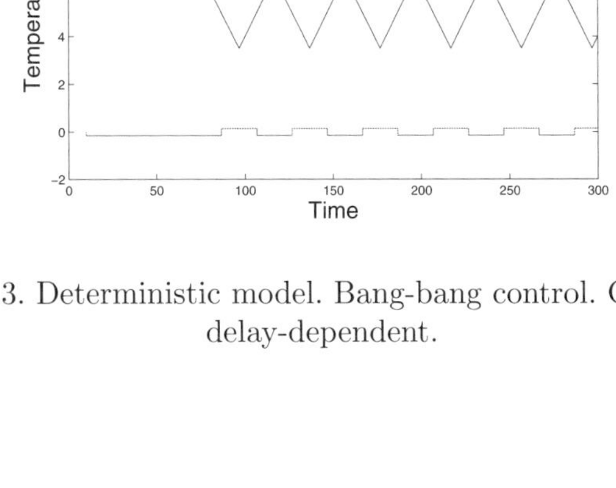

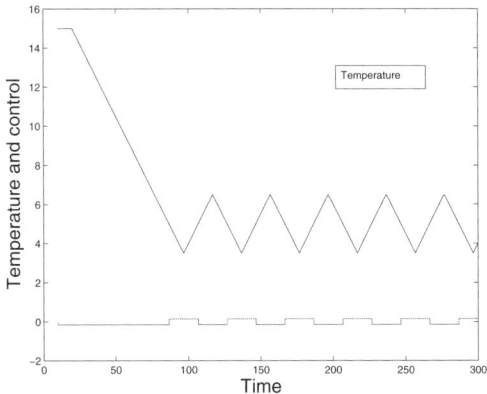

Figure 1.3. Deterministic model. Bang-bang control. Control not delay-dependent.

In the figure, the control is rapidly oscillating between the two allowed values $\pm.15$. The convergence is faster as the gain increases.

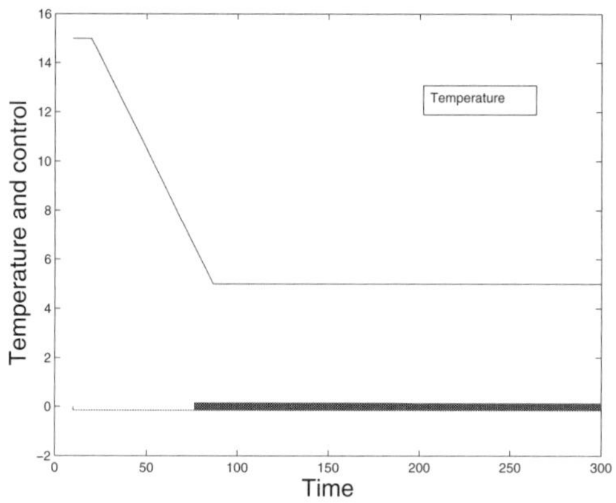

Figure 1.4. Deterministic model. Bang-bang control. Control delay-dependent.

The past-dependent control is also better at tracking time-varying temperature goals. The behavior with added noise follows a similar (perhaps even a worse) comparison. Let $\sigma = .4$. Figure 1.5 uses the control of Figure 1.1. The oscillations increase as time increases. Figure 1.6 uses the control of Figure 1.2. The behavior is that of a well-controlled system subject to uncontrollable noise. The driving noise processes in Figures 1.5 and 1.6 are the same.

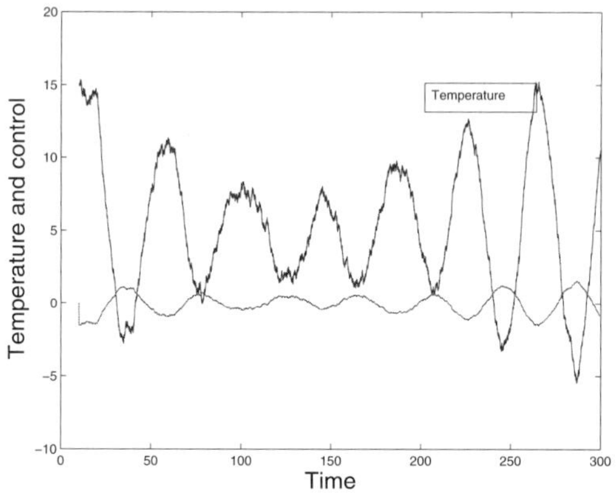

Figure 1.5. $\sigma = .4$. Control not delay-dependent.

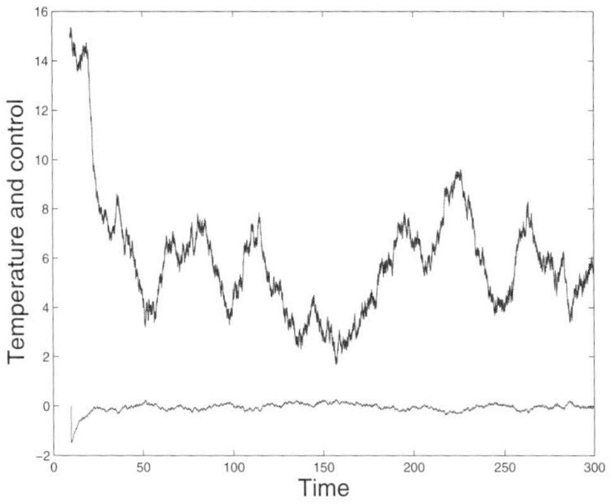

Figure 1.6. $\sigma = .4$. Control delay-dependent.

1.2 An Example from Internet Regulation

Due to the finite speed of electromagnetic signaling, delays are a common part of many telecommunications systems and can have significant effects on performance. One important example is the AIMD (additive increase multiplicative decrease) model that arises in (FTP) congestion control of internet traffic over long distances. The reference [1] studied the heavy traffic asymptotics [56] of many AIMD connections sharing a common router in the presence of other traffic, called "mice." These mice are in the system for a short time and are uncontrolled. The system was scaled by speed (bandwidth) and average number of sources, both proportional to a parameter n. With appropriate scalings of the packet rate and buffer size, an approximating controlled delayed-diffusion model was derived. The randomness due to the mice and/or number of connections appears in the limit. As often with heavy traffic models, the limit contains the main structural and parametric dependencies and is much simpler than the original system.

To improve the performance, another control, to be called the *preemptive control* is used. This control selects packets at random to be "marked" as they enter the buffer. The chance of being selected depends on the buffer state and/or source (input) rate, and is a control function to be chosen.[1]

[1] Early marking or discarding has become very popular since it was deployed in the RED (Random Early Detection) buffer management scheme [25]. It was designed to counter the synchronization in the source rate adjustments that can occur when there are many sources that send data to a common router buffer, and the buffer overflows. Then, after a communications delay, many of the sources would reduce their rates more or less simultaneously. Then, again after a delay, when it appears that the packets are getting through, they would increase their rates,

The selection probability will increase when the system nears a dangerous operating point, whose value would depend on the system delays. The selected packets are not dropped, as they would be if they were due to buffer overflows. But a signal is sent to the source indicating the marking, and the source responds by reducing its rate. Thus the effect on the source rate is similar to that of lost packets. This control, which anticipates the possibility of lost packets due to buffer overflows in the near future, can reduce the average queueing delay and rate of overflow considerably. Numerical data support this assertion. In a sense, the use of the preemptive control creates "false" rejections, which are spread out, thereby helping to reduce the oscillations that are caused by bursts of lost packets.

A simple form of the final model in [1] is

$$d\rho(t) = cdt - v_2\left[\frac{\kappa}{v_1 + a_m} dU(t - \bar{\theta}) + \kappa_1 u(t - \bar{\theta})dt\right], \tag{2.1}$$

$$x(t) - x(0) = \int_0^t [\rho(s) - b]\, ds + \sigma w_m(t) + L(t) - U(t). \tag{2.2}$$

Here $w_m(\cdot)$ is a standard Wiener process and $\sigma w_m(\cdot)$ represents the noise due to the presence of the mice. The variable $\rho(\cdot)$ represents the centered and scaled rate at which controlled packets are put into the system, $x(\cdot)$ represents the scaled content of a buffer, and $\bar{\theta}$ is the round trip transportation delay. The nondecreasing process $U(\cdot)$ represents the scaled buffer overflow and can increase only when the buffer is full. The nondecreasing process $L(\cdot)$ can increase only when the buffer is empty and assures that it does not become negative. The $b, c, \sigma, \kappa, \kappa_1, v_1, v_2$, and a_m are constants. The variables are measured at the buffer, and the control is determined there. The overflow and selections by the preemptive control are signaled to the source, which receives the information after a delay and adjusts its rate accordingly. The effect of a control signal sent from the buffer reaches the buffer after a delay of $\bar{\theta}$. Note the delayed reflection term $U(t - \bar{\theta})$. Such delayed reflection terms are common occurrences in models of communications systems, but have not previously been treated in the literature on delay systems.

A revised model, a tentative control and simulations. Since the model is a heavy traffic limit, the variables are centered and scaled, so ρ can be either positive or negative. Nevertheless the form (2.1) and (2.2) makes sense even for a non-heavy-traffic model, where ρ is constrained to be nonnegative. As long as the queue is not empty, the processor is working full time and the total system throughput is as high as possible. Then the main issues are average queueing delay, losses due to buffer overflow, and the mean source rate.

Consider the following modification of (2.1):

---

all acting at more or less the same time. This results in high oscillation and poor overall utilization.

$$d\rho(t) = cdt - v_0\rho(t)dU(t-\bar{\theta}) - u(t-\bar{\theta})dt + dL_\rho(t), \quad \rho(t) \geq 0, \tag{2.3}$$

where the term $\rho(t)dU(t-\bar{\theta})$ represents the (delayed) multiplicative decrease due to buffer overflow. The nondecreasing reflection term $L_\rho(t)$ can increase only when $\rho(t) = 0$ and serves to keep the source rate from going negative. These variables would normally be scaled. With no preemptive control, the system will overflow, and the source rate will be regulated only by the effects of the delayed loss term $\rho(t)dU(t-\bar{\theta})$. For the model (2.2) and (2.3), let $c = 1, b = 1, \bar{\theta} = 1, \sigma^2 = 1, v_0 = 1$, and let the scaled buffer size be $B = 7$. Figure 2.1 is a typical plot of the queue size and source rate when there is no preemptive control, with the queue being represented by the upper curve. The driving noise processes for all cases for this example are the same. The comparisons are similar with other choices of the noise processes.

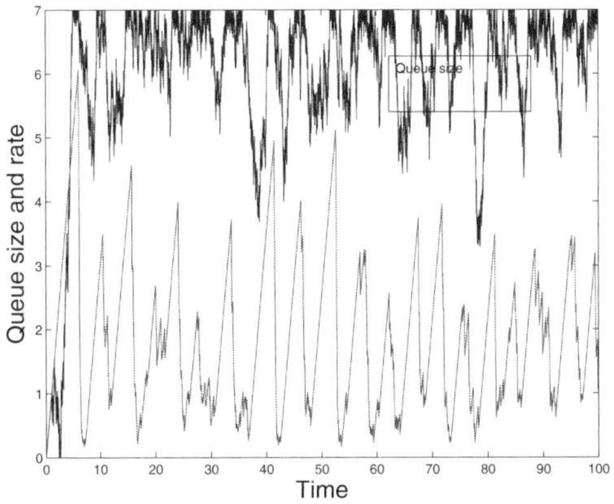

Figure 2.1. $\sigma = 1$. Uncontrolled AIMD model.

The queue is never empty, so the processor output rate cannot be improved. The average queue length is 6.0, the average source rate is 1.9, and the average buffer overflow loss per unit time is 0.867, which is exceedingly high, especially considering that the overflow packets need to be retransmitted. By increasing the penalty parameter v_0, the queue overflow behavior can be improved, but the general behavior is similar. Reducing the value of the continuous rate increase parameter c helps. Figure 2.2 illustrates the case of Figure 2.1, with c reduced to 0.1. The average queue length is 3.48, the average source rate is 1.12, and the average overflow per unit time is .17, which is still large.

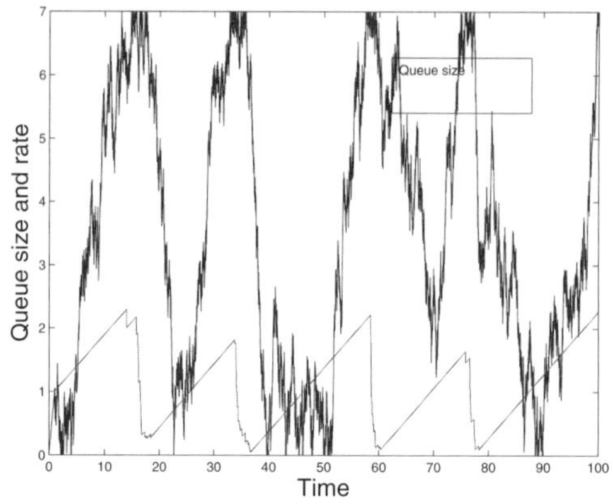

Figure 2.2. $\sigma = 1$. Uncontrolled AIMD model. Smaller rate of increase

The figure illustrates a familiar behavior. While the average source rate ρ is smaller than for the case of Figure 2.1, the queue is rarely empty and so the processor is working essentially full time, and less (although still too much) of the uncontrolled traffic is lost.

Controls. Let the (nonnegative) control be bounded above by $\bar{u}$. Suppose that the cost function is either discounted or ergodic, with cost rates that are linear in the overflow and queue size, and with no penalty on the preemptive control. When there is no delay, the optimal controls are of the "bang-bang" type. They would take the maximum value when some function of the rate and buffer occupancy is above a threshold value, and be zero otherwise. Computations show that this function can be well approximated by an affine function,[2] and the cost does not change much when the region in which the control is active is approximated by a rectangle. When there are delays the control would still be "bang-bang," but the form of the functions that determine the active region (in the infinite-dimensional) state space is not known. Part of the object of the book is to develop feasible ways of getting information about its structure. In this example, our only concern is to demonstrate the value of including information on past control actions in the computation of the present control value, analogously to what was done in the first example.

With a dynamic programming formulation, the control at time t would depend on the full system state, which is the pair $(x(t), \rho(t))$, and the control memory segment $\{u(t+\theta), -\bar{\theta} \leq \theta \leq 0\}$. Since, without the constraints that are handled by the reflection terms, the system is linear, and the buffer overflow should be small with any reasonable control, one could follow a common current procedure by dropping the bound on the control and the upper and lower bounds on the queue, use a quadratic cost function, and approximate

[2] See the figure in [1].

the consequent optimal linear control. One would still have to experiment with this cost function to assure that the desired tradeoffs among the quantities of interest are achieved for the system with a bounded queue and control.

Continuing, let $\bar{u} = 2.5$ be an upper bound for the control. Now, taking a clue from the form of the optimal controls for the no-delay problem, for parameters ρ_0, x_0 that are to be chosen let the control be zero if $\rho(t) \leq \rho_0$ or $x(t) \leq x_0$. Otherwise, we use the form

$$u(t) = \max\{0, x(t) + \rho^{1.5}(t) - 3S(t)\} \wedge \bar{u}, \tag{2.4}$$

where $S(t)$ was defined in the previous example.[3] The value of the control at a time t is the average rate of packet selection at that time.

This form was found to be useful via experimentation, but no claim is made concerning any optimality properties. Figure 2.3 plots the queue and source rate values for the parameters that were used in Figure 2.1. It is seen that, for the plotted trajectory, the queue rarely overflows and is rarely empty. The mean queue length is 3.39, the mean source rate is 1.025, and the mean overflow per unit time is .02, a vast improvement over the uncontrolled model and that for the model with the reduced rate increase $c = .1$ as well. The role of $S(t)$ is the same as in Example 1, to help avoid overcontrol by accounting for the control values chosen on the time interval $[t-\bar{\theta}, t)$ in computing the control value at time t. It is likely that the control can be modified to reduce the oscillations, but it is not clear whether that will help the overall performance. The variations in the queue process are influenced by the driving noise as well as by variations in the source rate process, so that process will always be random if $\sigma \neq 0$.

Let $\sigma = 0$. Then for the uncontrolled case of Figure 2.1, the average queue length is close to the maximum (7), and the source and overflow rates converge to 1.68 and .62, resp., quite bad. For the case of Figure 2.2, the average queue size is close to the maximum, and the source and overflow rates converge to 1.097 and .09, resp. For the controlled model of Figure 2.3 (but with $\sigma = 0$), the overflow rate is zero and the queue length and source rate converge to 2.96 and 1.02, respectively, a vast improvement over the above uncontrolled cases.

---

[3] A complication in applications is that, in general, the controllable source rates would not be known at the controller; only the queue lengths and the total queue input rates, which is the sum of the controllable and uncontrollable source rates. However, if a "locally" smoothed queue input rate is used, then the noise due to the uncontrollable sources is partly averaged out, yielding an approximation to the controllable source rate.

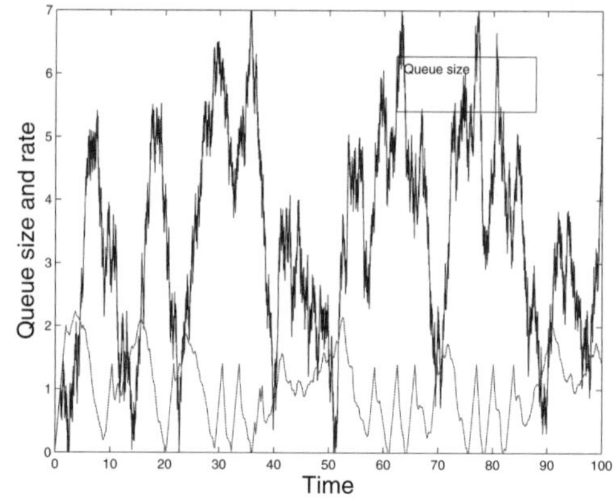

Figure 2.3. $\sigma = 1$. AIMD model with control.

Suppose that (2.3) is replaced by

$$d\rho(t) = cdt - v_0\rho(t)dU(t-\bar{\theta}) - \rho(t)u(t-\bar{\theta})dt + dL_\rho(t), \qquad (2.5)$$

where the effect of the preemptive control is proportional to the source rate. The result is plotted in Figure 2.4. The performance is close to that of the case of Figure 2.3, but the oscillations in the source rate are reduced. The average queue length is 3.2, and the average source and overflow rates are 1.02 and .02, resp.

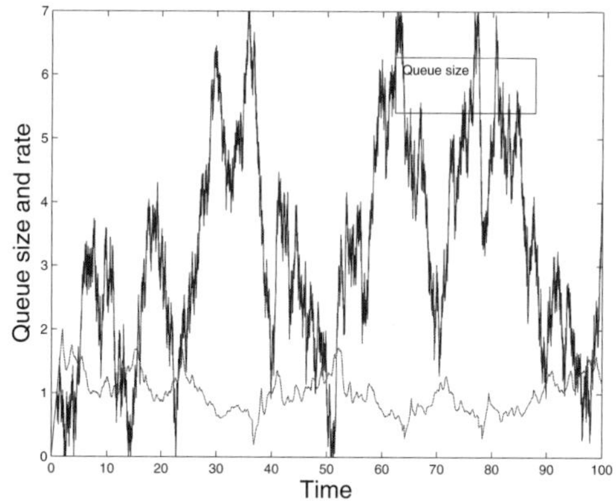

Figure 2.4. $\sigma = 1$. AIMD model with proportional control (2.5)

1.3 Models With Finite-Dimensional Equivalences

The numerical problem would be considerably simpler if there was a state transformation that would yield an equivalent system that is finite-dimensional.

Such models do exist, but the class is quite narrow. One such class is in [5, 33, 62] for an model with no boundary constraints:

$$dx(t) = b_1(x(t), y(t), u(t))dt + b_2(x(t), y(t))x(t-\bar{\theta})dt + \sigma(x(t), y(t), u(t))dw(t), \tag{3.1}$$

where, for a real number λ and real-valued $b_3(\cdot)$,

$$y(t) = \int_{-\bar{\theta}}^{0} e^{\lambda\theta} b_3(x(t+\theta))ds. \tag{3.2}$$

The cost function is defined over a finite interval and depends only on $x(\cdot)$ and $y(\cdot)$. A main assumption in [5] is that the system of PDEs defined by

$$e^{\bar{\theta}\lambda} T_x(x, y) b_2(x, y) = T_y(x, y) \tag{3.3}$$

has a solution, where T_x denotes the Jacobian of T with respect to x and $T(x, y)$ has the dimension of x. Then, under appropriate conditions on the functions in (3.1), (3.2), and in the cost rate, the system with state variable $Q(t) = T(x(t), y(t))$ has a finite-dimensional dynamical representation. See the references for full details. Applications for which explicit solutions to (3.3) are available are given for a financial-investment problem and for a problem of admission control arising in communications. This model is similar to that used in [33, 62]. An application to an investment-consumption problem is developed in [33], and necessary and sufficient conditions for this class of processes to have a finite-dimensional Bellman equation are given in [62].

Another example takes the (real-valued) form

$$\begin{aligned} dx(t) &= \left[a_1 x(t) + a_2 y(t) + a_3 x(t-\bar{\theta})\right] dt \\ &\quad + \sigma \left[x(t) + a_3 e^{\lambda\bar{\theta}} y(t)\right] dw(t) - d\gamma(t), \end{aligned}$$

where $\gamma(\cdot)$ is the total asset that has been consumed or harvested by time t. The value function (to be maximized) is over a long finite interval and has the particular form, with $x = x(t)$ and $y = y(t)$,

$$W(x, y, t) = E_{x,y}^{\gamma} \int_t^T e^{-\rho s} \left[\left(x(s) + a_3 e^{\lambda\bar{\theta}} y(s)\right)^k ds + d\gamma(s)\right].$$

If the parameters of the problem satisfy particular narrow relations, then the optimal control can be solved for explicitly. In this example [33] the control is singular. I.e., there is a boundary line of the form $v^* = c_1 x(t) + c_2 y(t)$, for some constant v^*, such that the control acts only on the boundary and keeps the pair $(x(t), y(t))$ on one side. Singular controls are common occurrences in the no-delay analog of such problems. Although the models and the allowed parameter sets are very special, the results are interesting since in principle they allow a comparison of the optimal cost values with those for controls that ignore the delay. The infinite-dimensional memory problem is not necessarily completely eliminated. Since the control is to be applied to the system (3.1) where $x(\cdot)$ is observed, to compute the control at time t in an application one needs to compute $y(t)$. If we cannot compute $y(\cdot)$ from a dynamical system whose inputs at time t are only the control $u(t)$ and $x(t)$, then we will have to keep the memory segment $\{x(t+\theta), -\bar{\theta} \leq \theta \leq 0\}$. This is still simpler than if the control depended on the full memory segment rather than on some finite-dimensional functional of it.

2

Weak Convergence and Martingales

2.0 Outline of the Chapter

This chapter contains a brief review of two of the main mathematical methods for dealing with the convergence of a sequence of approximations to a stochastic process or for showing that a sequence of stochastic processes has a limit and for characterizing it. The first method is the theory of weak convergence of a sequence of probability measures. The theory, which is an extension to a sequence of random processes of the theory of convergence in probability of a sequence of random variables, provides powerful tools for approximation and limit theorems. Once one knows that the sequence of processes of concern has a limit, that limit must be characterized. The methods of the so-called martingale problem are a standard and powerful approach to doing such a characterization, when the limit is a diffusion-type process.

The numerical approximations of concern will be representable as controlled Markov chains with multistep memory. The convergence and approximation theorems of the later chapters, which are based on weak convergence theory and the methods of the martingale problem, show that the expectations of a large set of functionals of these chains converge to the values for the original process, as the approximation parameter goes to zero. In particular, the optimal cost values converge to the optimal value for the original controlled process of interest. Also, suitable interpolations of the sequence of approximating chains, under their optimal controls, converges to an optimal limit process. In addition, for numerical purposes one often approximates the original model and uses that for the numerical computations. Then one must show that these approximations do indeed provide results that are close to those for the original model. The same methods are employed for these purposes. Only an outline of the results that are of main use to us will be given. The comprehensive references [8, 23] contain full details and much additional information. The references [55, 58, 61, 56] contain many applications of these methods to control and communications systems or to numerical approximations.

H.J. Kushner, *Numerical Methods for Controlled Stochastic Delay Systems*,
doi: 10.1007/978-0-8176-4621-9_2,

2.1 Weak Convergence

Let $I\!R^k$ denote Euclidean r-space with canonical value $x = (x_1, \ldots, x_k)$, and let $\{X_n\}$ be a sequence of $I\!R^k$-valued random variables on a probability space $(\Omega, \mathcal{F}, P)$. If there is an $I\!R^k$-valued random variable X such that $Ef(X_n) \to Ef(X)$ for each bounded and continuous real-valued function $f(\cdot)$ on $I\!R^k$, then X_n *converges to* X *in distribution.* The sequence $\{X_n\}$ is said to be *tight* or, equivalently said, *bounded in probability* if

$$\lim_{K\to\infty} \sup_n P\{|X_n| \geq K\} = 0. \tag{1.1}$$

An equivalent definition is that for each small $\mu > 0$ there are finite M_μ and K_μ such that $P\{|X_n| \geq K_\mu\} \leq \mu$ for $n \geq M_\mu$. Convergence in distribution is also called *weak convergence*. Tightness is a necessary and sufficient condition that any subsequence of $\{X_n\}$ have a further subsequence that converges in distribution [10, 23].

Let $\{\xi_n\}$ be a sequence of mutually independent and identically distributed real-valued random variables, with mean zero and unit variance and $w(\cdot)$ a real-valued Wiener process with unit variance parameter. For $t > 0$ define

$$w^n(t) = \frac{1}{\sqrt{n}} \sum_{i=1}^{[nt]} \xi_i, \tag{1.2}$$

where $[nt]$ denotes the integer part of nt. Then the central limit theorem says that $w^n(t)$ converges in distribution to a normally distributed random variable with mean zero and variance t.

For an integer k, and $0 = t_0 < t_1 < \cdots < t_{k+1}$, the multivariate central limit theorem [10] says that $\{w^n(t_{i+1}) - w^n(t_i),\ i \leq k\}$ converges in distribution to $\{w(t_{i+1}) - w(t_i), i \leq k\}$. Now consider $w^n(\cdot)$ to be a random process with paths that are constant on the intervals $[i/n, (i+1)/n)$. It is then natural to ask whether the sequence of processes $w^n(\cdot)$ converges to $w(\cdot)$ in a stronger sense. For example, will the distribution of the maximum $\max\{w^n(t) : t \leq 1\}$ converge in distribution to $\max\{w(t) : t \leq 1\}$? Donsker's theorem states that $F(w^n(\cdot))$ converges in distribution to $F(w(\cdot))$ for a large class of functionals $F(\cdot)$ [7, 23], for example for measurable $F(\cdot)$ that depend on only a finite segment of the path and are continuous almost everywhere with respect to the measure of $w(\cdot)$. This is an example of the theory of weak convergence.

The two main steps in getting the limit theorems for random processes are analogous to what is done for proving the central limit theorem: First show that there are appropriately convergent subsequences and then identify the limits. For vector-valued random variables, the necessary and sufficient condition (1.1) for the first step says that, neglecting an n-dependent set of small (uniformly in n) probability, the values of the random variables X_n are confined to some compact set. When random processes replace random variables, there will be an analogous condition ensuring that the paths are in a compact set with a "high probability."

2.1.1 Basic Theorems of Weak Convergence

Definitions. Let S denote a metric space with metric $\rho(\cdot)$ and $C(S)$ the set of real-valued continuous functions on S, with $C_b(S)$ being the subset of bounded functions. Let $\mathcal{B}(S)$ denote the collection of Borel subsets of S. Let $\mathcal{P}(S)$ denote the space of probability measures on $(S, \mathcal{B}(S))$. Let $X_n, n < \infty$, and X be S-valued random variables, with distributions $P_n, n < \infty$, and P, respectively. The sequence $\{X_n, n < \infty\}$ is said to *converge in distribution* to X if $Ef(X_n) \to Ef(X)$ for all $f \in C_b(S)$ or, equivalently written, if $\int_S f(s)P_n(ds) \to \int_S f(s)P(ds)$. This is called *weak convergence* and written as $P_n \Rightarrow P$. We will often say that the sequence of random variables X_n converges weakly to X, and denote this by $X_n \Rightarrow X$ as well. The X will be said to be the *weak-sense limit.*

For $\lambda \in \Lambda$, an arbitrary index set, let $P_\lambda \in \mathcal{P}(S)$. The set $\{P_\lambda,\ \lambda \in \Lambda\}$ is called *tight* if for each $\varepsilon > 0$ there is a compact set $K_\varepsilon \subset S$ such that

$$\inf_{\lambda \in \Lambda} P_\lambda(K_\varepsilon) \geq 1 - \varepsilon. \tag{1.3}$$

If P_λ is the measure defined by an S-valued random variable X_λ, then we will also say that $\{X_\lambda, \lambda \in \Lambda\}$ is tight. If all of the X_λ are defined on the same probability space, then (1.3) is equivalent to

$$\inf_{\lambda \in \Lambda} P\left\{X_\lambda \in K_\varepsilon\right\} \geq 1 - \varepsilon. \tag{1.4}$$

The Prohorov metric. Let $P_i \in \mathcal{P}(S), i = 1, 2$. For $A \in \mathcal{B}(S)$, define the set $A^\varepsilon = \{s' : \rho(s', s) < \varepsilon \text{ for some } s \in A\}$. Then the *Prohorov metric* $\pi(\cdot)$ on $\mathcal{P}(S)$ is defined by

$$\pi(P_1, P_2) = \inf\left\{\varepsilon > 0 : P_1(A) \leq P_2(A^\varepsilon) + \varepsilon \text{ for all closed } A \in \mathcal{B}(S)\right\},$$

and is always used on the space $\mathcal{P}(S)$. The following two theorems are fundamental.

Theorem 1.1. [23, page 101.] *If S is complete and separable, then $\mathcal{P}(S)$ is complete and separable.*

Theorem 1.2. [23, Theorem 3.2.2.] *If S is complete and separable, then a set $\{P_\lambda, \lambda \in \Lambda\} \subset \mathcal{P}(S)$ has compact closure if and only if $\{P_\lambda, \lambda \in \Lambda\}$ is tight.*

Suppose that S is complete and separable and that a given sequence of probability measures has compact closure (Prohorov metric). Theorem 1.2 then implies the existence of a convergent subsequence [19, Theorem 13, page 21]. The theorem gives a practical method for verifying the compact closure property, as tightness is also a property of the random variables (or processes) associated with the P_λ. These random variables typically have explicit representations (for example, they might be solutions to a stochastic differential equation) that can be used to verify the tightness property. A sequence of vector-valued random variables is tight if the sequence of each of its components is tight, as asserted in the next result.

Corollary 1.3. *Let S_1 and S_2 be complete and separable metric spaces, and define $S = S_1 \times S_2$ with the usual product space topology. For $\{P_\lambda, \lambda \in \Lambda\} \subset \mathcal{P}(S)$, let $P_{\lambda,i}$ be the marginal distribution of P_λ on S_i. Then $\{P_\lambda, \lambda \in \Lambda\}$ is tight if and only if $\{P_{\lambda,i}, \lambda \in \Lambda\}$, $i = 1, 2$, are tight.*

The next theorem contains some statements that are equivalent to weak convergence. Let ∂B be the boundary of the set $B \in \mathcal{B}(S)$.

Theorem 1.4. [23, Theorem 3.3.1.] *Let S be a metric space and let P_n, $n < \infty$, and P be elements of $\mathcal{P}(S)$. Then statements (i)–(iv) below are equivalent and are implied by (v). If S is separable, then (i)–(v) are equivalent:*

$$
\begin{aligned}
&(i)\ P_n \Rightarrow P,\\
&(ii)\ \limsup_n P_n(F) \le P(F) \text{ for closed sets } F,\\
&(iii)\ \liminf_n P_n(O) \ge P(O) \text{ for open sets } O,\\
&(iv)\ \lim_n P_n(B) = P(B) \text{ if } P(\partial B) = 0,\\
&(v)\ \pi(P_n, P) \to 0.
\end{aligned}
$$

The theorem implies that, for separable S, convergence in the Prohorov metric is equivalent to weak convergence. Part (iv) implies the following important extension of the class of functionals that converge in distribution.

Theorem 1.5. [7, Theorem 5.1.] *Let S be a metric space, and let $P_n, n < \infty$, and P be probability measures on $\mathcal{P}(S)$ satisfying $P_n \Rightarrow P$. Let $f(\cdot)$ be a real-valued measurable function on S and define D_f to be the measurable set of points at which $f(\cdot)$ is not continuous. Let X_n and X be random variables that induce the measures P_n and P on S, respectively. Then $f(X_n) \Rightarrow f(X)$ whenever $P\{X \in D_f\} = 0$.*

The Skorokhod representation. Suppose that $X_n \Rightarrow X$, where the X_n and X might be defined on different probability spaces. The probability spaces are unimportant, as weak convergence is a statement on the measures of the random variables. But for the purpose of characterizing the weak-sense limit X, it can be very useful to have all processes defined on the same space and weak convergence replaced by probability one convergence. This can be done without changing the distributions of the X_n or X. The result is known as the *Skorokhod representation* [23].

Theorem 1.6. [23, Theorem 3.1.8.] *Let S be a separable metric space, and assume that $P_n \in \mathcal{P}(S)$ converges weakly to $P \in \mathcal{P}(S)$ as $n \to \infty$. Then there exists a probability space $(\tilde{\Omega}, \tilde{\mathcal{F}}, \tilde{P})$ on which there are defined random variables $\tilde{X}_n, n < \infty$, and $\tilde{X}$ such that for all Borel sets B and all $n < \infty$,*

$$\tilde{P}\left\{\tilde{X}_n \in B\right\} = P_n(B), \qquad \tilde{P}\left\{\tilde{X} \in B\right\} = P(B), \tag{1.5}$$

and such that

$$\tilde{X}_n \to \tilde{X} \text{ with probability one.} \tag{1.6}$$

2.1.2 The Function Spaces $D(S; I)$

For a complete and separable metric space S, let $D(S; I)$ denote the set of S-valued functions on the interval I that are right continuous and have left-hand limits. Let $C(S; I)$ denote the subset of continuous functions. The interval I will be either $[0, \infty), [t_1, \infty)$ or $[t_1, t_2]$ for some $t_1 < t_2$. Even when the weak-sense limit processes have continuous paths, it is usually easier to prove tightness and weak convergence using the path spaces $D(S; [0, \infty))$. If there are Poisson jumps in the system dynamics or if the control is of an impulsive or singular nature, then $D(S; [0, \infty))$ must be used. We next define the metric.

The Skorokhod metric [23, Chapter 3.5], [7, Chapter 3]. For $T > 0$, let Λ_T denote the space of continuous and strictly increasing functions from $[0, T]$ onto $[0, T]$. The functions in this set will be "allowable timescale distortions." For $\lambda(\cdot) \in \Lambda_T$ define

$$|\lambda| = \sup_{s<t} \left| \log \left\{ \frac{\lambda(t) - \lambda(s)}{t - s} \right\} \right|.$$

The Skorokhod metric $d'_T(\cdot)$ on $D(I\!\!R^k; [0, T])$ is defined by, for $\lambda(\cdot) \in \Lambda_T$,

$$d'_T(f(\cdot), g(\cdot)) = \inf\{\epsilon : |\lambda| \le \epsilon, \quad \sup_{0 \le s \le T} |f(s) - g(\lambda(s))| \le \epsilon, \text{ for some } \lambda(\cdot)\}. \tag{1.7}$$

On the space $D(I\!\!R^k; [0, \infty))$, the metric is defined by

$$d'(f(\cdot), g(\cdot)) = \int_0^\infty e^{-t} \min\left[1, d'_t(f(\cdot), g(\cdot))\right] dt. \tag{1.8}$$

Now let S be a complete and separable metric space with metric $\rho(\cdot)$. Then the Skorokhod metric on the spaces $D(S; [0, T])$ is defined by the $d'_T(\cdot)$ above, but with $\rho(f(s), g(\lambda(s)))$ used in place of $|f(s) - g(\lambda(s))|$, where both $f(\cdot)$ and $g(\cdot)$ are now points in $D(S; [0, T])$. Define the space $D(S; [0, \infty))$ analogously. If S is complete and separable, then so are $D(S; [0, T])$ and $D(S; [0, \infty))$ [23].

If $f_n(\cdot) \to f(\cdot)$ in $d_T(\cdot)$ where $f(\cdot)$ is continuous, then the convergence must be uniform on $[0, T]$. If there are $\eta_n \to 0$ such that the discontinuities of $f_n(\cdot)$ are less than η_n in magnitude and if $f_n(\cdot) \to f(\cdot)$ in $d_T(\cdot)$, then the convergence is uniform on $[0,T]$ and $f(\cdot)$ must be continuous. Because of the "timescale distortion" that is involved in the definition of the metric $d_T(\cdot)$, we can have (loosely speaking) convergence of a sequence of discontinuous functions where there are only a finite number of discontinuities, where both the locations and the values of the discontinuities converge, and a type of "equicontinuity" condition holds between the discontinuities. See [7, 23] for full detail.

A criterion for tightness in $D(S;[0,T])$ and $D(S;[0,\infty))$. The following criterion for tightness will be used. Recall that for a filtration $\{\mathcal{F}_t, t \geq 0\}$, the random time τ is an $\mathcal{F}_t$-stopping time if $\{\tau \leq t\} \in \mathcal{F}_t$ for all $t \in [0,\infty)$.

Theorem 1.7. [49, Theorem 2.7b.] *Let $x^n(\cdot)$ be processes with paths in $D(S;[0,\infty))$, where S is a complete and separable metric space with metric $\rho(\cdot)$. For each $\delta > 0$ and rational $t < \infty$, let there be a compact set $S_{\delta,t} \subset S$ such that*

$$\sup_n P(x^n(t) \notin S_{\delta,t}) \leq \delta. \tag{1.9}$$

Let $\mathcal{F}_t^n$ be the σ-algebra determined by $\{x^n(s),\ s \leq t\}$ and let $\mathcal{T}_n(T)$ be the set of $\mathcal{F}_t^n$-stopping times that are no bigger than T. Suppose that

$$\lim_{\delta \to 0} \limsup_n \sup_{\tau \in \mathcal{T}_n(T)} E \min\{1, \rho(x^n(\tau+\delta), x^n(\tau))\} = 0 \tag{1.10}$$

for each $T < \infty$. Then $\{x^n(\cdot),\ n < \infty\}$ is tight in $D(S;[0,\infty))$.

Let $C(G;[a,b])$ denote the space of S-valued continuous functions on the interval $[a,b]$ with the sup norm topology, where G is a compact subset of a Euclidean space. If the interval $[a,b]$ is unbounded, then the local sup norm topology is used. The next theorem gives a necessary and sufficient condition for tightness of sequence of G-value continuous processes.

Theorem 1.8. [8, Theorem 7.3.] *The sequence of G-valued processes $x^n(\cdot)$ is tight in $C(G;[0,1])$ if and only if: For each $\epsilon > 0$ and $\eta > 0$, there is a $\delta > 0$ and an $n_0 < \infty$ such that, for $n \geq n_0$,*

$$P\left\{\sup_{i\delta<1} \sup_{s \leq \delta} |x^n(s) - x^n(i\delta)| \geq \epsilon\right\} \leq \eta.$$

2.2 Martingales and the Martingale Method

2.2.1 Martingales

Definitions. Let $(\Omega, \mathcal{F}, P)$ denote a probability space. It will always be assumed that $\mathcal{F}$ is *complete;* i.e., it contains all subsets of P-null sets.

Let $\overline{\mathcal{F} \times \mathcal{B}([0,\infty))}$ denote the completion of the product σ-algebra with respect to the product measure, with Lebesgue measure used on $\mathcal{B}([0,\infty))$. A function $\phi(\cdot)$ on $\Omega \times [0,\infty)$ and with values $\phi(\omega,t)$ in some metric space S is said to be a *measurable process* if it is a measurable mapping from $(\Omega \times [0,\infty), \overline{\mathcal{F} \times \mathcal{B}([0,\infty))})$ to $(S, \mathcal{B}(S))$. All processes are assumed to be measurable and separable. It is always assumed that S and $C(S;[0,\infty))$ and $D(S;[0,\infty))$ are complete and separable metric spaces ([7, 23]).

A family of σ-algebras $\{\mathcal{F}_t, t \geq 0\}$ is called a *filtration* on this probability space if $\mathcal{F}_s \subset \mathcal{F}_t \subset \mathcal{F}$ for all $0 \leq s \leq t$. We will always assume that the $\mathcal{F}_t$ are complete in that $\mathcal{F}_t$ contains all the subsets of null sets in $\mathcal{F}$. If A is a collection of random variables defined on $(\Omega, \mathcal{F}, P)$, then we use $\mathcal{F}(A)$ to denote the σ-algebra generated by A. Let $E_{\mathcal{F}_t}$ and $P_{\mathcal{F}_t}$ denote the expectation and probability, respectively, conditioned on the σ-algebra $\mathcal{F}_t$.

Let $M(\cdot)$ be a stochastic process defined on $(\Omega, \mathcal{F}, P)$ with filtration $\{\mathcal{F}_t, t \geq 0\}$. If $M(t)$ is $\mathcal{F}_t$-measurable for each t, then $M(\cdot)$ is said to be $\mathcal{F}_t$-adapted. Let $M(\cdot)$ be $\mathcal{F}_t$-adapted and take values in the path space $D(\mathbb{R}^k;[0,\infty))$. Then $M(\cdot)$ is said to be an *$\mathcal{F}_t$-martingale* if $E|M(t)| < \infty$ for all $t \geq 0$ and

$$E_{\mathcal{F}_t} M(t+s) = M(t) \text{ w.p.1 for all } s,t \geq 0. \tag{2.1}$$

If the filtration is unimportant or obvious, then we will simply say that $M(\cdot)$ is a martingale. If $M(\cdot)$ is an $\mathcal{F}_t$-martingale, then it is also an $\mathcal{F}(M(s), s \leq t)$-martingale. We say that $\{\mathcal{F}(M(s), s \leq t), t \geq 0\}$ is the filtration generated by $M(\cdot)$.

Martingales are a fundamental tool in stochastic analysis. Processes can often be decomposed into a sum of a process of bounded variation and a martingale. This decomposition can be used to facilitate the analysis, as the bounded-variation term is often relatively easy to handle, and there are many useful techniques for the analysis of martingales. The following inequalities will be useful. Let $M(\cdot)$ be a real or vector-valued $\mathcal{F}_t$-martingale with paths in $D(\mathbb{R}^k;[0,\infty))$ for some $k \geq 1$. Then [10, 16, 42, 76] for any $c > 0$ and $0 \leq t \leq T$,

$$P_{\mathcal{F}_t}\left\{ \sup_{t \leq s \leq T} |M(s)| \geq c \right\} \leq E_{\mathcal{F}_t}|M(T)|^2/c^2 \text{ w.p.1}, \tag{2.2}$$

$$E_{\mathcal{F}_t} \sup_{t \leq s \leq T} |M(s)|^2 \leq 4E_{\mathcal{F}_t}|M(T)|^2 \quad \text{w.p.1}. \tag{2.3}$$

Stopping time. Let $\{\mathcal{F}_t, t \geq 0\}$ be a filtration. If $M(\cdot)$ is an $\mathcal{F}_t$-martingale and τ is an $\mathcal{F}_t$-stopping time, then the "stopped" process defined by $M(t \wedge \tau)$ is also an $\mathcal{F}_t$-martingale [10, 76]. Let $\mathcal{F}_\tau$ denote the "stopped" σ-algebra that is composed of the sets $A \in \mathcal{F}$ such that $A \cap \{\tau \leq t\} \in \mathcal{F}_t$ for all t.

2.2.2 Verifying That a Process Is a Martingale

We now give a method that will be useful in showing that a process is a martingale. It is only a rewording of the definition of a martingale in terms of conditional expectations.

Let Y be a vector-valued random variable with $E|Y| < \infty$, and let $V(\cdot)$ be a process with paths in $D(S;[0,\infty))$, where S is a complete and separable metric space. Suppose that for some given $t > 0$, each integer p and each set of real numbers $0 \le s_i \le t$, $i = 1,\ldots,p$, and each bounded and continuous real-valued function $h(\cdot)$, $Eh(V(s_i), i \le p)Y = 0$. This fact and the arbitrariness of p, s_i, t, and of the function $h(\cdot)$ imply that

$$E[Y|V(s), s \le t] = 0$$

with probability one [10].

Next, let $U(\cdot)$ be a random process with $E|U(t)| < \infty$ for each t, with values in $D(S;[0,\infty))$, and such that for all p, $h(\cdot)$, $s_i \le t, i \le p$, as given above, and a given real $\tau > 0$,

$$Eh(U(s_i), i \le p)\,[U(t+\tau) - U(t)] = 0. \tag{2.4}$$

Then $E[U(t+\tau) - U(t)|U(s), s \le t] = 0$. If this holds for all t and $\tau > 0$, then by the definition (2.1) of a martingale, $U(\cdot)$ is a martingale with respect to the filtration generated by $U(\cdot)$. It is often more convenient to work with the following more general setup.

Theorem 2.1. *Let $U(\cdot)$ be a random process with paths in $D(\mathbb{R}^k;[0,\infty))$ and with $E|U(t)| < \infty$ for each t. Let $V(\cdot)$ be a process with paths in $D(S;[0,\infty))$, where S is a complete and separable metric space. Let $U(t)$ be measurable on the σ-algebra $\mathcal{F}_t^V$ determined by $\{V(s), s \le t\}$. Suppose that for each real $t \ge 0$ and $\tau \ge 0$, each integer p, and each set of real numbers $s_i \le t$, $i = 1,\ldots,p$, and each bounded and continuous real-valued function $h(\cdot)$,*

$$Eh(V(s_i), i \le p)\,[U(t+\tau) - U(t)] = 0. \tag{2.5}$$

Then $U(\cdot)$ is an $\mathcal{F}_t^V$-martingale.

An application. A sufficient condition for a Wiener process. The numerical approximations can be represented as processes that have the drift of the original diffusion and are driven by martingales. For the convergence proofs one needs to prove that these martingales converge to a Wiener process. The following result is useful for this purpose.

A process $v(\cdot)$ is said to be nonanticipative with respect to a Wiener process $w(\cdot)$ if $w(\cdot)$ is a martingale with respect to the filtration generated by $(v(\cdot), w(\cdot))$. Equivalently, for all t, $w(t+\cdot) - w(t)$ is independent of $\{v(s), w(s), s \le t\}$. Let $x(\cdot)$ and $z(\cdot)$ be $\mathbb{R}^r$-valued continuous processes with $z(\cdot)$ having bounded variation (w.p.1) on any bounded time interval. Let $b(\cdot), \sigma(\cdot), x(\cdot)$ be measurable processes and define $a(t) = \sigma(t)\sigma'(t) = \{a_{i,j}(t); i,j\}$. For $f(\cdot)$ a real-valued function with compact support that is continuous and bounded together with its first and second derivatives, define

$$\mathcal{L}f(x(t)) = f_x'(x(t))b(t) + \frac{1}{2}\sum_{i,j} a_{i,j}(t) f_{x_i x_j}(x(t)).$$

Let $h(\cdot)$ be a bounded and continuous function of its arguments, and for an integer k, and nonnegative t, T, let $0 \le t_1 \le \cdots \le t_k < t < t+T$, and let $f(\cdot)$ be as above. Suppose that for all such $h(\cdot), f(\cdot), k, t, T, t_i$, we have

$$\begin{aligned} &Eh\left(x(t_i), b(t_i), \sigma(t_i), z(t_i), i \le k\right) \\ &\quad \times \left[f(x(t+T)) - f(x(t)) - \int_t^T (\mathcal{L}f(x(s))ds + f_x'(x(s))dz(s)) \right] = 0. \end{aligned} \tag{2.6}$$

Then there is a standard $I\!R^r$-valued Wiener process on the same probability space (perhaps augmenting the space by adding an "independent" Wiener process) such that [42, Chapter 5, Proposition 4.6] $x(\cdot), z(\cdot), b(\cdot), \sigma(\cdot)$ are nonanticipative with respect to $w(\cdot)$, and

$$x(t) = x(0) + \int_0^t b(s)ds + \int_0^t \sigma(s)dw(s) + z(t).$$

3

Stochastic Delay Equations: Models

3.0 Outline of the Chapter

In this chapter, we will describe most of the models that will be of interest in the sequel. The subject of delay equations is vast, whether deterministic, stochastic, or controlled or not; for example, see [27, 44, 45, 68, 74, 73, 77]. The behavior can be quire bizarre, as seen in the examples in [74]. Our main concern is with models for which practical convergent numerical algorithms are possible.

Numerical approximations usually require that the process be confined to some compact set G, either directly or after a state space transformation. This motivates the models that we use. Two types of physical models are discussed. With the first, in Section 1, the process is either stopped on first reaching the boundary of a compact set G or else it is confined to G via its own dynamics. Thus the behavior on the boundary of G is irrelevant. A process might be stopped on first reaching the boundary either because that is the problem of interest or to obtain a bounded state space for purely numerical purposes. The models of interest are defined, with path and/or control delays. Standard issues, such as relaxed controls and weak-sense uniqueness, are reviewed. We start with the simplest model (1.1), where only the path is delayed, and then discuss the models in the order of increasing complexity.

Models for many physical problems have reflecting boundaries. They occur naturally in problems arising in queueing and communications systems [56], where the state space is often bounded owing to the finiteness of buffers and nonnegativity of their content, and the internal routing and buffer overflows determine the reflection directions on the boundary. One common way of "numerically" bounding a state space is to impose a reflecting boundary, if one does not exist already. In this case, one selects the region so that the boundary plays a minor role. Section 2 introduces the standard reflecting diffusion model, also known as the Skorokhod problem, where the process is confined to G by means of a reflection process on the boundary. Estimates of the moments of the reflection terms are given as well as criteria for the

H.J. Kushner, *Numerical Methods for Controlled Stochastic Delay Systems*,
doi: 10.1007/978-0-8176-4621-9_3,

tightness of a sequence of solutions. The outline is brief and additional material can be found in [56] and in the other references cited in Section 2. If the path, control (and possibly the reflection terms as well) are delayed, then the model (2.11) is useful and it is the form that will be used in Chapter 9, which concerns an approach that is designed to deal with the memory requirements of the numerical procedures when the path, control, and possibly the reflection processes are delayed. The so-called neutral equation model is in Subsection 2.3. The system models that are discussed are the ideal ones. Because the state spaces for such problems are infinite-dimensional, approximations need to be made before the numerical problem can be considered, and some such approximations will be discussed in Chapter 4.

In order not to overcomplicate the development and to focus on the issues that are of main concern in the delay case, we omit treatment of many interesting cases, such as where the control occurs in the variance term or where there are Poisson-type jumps, controlled or not. These are fully dealt with in [58] for the no-delay case, and that development can be carried over to the delay case in the same manner that we carry over the cases that are treated. The Girsanov transformation method, discussed in Section 3, is a powerful tool for introducing nonlinear control while maintaining uniqueness. It will be very useful for the treatment of the ergodic cost problem in Chapter 5. Typical cost functionals are discussed in Section 4. The existence of an optimal control is shown in Section 5. The proofs are of interest also because they provide a template for the proofs of convergence of the numerical approximations in Chapters 7–9. The singular and impulsive control problems are considered in Section 6. The models are described and existence of an optimal control proved. Singular controls occur frequently in routing and admission control in communications problems, where it is usually determined by a "free boundary;" that is where there is a set in the state space with the control acting when the process reaches its boundary, and attempts to keep the process from exiting the set.

We will suppose that the maximum delays are the same for the path and control. This will not be the case in general, where the maximum delay will depend on the component of the path or control. But the convention greatly simplifies the notation, and it is trivial to adjust any of the results to the case where the maximum delays are different.

We have supposed that the delays are not random and are known. Suppose that the delay is random, say a finite-state Markov chain, but its value is always known. (See, e.g., [79] for an example arising in communications.) In this case, we simply augment the system state by the state of this Markov process, assuming that its value is always known. The memory segment will be that for the maximum delay.

3.1 The System Model: Boundary Absorption

Two types of delayed stochastic differential equations will be used as system models, depending on the boundary conditions. In this section, we discuss a simple model for use when the control stops on first leaving a predetermined set or hitting its boundary. Because interest in the process stops on first contact with the boundary, we need not specify reflection or other boundary processes. In the next section, we discuss the models when there are reflecting boundaries.

Terminology. The path of the stochastic differential equation of concern takes values in $I\!R^r$, Euclidean r-space, and the solution is denoted by $x(\cdot)$. In this section, the process is stopped on first hitting the boundary of a compact set $G \subset I\!R^r$. The control $u(t)$ takes values in a compact set U. An *admissible control* is an (ω, t)-measurable U-valued function that is nonanticipative with respect to the driving Wiener process $w(\cdot)$. The maximum delay is denoted by $0 < \bar{\theta} < \infty$. Define $\bar{x}(t)$ to be the path memory segment $\{x(t+\theta), -\bar{\theta} \leq \theta \leq 0\}$. The value of the function $\bar{x}(t)$ at θ will be written as $x(t, \theta)$. The dynamical terms depend on the segments $\bar{x}(t)$ of the $x(\cdot)$-process over an interval of length $\bar{\theta}$, the state of the process, and we need to define the space of such segments. In work on the mathematics of delay equations, it is common to use either the path space $C(G; [-\bar{\theta}, 0])$ of continuous G-valued functions on the interval $[-\bar{\theta}, 0]$ (with the sup norm topology) or $L_2(G; [-\bar{\theta}, 0])$, the space of square integrable functions on $[-\bar{\theta}, 0]$. The Skorokhod space $D(G; [-\bar{\theta}, 0])$ is more appropriate for the work on the convergence of the numerical algorithms and for the results concerning the existence of an optimal control later in this chapter, as well as for the approximation results in Chapter 4, which use weak convergence methods, and its use involves no loss of generality. If the model is extended to include a Poisson-type jump term or has singular or impulsive controls, then the use of $D(G; [-\bar{\theta}, 0])$ is indispensable.

Note that if $f^n(\cdot)$ converges to $f(\cdot)$ in $D(G; [t_1, t_2])$ and $f(\cdot)$ is continuous, then the convergence is uniform on any bounded subinterval.

We will use $\hat{x}$, with values $\hat{x}(\theta)$, $-\bar{\theta} \leq \theta \leq 0$, to denote the canonical point in $D(G; -[\bar{\theta}, 0])$. A shortcoming of the Skorokhod topology is that the function $f(\cdot)$ defined by $f(\hat{x}) = \hat{x}(\theta_0)$, for any fixed $\theta_0 \in [-\bar{\theta}, 0]$, is not continuous (it is measurable). But it is continuous at all points $\hat{x}$ that are continuous functions. In our case, all the solution paths $x(\cdot)$ will be continuous.

Delay in the path only. When the path only is delayed, the process model is the controlled diffusion

$$dx(t) = b(\bar{x}(t), u(t))dt + \sigma(\bar{x}(t))dw(t), \tag{1.1}$$

with initial condition $\hat{x} = \{x(s), -\bar{\theta} \leq s \leq 0\}$. We will use the following assumptions. The last sentence of (A1.1) (and its analog in the other conditions) is not needed in this chapter. But we write it for use in the later chapters.

A1.1. *$b(\cdot)$ is bounded and measurable and is continuous on $D(G;[-\bar{\theta},0]) \times U$ at each point $(\hat{x}(\cdot),\alpha)$ such that $\hat{x}(\cdot)$ is continuous.*[1] *More generally, suppose that if $f^n(\cdot) \to f(\cdot)$ in $D(G;[-\bar{\theta},0])$ and that the values of $f^n(\cdot)$ at the points of discontinuity of $f(\cdot)$ converge to those of $f(\cdot)$, then $b(f^n,\alpha) \to b(f,\alpha)$ uniformly in α.*

A1.2. *The matrix-valued function $\sigma(\cdot)$ is bounded and measurable and is continuous on $D(G;[-\bar{\theta},0])$ at each point $\hat{x}(\cdot)$ that is continuous. More generally, suppose that if $f^n(\cdot) \to f(\cdot)$ in $D(G;[-\bar{\theta},0])$ and that the values of $f^n(\cdot)$ at the points of discontinuity of $f(\cdot)$ converge to those of $f(\cdot)$, then $\sigma(f^n) \to \sigma(f)$. The function $k(\cdot)$ that will appear in the cost function satisfies the condition on $b(\cdot)$.*

Assumption (A1.1) covers the common case where $b(\bar{x}(t),\alpha) = \sum_i b_i(x(t-\theta_i),\alpha), 0 \le \theta_i \le \bar{\theta}$, where the $b_i(\cdot)$ are continuous.

Delays in the path and control. We will also consider the problem where the control as well as the path is delayed. Let $\mathcal{B}(U;[-\bar{\theta},0])$ denote the space of measurable functions on $[-\bar{\theta},0]$ with values in U, and let $\hat{u}$, with values $\hat{u}(\theta), -\bar{\theta} \le \theta \le 0$, denote a canonical element of $\mathcal{B}(U;[-\bar{\theta},0])$. Then the dynamical term $b(\cdot)$ becomes a function of both $\hat{x},\hat{u}$. As noted in the Chapter Outline, we will suppose that the maximum delays are the same for the path and control. This will not be the case in general, where the delay will depend on the component of the path or control. But the convention greatly simplifies the notation, and it is trivial to adjust any of the results to the case where the maximum delays are different.

Depending on the applications of interest, there are a variety of choices for the way that the control appears in $b(\cdot)$. We will use the following common assumption, where $\bar{u}(t)$ denotes the "memory-segment" function $\{u(t+\theta),\theta \in [-\bar{\theta},0]\}$.

A1.3. *Let $\mu_c(\cdot)$ be a bounded measure on the Borel sets of $[-\bar{\theta},0]$ and let $b(\cdot)$ be a bounded measurable function on $D(G;[-\bar{\theta},0]) \times U \times [-\bar{\theta},0]$. The function $b(\hat{x},\alpha,v)$ is continuous in $(\hat{x},\alpha,v)$ at each point $\hat{x}$ that is a continuous function. More generally, suppose that if $f^n(\cdot) \to f(\cdot)$ in $D(G;[-\bar{\theta},0])$ and that the values of $f^n(\cdot)$ at the points of discontinuity of $f(\cdot)$ converge to those of $f(\cdot)$, or else $\sup_{-\bar{\theta}\le\theta\le 0} |f^n(\theta) - f(\theta)| \to 0$, then $b(f^n,\alpha,\theta) \to b(f,\alpha,\theta)$ uni-*

[1] The space $D(G;[-\bar{\theta},0])$ is used because we will be using weak convergence methods, but also because the various approximations to $\bar{x}(t)$ will be discontinuous, although their limits will be continuous. The continuity of $\hat{x}$ implies that if $\hat{x}_n \to \hat{x}$ in $D(G;[-\bar{\theta},0])$ and $\hat{x}$ is continuous then $b(\hat{x}_n,\alpha) \to b(\hat{x},\alpha)$ uniformly in α. The function $b(\cdot)$ would not generally be continuous on $D(G;[-\bar{\theta},0])$. For example, let $\bar{\theta} = 1$ and $b(\hat{x}) = \hat{x}(-.5)$. Define $\hat{x}_n(\theta) = 1$ for $\theta \in [-.5+1/n,0]$ and equal to zero otherwise. Then, in $D(G;[-\bar{\theta},0])$, $\hat{x}_n$ converges to the function with value unity at $\theta \ge -.5$. But $b(\hat{x}_n) = 0$ for all n.

formly in α, θ. The function $k(\cdot)$ that will appear in the cost function satisfies the condition on $b(\cdot)$.

Define

$$\bar{b}(\bar{x}(t), \bar{u}(t)) = \int_{-\bar{\theta}}^{0} b(\bar{x}(t), u(t+\theta), \theta)\mu_c(d\theta). \tag{1.2}$$

Then the system is

$$dx(t) = \bar{b}(\bar{x}(t), \bar{u}(t))dt + \sigma(\bar{x}(t))dw(t), \tag{1.3}$$

with initial condition $(\hat{x}, \hat{u}) = \{\hat{x}(\theta), \hat{u}(\theta), -\bar{\theta} \le \theta \le 0\}$.

An example of the general form covered by (A1.3) is, for $0 \le \theta_i \le \bar{\theta}$,

$$dx(t) = x(t)x(t-\theta_1)u(t-\theta_2)dt + u^2(t-\theta_3)dt + b_0(\bar{x}(t))dt + \sigma(\bar{x}(t))dw,$$

in which case $\mu_c(\cdot)$ is concentrated on the three points $\{0, -\theta_2, -\theta_3\}$. What is not covered are "cross" terms in the control such as $u(t-\theta_1)u(t-\theta_2)$, where $\theta_1 \ne \theta_2$.

Relaxed controls. Given a control $u(\cdot)$, for a real Borel set T and a Borel set $A \subset U$, let $r(A,T)$ denote the Lebesgue measure of the subset of T on which the control takes values in A. The (random) measure $r(\cdot)$ is equivalent to the control $u(\cdot)$ in that one can be obtained from the other. A relaxed control is an extension of this idea. When proving approximation and limit theorems, it is common practice to work in terms of relaxed controls, as they are better behaved than ordinary controls in that $r(A,[0,t])$ is continuous in t.

A relaxed control $r(\cdot)$ [58] is a measure on the Borel sets of $U \times [0,\infty)$, with $r(A \times [0,\cdot])$ being measurable and nonanticipative with respect to $w(\cdot)$ for each Borel $A \in U$, and satisfying $r(U \times [0,t]) = t$. It must then be Lipschitz continuous, with Lipschitz constant ≤ 1. Write $r(A,t) = r(A \times [0,t])$. The left-hand derivative[2] $r'(d\alpha,t) = \lim_{0<\delta\to 0}[r(d\alpha,t) - r(d\alpha,t-\delta)]/\delta$ is defined for almost all (ω,t). By the definitions, $r(d\alpha\, ds) = r'(d\alpha,s)ds$. For $-\bar{\theta} \le \theta \le 0$, we write $r(d\alpha, ds+\theta)$ for $r(d\alpha, s+ds+\theta) - r(d\alpha, s+\theta)$. The weak topology is used on the relaxed controls. Thus $r^n(\cdot)$ converges to $r(\cdot)$ if and only if $\int\int \phi(\alpha,s)r^n(d\alpha\, ds) \to \int\int \phi(\alpha,s)r(d\alpha\, ds)$ for all continuous functions $\phi(\cdot)$ with compact support. With this topology, the space of relaxed controls is compact, which is one of its main advantages. An ordinary control $u(\cdot)$ can be written as the relaxed control $r(\cdot)$ defined by its derivative $r'(A,t) = I_{\{u(t)\in A\}}$, where I_K is the indicator function of the set K. Then, as noted above, $r(A,t)$ is the amount of time that the control takes values in the set A by time t.

Rewriting (1.1) in terms of relaxed controls yields

---

[2] In [58] m_t or r_t were used to denote the derivative. But this notation would be confusing in the context of the notation required to represent the various delays in this book.

$$\begin{aligned} x(t) &= x(0) + \int_0^t \int_U b(\bar{x}(s), \alpha) r(d\alpha\, ds) + \int_0^t \sigma(\bar{x}(s)) dw(s) \\ &= x(0) + \int_0^t \int_U b(\bar{x}(s), \alpha) r'(d\alpha, s) ds + \int_0^t \sigma(\bar{x}(s)) dw(s)). \end{aligned} \tag{1.4}$$

In relaxed control form, the integral of (1.2) on $[0, t]$ is

$$\begin{aligned} &\int_0^t \int_{-\bar{\theta}}^0 \int_U b(\bar{x}(s), \alpha, \theta) r'(d\alpha, s+\theta) \mu_c(d\theta) ds \\ &\quad = \int_{-\bar{\theta}}^0 \left[\int_0^t \int_U b(\bar{x}(s), \alpha, \theta) r(d\alpha, ds+\theta) \right] \mu_c(d\theta). \end{aligned} \tag{1.5}$$

Define the control memory segment

$$\bar{r}(t) = \{ r(s) - r(t-\bar{\theta}), t - \bar{\theta} \le s \le t \}$$

with canonical value $\hat{r}$, and set

$$\bar{b}(\bar{x}(t), \bar{r}(t)) = \int_{-\bar{\theta}}^0 \int_U b(\bar{x}(t), \alpha, \theta) r'(d\alpha, t+\theta) \mu_c(d\theta). \tag{1.6}$$

Then the full system equation (1.3) becomes

$$\begin{aligned} x(t) &= x(0) + \int_{-\bar{\theta}}^0 \left[\int_0^t \int_U b(\bar{x}(s), \alpha, \theta) r(d\alpha, ds+\theta) \right] \mu_c(d\theta) + \int_0^t \sigma(\bar{x}(s)) dw(s) \\ &= x(0) + \int_0^t \bar{b}(\bar{x}(s), \bar{r}(t)) ds + \int_0^t \sigma(\bar{x}(s)) dw(s), \end{aligned} \tag{1.7}$$

and the initial condition is $(\bar{x}(0), \bar{u}(0))$. For ordinary controls, (1.6) can be written in the simpler form (1.2).

For use in Chapter 4 and later, we introduce an alternative way of writing (1.6). Define the function $\tilde{r}'(d\alpha, t, \theta) = r'(d\alpha, t+\theta)$. Then (1.6) can be written as

$$\bar{b}(\bar{x}(t), \bar{r}(t)) = \int_{-\bar{\theta}}^0 \int_U b(\bar{x}(t), \alpha, \theta) \tilde{r}'(d\alpha, t, \theta) \mu_c(d\theta). \tag{1.8}$$

This form will be useful because when we will be approximating the control memory segment in Chapter 4, it is the function $r'(d\alpha, t+\theta)$ that will be approximated for each t and the result will not necessarily be representable as a function of α and the sum $t+\theta$.

Random delays. In principle, there is no difficulty in letting the delays be random. The issues are of a practical nature. Suppose that the delay is random in that there are real-valued nonanticipative processes $\theta_i(\cdot), i \le L$, with $0 \le \theta_i(t) \le \theta$, such that the dynamical terms are $b(x(t - \theta_i(t)), i \le L, u(t))$ and $\sigma(x(t - \theta_i(t)), i \le L)$. Such systems are discussed in [73, pages

167–186]. Suppose that $\theta(\cdot)$ is a finite-state Markov chain, the simplest model. The randomness complicates the numerical problem because it adds a new state variable, the state of the chain. If the state of the chain is known, then numerical algorithms can be developed along the lines of the later chapters. One needs to keep track of the maximum memory segment, assumed to be of length $\bar{\theta}$. If the state of the chain is not known, then we have the substantially more difficult problem of "partial information," with which we do not deal.

Strong and weak-sense solutions. For notational convenience only, we will suppose that the initial conditions $\bar{x}(0)$ and $\bar{u}(0)$ are not random. A strong-sense solution to (1.1), (1.3), (1.4), or (1.7), is a process $x(\cdot)$ such that, for each t, $x(t)$ is measurable function of $\{w(s), u(s), s \le t\}$ or of $\{w(s), r(s), s \le t\}$, according to the case. The definitions are the same for the reflected diffusion model discussed in the next section.

Consider the model (1.1) or (1.4). If $w(\cdot)$ is a Wiener process on $[0,\infty)$ and $r(\cdot)$ is a relaxed control on the same probability space and it is defined on $[0,\infty)$, and is nonanticipative with respect to $w(\cdot)$, then we say that the pair is admissible or, if $w(\cdot)$ is understood, that $r(\cdot)$ is admissible. Suppose that, given an admissible pair $(w_1(\cdot), r_1(\cdot))$ and initial condition $\hat{x}$, there is a probability space on which is defined a set $(x(\cdot), w(\cdot), r(\cdot))$ solving (1.4) with $\bar{x}(0) = \hat{x}$, where $(x(\cdot), r(\cdot))$ is nonanticipative with respect to the Wiener process $w(\cdot)$, $(w(\cdot), r(\cdot))$ has the same probability law as $(w_1(\cdot), r_1(\cdot))$, and the probability law of the solution set does not depend on the probability space. Then we say that there is a unique solution to (1.1) in the weak sense [42]. The definition is the same with model (1.7), except that the initial control condition $\hat{u}$ or $\hat{r}$ needs to be specified. Because all controls will be admissible, the qualifier will often be dropped. We always assume the following condition.

A1.4. *There is a weak-sense unique weak-sense solution to* (1.4) *or* (1.7), *whichever is appropriate, for each admissible pair* $(w(\cdot), r(\cdot))$ *and initial data.*

The techniques that are used to prove existence and uniqueness for the no-delay problem can be adapted to the delay problem. For example, use (A1.1), (A1.2), and the Lipschitz condition

$$|\sigma(\hat{x}) - \sigma(\hat{y})| + |b(\hat{x}, \alpha, v) - b(\hat{y}, \alpha, \theta)| \le K \sup_{-\bar{\theta} \le \theta \le 0} |\hat{x}(\theta) - \hat{y}(\theta)|,$$

and a standard Picard iteration. In fact, this ensures that there is a strong-sense solution for any admissible pair $(w(\cdot), r(\cdot))$. See also [67, Section 1.7] and [73, 74] for a more complete development for the uncontrolled problem. Alternatively, one can use the Girsanov measure transformation methods [56, 58] that will be discussed in Section 3.

3.2 Reflecting Diffusions: The Skorokhod Problem

3.2.1 The Reflected Diffusion

In this section, we describe a standard model of a reflecting diffusion. The process $x(\cdot)$ will be confined to a convex polyhedron $G \in I\!R^r$, with a nonempty interior and boundary ∂G, by means of the boundary reflection process $z(\cdot)$. The conditions on the reflection directions on the boundary are stated below. Let $y_i(\cdot)$ denote the nondecreasing process that is the component of $z(\cdot)$ that is due to reflection from the ith face of G. In problems arising in communications theory, it is usually buffer overflows that are penalized, and on the boundaries associated with overflows, the reflection directions are inward normals to the boundary. The reflected diffusion models (2.1) or (2.3) below and their relaxed control forms are known as the Skorokhod problem. For a detailed discussion of the Skorokhod problem and the assumptions (A2.1) and (A2.2), see [56, Chapter 3] and [20, 21]. In this subsection we consider the cases where the path and/or the control is delayed. The next subsection gives an alternative model that can be used when part of the reflection or Wiener process also appears in a delayed form.

Assumptions on the state space G. Assumptions (A2.1) and (A2.2) are the ones used in [58] (see Section 5.7 of this reference), and are standard in the treatment of general reflecting diffusions [20, 21], [56, Section 3.5]. They might look complicated, but are quite reasonable, as seen from the discussion in [56, Chapter 3], which relates them to conditions that arise in various applications in queueing and communications.

A2.1. *The state space G is the intersection of a finite number of closed half spaces in Euclidean r-space $I\!R^r$ and is the closure of its interior. Let ∂G_i, $i = 1, \ldots,$ denote the faces of G, and n_i the interior normal to ∂G_i. Interior to ∂G_i, the reflection direction is denoted by the unit vector d_i, and $\langle d_i, n_i \rangle > 0$ for each i. The possible reflection directions at points on the intersections of the ∂G_i are in the convex hull of the directions on the adjoining faces. No more than r constraints are active at any boundary point.*

Let $d(x)$ denote the convex hull of the set of reflection directions at the point $x \in \partial G$, whether it is a singleton or not.

A2.2. *For each $x \in \partial G$, define the index set $I(x) = \{i : x \in \partial G_i\}$. Suppose that $x \in \partial G$ lies in the intersection of more than one boundary; that is, $I(x)$ has the form $I(x) = \{i_1, \ldots, i_k\}$ for some $k > 1$. Let $N(x)$ denote the convex hull of the interior normals $n_{i_1}, \ldots, n_{i_k}$ to $\partial G_{i_1}, \ldots, \partial G_{i_k}$, resp., at x. Then there is some vector $v \in N(x)$ such that $\gamma' v > 0$ for all $\gamma \in d(x)$.* (See Figure 2.1 for an illustration of this condition.)

There is a neighborhood $N(G)$ and an extension of $d(\cdot)$ to $\overline{N(G)} - G$ that is upper semicontinuous in the following sense: For each $\epsilon > 0$, there is $\rho > 0$ that *goes to zero as* $\epsilon \to 0$ *and such that if* $x \in N(G) - G$ *and distance*$(x, \partial G) \leq \rho$, *then* $d(x)$ *is in the convex hull of the directions* $\{d(v); v \in \partial G, distance(x, v) \leq \epsilon\}$.

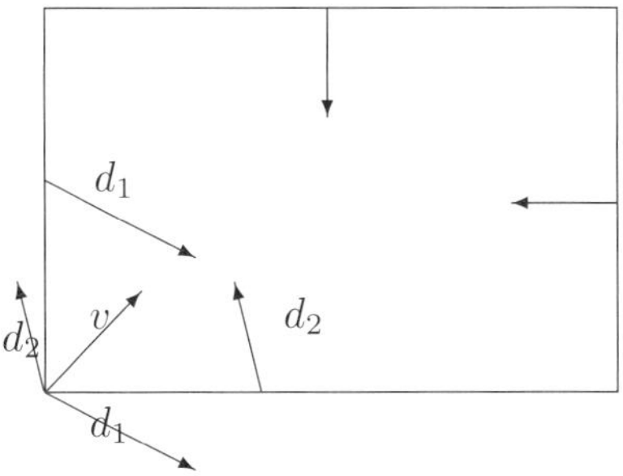

Figure 2.1. Illustration of (A2.2).

With only the path delayed, the model is the reflected form of (1.1):

$$dx(t) = b(\bar{x}(t), u(t))dt + \sigma(\bar{x}(t))dw(t) + dz(t), \tag{2.1}$$

with relaxed control representation

$$dx(t) = \int_U b(\bar{x}(t), \alpha)r'(d\alpha, t)dt + \sigma(\bar{x}(t))dw(t) + dz(t). \tag{2.2}$$

When both the path and control are delayed, the model is the reflected form of (1.2):

$$dx(t) = \bar{b}(\bar{x}(t), \bar{u}(t))dt + \sigma(\bar{x}(t))dw(t) + dz(t), \tag{2.3}$$

with relaxed control form, analogously to (1.7),

$$x(t) = x(0) + \int_0^t \bar{b}(\bar{x}(s), \bar{r}(s))ds + \int_0^t \sigma(\bar{x}(s))dw(s) + z(t), \tag{2.4}$$

where $\bar{b}(\cdot)$ was defined in either (1.2) or (1.6), according to the case. The analog of (A1.4) is:

A2.3. *There is a unique weak-sense solution to* (2.4) *for each initial admissible relaxed control and initial data.*

Note on the solution. Let $|z|(t)$ denote the variation of $z(\cdot)$ on the interval $[0, t]$. By a solution to any of (2.1)–(2.4), we mean the following. $z(\cdot)$ is the reflection process and satisfies the following conditions: $|z|(t) < \infty$ with probability one (w.p.1) for all t, and there is a measurable function $\gamma(\cdot)$ with $\gamma(t) \in d(x(t))$ w.p.1 such that $z(t) = \int_0^t \gamma(s)d|z|(s)$. This says only that the reflection process can change only when $x(t)$ is on the boundary, and the increments are in a correct reflection direction. We can write $z(t) = \sum_i d_i y_i(t)$, where $y_i(\cdot)$ is a nondecreasing process that can increase only when $x(t)$ is on the ith face of G. See [56, 58] for more detail on controlled reflected diffusions.

Remarks on the assumptions. The extension of $d(x)$ to an outer neighborhood of ∂G that is required in the last part of (A2.2) is needed as the Markov chain processes associated with the numerical procedures are defined on a discrete grid and might leave G before being reflected back in. The extension can always be constructed, so the requirement is insignificant. Under (A2.1)–(A2.2), the choice of the reflection direction (in the allowed convex sets) on the corners and edges of G has no effect on the process. If the state space is being bounded for purely numerical reasons, then the reflections are introduced only to give a compact set G, which should be large enough so that the effects on the solution in the region of main interest are small. A common choice is a hyperrectangle with interior normal reflection directions. The condition (A2.2) implies (see [20, 56]), the so-called "completely-S" condition, the fundamental boundary condition for the modeling of stochastic networks, [29, 56, 84], and which ensures that $z(\cdot)$ has bounded variation w.p.1.

Estimates of the reflection term $z(\cdot)$. Estimates of the variation of $z(\cdot)$ will be needed in many of the proofs and to ensure that the cost functions are well defined. The basic results are covered by the following theorem, which is a rewording of [58, Theorem 1.1, Chapter 11], except for the assertion concerning (2.7). This assertion is proved by working recursively on intervals $(0, \theta_0], (\theta_0, 2\theta_0], \ldots$, using the previous part of the theorem, until the interval $(0, T]$ is covered, where θ_0 is defined in the theorem. With $p(\cdot)$ omitted, the bounds on the expectations in (2.5) and (2.6) depend on $E \sup_{t \le T} |X(0) + \int_0^t f(s)ds + \int_0^t \sigma(s)dw(s)|^2$, and the appropriate "recursive"adjustments are made when $p(\cdot) \neq 0$.

Lemma 2.1. *Assume* (A2.1)–(A2.2). *Let $f(\cdot), \sigma(\cdot)$ be measurable and nonanticipative processes of the appropriate dimension, and bounded in norm by some constant $K < \infty$. Define*

$$dX(t) = f(t)dt + \sigma(t)dw(t) + dZ(t), \quad X(0) \in G,$$

where $Z(\cdot)$ is the reflection term. Let $|Z|(t)$ denote the variation of $Z(\cdot)$ on the interval $[0, t]$. Then

$$\lim_{T \to 0} \sup_{X(0), f, \sigma} E|Z|^2(T) = 0. \tag{2.5}$$

For each $T < \infty$,

$$\sup_{X(0), f, \sigma} E|Z|^2(T) < \infty. \tag{2.6}$$

Let $Y_i(\cdot)$ denote the component of the reflection process that is due to reflection on the ith face, with corners and edges assigned in any way at all to the adjacent faces, and define $Y(\cdot) = \{Y_i(\cdot), i\}$. Then (2.5) and (2.6) hold for $Y(\cdot)$ replacing $Z(\cdot)$. Let $p(\cdot)$ be a bounded function of θ and x, with θ-support concentrated on $[-\bar{\theta}, 0]$ and that takes the value zero on $-\theta_0 \le \theta \le 0$, where $\theta_0 > 0$. For $X(0) \in G$, redefine $X(\cdot)$ by

$$dX(t) = f(t)dt + \sigma(t)dw(t) + dt \int_{\theta=-\bar{\theta}}^{0} p(X(t+\theta), \theta) d_\theta Y(t+\theta) + dZ(t). \quad (2.7)$$

Then (2.5) and (2.6) continue to hold.

Suppose that, for $\Delta > 0$, on the intervals $[n\Delta, n\Delta + \Delta)$ the system evolves as (2.7) without the reflection term, and the reflection comes in at times $n\Delta$ if $X(n\Delta-) \notin G$. Then the lemma holds if $\lim_{T,\Delta \to 0}$ replaces $\lim_{T\to 0}$ in (2.5).

Tightness of a sequence of solutions to the Skorokhod problem. The following result will be useful to establish tightness in $D(G; [0, \infty))$ of a sequence of processes and is a special case of [56, Theorem 3.6.1]. See also [58, Theorem 11.1.2]. Some details of the proof will be given in Theorem 5.5, which gives a generalization that will be useful for the proofs of convergence of the numerical procedures in Chapters 7–9.

Lemma 2.2. *Assume* (A2.1) *and* (A2.2). *Assume the form*

$$x^n(t) = x^n(0) + F^n(t) + z^n(t), \quad x(t) \in G, \quad (2.8)$$

where $z^n(\cdot)$ is the reflection process and $F^n(\cdot)$ is asymptotically continuous in the sense that for each $\nu > 0$ and $T > 0$,

$$\lim_{\delta\to 0} \limsup_{n\to\infty} P\left\{ \sup_{t\le T} \sup_{s\le\delta} |F^n(t+s) - F^n(t)| \ge \nu \right\} = 0. \quad (2.9)$$

Then $\{z^n(\cdot), y^n(\cdot)\}$ is tight in $D(G; [0, \infty))$, the limit $(x(\cdot), z(\cdot))$ of any weakly convergent subsequence is continuous with probability one, and $z(\cdot)$ is a reflection process for $x(\cdot)$.

The following extension of Lemma 2.2 will be useful in showing the convergence of the numerical approximations, and the proof is that of Lemma 2.2. The lemma is designed to handle problems where the "reflection" might occur outside of but close to the boundary, at a maximum distance that goes to zero as $n \to \infty$.

Lemma 2.3. *Assume* (A2.1), (A2.2), *the form* (2.8), *and the condition* (2.9). *The functions $F^n(\cdot)$ are assumed to be piecewise-constant. Suppose that $F^n(\cdot)$ jumps at t, with $x(t-) + dF^n(t) \notin G$. Let the reflection direction that takes* *the path back into G take the value at the point $x(t-) + dF^n(t)$, as defined by the last paragraph of* (A2.2). *Then the conclusions of Lemma 2.2 hold.*

Assumptions (A2.1) and (A2.2) imply that, for large n, the $z^n(\cdot)$ in (2.8) can be written as[3]

$$z^n(t) = \sum_i d_i y_i^n(t), \tag{2.10}$$

where the $y_i^n(\cdot)$ are nondecreasing and can increase only at t where $x^n(t)$ is within a distance of ∂G_i that goes to zero as $n \to \infty$. The representation (2.10) is not necessarily unique as the assignment when the path is sent back to an edge or corner is not specified. It is not necessary to specify it. But, if desired, it can be specified in any measurable way.

Hölder continuity of the sample paths.

Lemma 2.4. [42, Theorem 2.9.25.] *The sample paths of a Wiener process on any interval $[0, T]$ are Hölder continuous with an exponent slightly less than 1/2, with an arbitrarily high probability, in the sense that, for any $\rho > 0$*

$$P\left\{ \lim_{\delta \to 0} \sup_{0<s<t<1, t-s \le \delta} \frac{|w(t) - w(s)|}{\delta^{.5-\rho}} > 0 \right\} = 0.$$

By a time change argument, this extends to any stochastic integral with a bounded integrand. Hence it holds for the unreflected process.

The first sentence in the next theorem is the remark in [8, bottom of page 124]. The second sentence is a consequence of the first and the criterion in Theorem 2.1.7. The theorem implies that for each $\epsilon > 0$ there is a compact set $K_\epsilon \subset C(G; [0, 1])$ such that for all $T < \infty$,

$$\inf_{\bar{x}(0), \bar{u}(0)} P\{(x(T + s), s \in [0, 1]) \in K_\epsilon\} \ge 1 - \epsilon.$$

Theorem 2.5. *A sequence of continuous processes that is tight in $D(G; [0, 1])$ is also tight in $C(G; [0, 1])$. Let $x^n(\cdot)$ be a sequence of solutions to any of the stochastic differential equations that we have considered in this chapter. Then the sequence $\{x^n(l + \cdot); l = 1, 2, \ldots, n < \infty\}$ on $[0, 1]$ is tight in $C(G; [0, 1])$.*

[3] For the model (2.8) and the models that arise as representations of the numerical approximations, the process might jump out of G by a small amount, say at most by ϵ_n that goes to zero as $n \to \infty$. Then the reflection back to G occurs at a point that is not in G. By the extension in (A2.2) the form (2.10) should be $\sum_i d_i^n(t) y_i^n(t)$, where $d_i^n(t) \to d_i$ as $n \to \infty$. But we use the form (2.10) for simplicity. The modifications in the development with the more general form are trivial.

3.2.2 Delayed Control, Reflection Term, and/or Wiener Process

In many applications delayed values of the reflection term and/or the Wiener process also appear in the dynamics. This causes serious problems with the memory requirements for the numerical algorithms. The following model is used for the numerical approximations developed in Chapter 9. Those methods help to reduce the memory requirements for such problems as well as for the problem with delayed controls. All functions of θ have value zero for $\theta < -\bar{\theta}$ and $\theta > 0$.

Definition. Let $\mu_a(\cdot)$ be a distribution function on the interval $[-\bar{\theta} - \epsilon_0, \epsilon_0]$, for any $\epsilon_0 > 0$, and with $\mu_a(\cdot)$ being left-continuous. The value of ϵ_0 is unimportant and is introduced only to simplify the notation for the increments at the end points of $[-\bar{\theta}, 0]$. The mass on the end intervals $[-\bar{\theta} - \epsilon_0, -\bar{\theta})$ and $(0, \epsilon_0]$ is zero. Define the differential $d\mu_a(\theta) = \mu_a(\theta + d\theta) - \mu_a(\theta)$.

The model is

$$\begin{aligned} dx(t) &= c(x(t), u(t))dt \\ &\quad + dt \int_{-\bar{\theta}}^{0} b(x(t+\theta), u(t+\theta), \theta) d\mu_a(\theta) + \sigma(x(t))dw(t) + dz(t) \\ &\quad + dt \int_{\theta=-\bar{\theta}}^{0} p(x(t+\theta), \theta) d_\theta y(t+\theta). \end{aligned} \tag{2.11}$$

The last integral is with respect to θ in the sense that

$$p(x(t+\theta), \theta) d_\theta y(t+\theta) = p(x(t+\theta), \theta)\left[y(t+\theta+d\theta) - y(t+\theta)\right].$$

We suppose that $z(t) = 0$ for $t \le 0$. As with (1.7) or (2.4), the initial condition is the pair $\hat{x} = \{x(s), -\bar{\theta} \le s \le 0\}$, $\hat{u} = \{u(s), -\bar{\theta} \le s \le 0\}$. One could incorporate the term containing $b(\cdot)$ into the term containing $c(\cdot)$. But we prefer to use both terms, with $b(\cdot)$ being used to represent the component of the drift with delayed arguments, and $c(\cdot)$ being used to represent the component without delays.

The relaxed control form of (2.11) is

$$\begin{aligned} dx(t) &= dt \int_U c(x(t), \alpha) r'(d\alpha, t) \\ &\quad + dt \int_{-\bar{\theta}}^{0} \int_U b(x(t+\theta), \alpha, \theta) r'(d\alpha, t+\theta) d\mu_a(\theta) + \sigma(x(t))dw(t) \\ &\quad + dz(t) + dt \int_{\theta=-\bar{\theta}}^{0} p(x(t+\theta), \theta) d_\theta y(t+\theta). \end{aligned} \tag{2.12}$$

For the boundary absorption case, where there is no reflection term, simply omit the terms containing $z(\cdot)$ and $y(\cdot)$.

The models (2.11) or (2.12) include drift terms such as $f_0(x(t-\theta_1)) + f_1(u(t-\theta_2))$, where the delays θ_1 and θ_2 are equal or unequal. We can also use the form

$$\sum_{i=1}^{L} \int_{-\bar{\theta}}^{0} b_i(x(t+\theta), u(t+\theta), \theta) d\mu_{ai}(\theta) \tag{2.13}$$

(with the analogous replacement in the relaxed control form (2.12)), where the $b_i(\cdot)$ and $\mu_{ai}(\cdot)$ satisfy the conditions to be imposed on $b(\cdot)$ and $\mu_a(\cdot)$, resp.

In models arising in communications where there is a delayed component of the reflection process, it arises due to a transportation delay. This motivates (A2.4) below. In any case, the assumption is needed for technical reasons, to ensure that the solution processes are well defined.

A2.4. *There is* $\theta_0 \in [-\bar{\theta}, 0)$ *such that* $p(x,\theta) = 0$ *for* $\theta \geq \theta_0$.

A2.5. *The functions* $b(\cdot), c(\cdot), p(\cdot), c(\cdot)$, *and* $\sigma(\cdot)$, *are bounded and continuous. The distribution function* $\mu_a(\cdot)$ *satisfies* $\mu_a(0+) - \mu_a(\theta) \to 0$ *as* $\theta \uparrow 0$. *The function* $k(\cdot)$ *that will appear in the cost function satisfies the condition on* $b(\cdot)$. *We suppose that* $z(t) = 0$ *for* $t \leq 0$.

A2.6. *There is a unique weak sense solution to* (2.12) *for each initial condition and admissible control.*

Comment on the delayed reflection term in (2.11). Consider a one-dimensional problem. Let $p(x,\theta) = 1/\delta$ on the interval $[-\Delta-\delta, -\Delta]$, $\Delta > 0, \delta > 0$, with value zero elsewhere. Then, with $y(t) = 0, t \leq 0$, we have

$$\int_0^t ds \int_{-\bar{\theta}}^{0} p(\theta) d_\theta y(s+\theta) = \frac{1}{\delta} \int_{t-\Delta-\delta}^{t-\Delta} y(s) ds \approx y(t-\Delta) \tag{2.14}$$

for small δ. In this way, point delays as well as distributed delays can be approximated.

Extensions. One can add the delayed Wiener process term

$$dt \int_{-\bar{\theta}}^{0} p_w(x(t+\theta), \theta) d_\theta w(t+\theta), \tag{2.15}$$

where $p_w(\cdot)$ satisfies the conditions on $p(\cdot)$. We can also use the form

$$\left[\int_{-\bar{\theta}}^{0} \sigma(x(t+\theta), \theta) \mu_\sigma(d\theta)\right] dw(t), \tag{2.16}$$

where $\mu_\sigma(\cdot)$ has a bounded derivative with respect to Lebesgue measure, for the noise term in (2.11) or (2.12), with little additional difficulty for the numerical problem. The term $c(x(t), u(t))$ can be replaced by the form $c(\bar{x}(t), u(t))$, which uses the path memory segment, but this would increase the memory requirements considerably. See Chapter 9, where the numerical approximations for (2.11) and (2.12) are done.

3.2.3 Neutral Equations

The numerical algorithms will also apply to models of the form

$$d\left[x(t) - F(\bar{x}(t))\right] = \int_U b(\bar{x}(t), \alpha) r'(d\alpha, t) dt + \sigma(\bar{x}(t)) dw(t) + dz(t),$$

where $F(\cdot)$ is a continuous $I\!R^r$-valued function on $C(G; [-\bar{\theta}, 0])$. Such forms are known as *neutral equations*. Rewrite in integral form:

$$\begin{aligned} x(t) - F(\bar{x}(t)) &= x(0) - F(\bar{x}(0)) \\ &+ \int_0^t \int_U b(\bar{x}(s), \alpha) r'(d\alpha, s) ds + \int_0^t \sigma(\bar{x}(s)) dw(s) + z(t). \end{aligned} \tag{2.17}$$

If the control is delayed as well, then use the form (2.4) with $\bar{b}(\cdot)$. The memory segments at time t are still on the interval $[t - \bar{\theta}, t]$. See [68, Chapter 6] for a discussion of the existence and uniqueness of solutions for the model without boundary reflections. For the deterministic form, various applications and an analysis of the stability problem can be found in [77], and a development of the general theory is in [27].

We will suppose that $F(\cdot)$ satisfies the following "gap" condition. The other conditions that apply to (2.3) or (2.4) are assumed as well. If the boundary is absorbing and not reflecting, then drop $z(\cdot)$ and the assumptions on the reflection directions.

A2.7. *There is* $0 < \bar{\theta}_1 < \bar{\theta}$ *such that* $F(\hat{x})$ *depends only on the part of* $\hat{x}$ *on* $[-\bar{\theta}, -\bar{\theta}_1]$.

3.2.4 Controlled Variance and Jumps

All of the models in [58] can be handled with delays in the dynamics. However, in this book, for the sake of simplicity, we will not deal with models subject to (controlled or uncontrolled) jump driving forces or with controlled variance. The main new issues when there are delays are the same as those for the models in the previous subsections. The algorithmic development and proofs of convergence would be adaptations of those in [58, Chapter 13] to the models with delays, along the lines of what is done in this book for the models of the previous subsections of this section. We will write simple forms of the possible models so that the full range of control problems can be seen.

Poisson jumps. Let $\{\mathcal{F}_t, t < \infty\}$ be a filtration on a probability space, with $w(\cdot)$ a standard vector-valued $\mathcal{F}_t$-Wiener process and $N(\cdot)$ an $\mathcal{F}_t$-Poisson random measure with intensity measure $h(dt\,d\rho) = \lambda dt \times \Pi(d\rho)$, where $\Pi(\cdot)$ is a measure on a compact set Γ. Let $q(\cdot)$ be a bounded measurable function. Then, with ordinary controls used, the analog of (1.1) is

$$\begin{aligned} x(t) = x(0) &+ \int_0^t b(\bar{x}(s), u(s))ds \\ &+ \int_0^t \sigma(\bar{x}(s))dw(s) + \int_0^t \int_\Gamma q(\bar{x}(s-), \rho)N(ds\,d\rho). \end{aligned} \tag{2.18}$$

There are obvious analogs of the other forms in Sections 1 and 2.

Controlled variance. The analog of (1.1) with controlled variance is

$$x(t) = x(0) + \int_0^t b(\bar{x}(s), u(s))ds + \int_0^t \sigma(\bar{x}(s), u(s))dw(s). \tag{2.19}$$

A reflecting boundary satisfying (A2.1) and (A2.2) and associated reflection term $z(\cdot)$ can be added in all cases and the controls can be delayed as well, as in (2.3) or (2.4).

3.3 The Girsanov Transformation

The Girsanov transformation method for obtaining weak-sense existence and uniqueness is widely used in control problems, particularly where we wish to use feedback controls $u(x)$ that are not Lipschitz continuous, or where $b(\cdot)$ is not Lipschitz continuous. See [42] and the references therein. Given a solution (either in the weak or strong sense) to the uncontrolled forms of any of the dynamical equations (1.3), (1.7), (2.3), (2.4), (2.11) or (2.12), the Girsanov transformation method defines another solution, with a controlled drift term that we are able to choose, by simply transforming the probability measure P on the original probability space to a new measure, denoted by $\tilde{P}$. The new drift term contains the desired control. The key is the construction of the Radon–Nikodym derivative of the new measure with respect to P. The procedure will be particularly useful in Chapter 5 for the ergodic cost problem.

As usual, let $(\Omega, \mathcal{F}, P)$ be the probability space, with filtration $\{\mathcal{F}_t, t \geq 0\}$ and $\mathcal{F} = \cup_t \mathcal{F}_t$. Let $w(\cdot)$ be an m-dimensional standard $\mathcal{F}_t$-Wiener process. Let $v(\cdot)$ be an m-dimensional bounded and adapted (to $\mathcal{F}_t$) process. Define

$$R(t) = \exp\left(\int_0^t v'(s)dw(s) - \frac{1}{2}\int_0^t |v(s)|^2 ds \right). \tag{3.1}$$

By Itô's Lemma,

$$R(t) = 1 + \int_0^t R(s)v'(s)dw(s).$$

In fact [34, 42, 80], for $s > 0, t \geq 0$,

$$ER(t) = 1, \quad E\left[R(t+s)\big|v(u), w(u), u \le t\right] = R(t), \text{ w.p.1.}$$

Thus the process $R(\cdot)$ is an $\mathcal{F}_t$-martingale.

For $0 < T < \infty$, define the probability measure $\tilde{P}_T$ on $(\Omega, \mathcal{F}_T)$ by the Radon–Nikodym derivative $d\tilde{P}_T/dP = R(T)$. That is,

$$\tilde{P}_T(A) = E\left[I_A R(T)\right], \text{ for } A \in \mathcal{F}_T, \tag{3.2}$$

where I_A denotes the indicator function of the event A and the expectation is with respect to P. Because $ER(T) = 1$, $\tilde{P}_T$ is indeed a probability measure. There is a unique extension of the $\tilde{P}_T$, $T < \infty$, to a measure $\tilde{P}$ on $(\Omega, \mathcal{F})$. The next theorem gives a key consequence.

Theorem 3.1. [42, 65, 80]. *Assume that $R(\cdot)$ defined by equation (3.1) is a martingale (which is assured here as $v(\cdot)$ is bounded). Then*

$$\tilde{w}(t) = w(t) - \int_0^t v(s)ds \tag{3.3}$$

is a standard m dimensional $\mathcal{F}_t$-Wiener process on the probability space $(\Omega, \mathcal{F}, \tilde{P})$.

Application. In control theory this result is often used in the following way. Let $w_1(\cdot)$ and $w_2(\cdot)$ denote the first m and last $k-m$ components of a k-dimensional Wiener process $w(\cdot)$, respectively, and define $x_1(\cdot)$ and $x_2(\cdot)$ similarly, with $x(\cdot) = (x_1(\cdot), x_2(\cdot))$. Let $\sigma_1(\cdot)$ be an $m \times m$ matrix-valued function with the property that the inverse $\sigma_1^{-1}(x)$ exists and is uniformly bounded in x. Suppose that the stochastic differential equation

$$\begin{aligned} dx_1(t) &= b_1(x(t))dt + \sigma_1(x(t))dw_1, \\ dx_2(t) &= b_2(x(t))dt + \sigma_2(x(t))dw_2, \end{aligned} \tag{3.4}$$

has a unique weak-sense solution with the given initial condition.[4] Let $\tilde{b}_1(\cdot)$ be a bounded, measurable, and $I\!R^m$-valued function. Define

$$q_1(\cdot) = \sigma_1^{-1}(x(\cdot))\tilde{b}_1(x(\cdot)).$$

Define $R(\cdot)$, $\tilde{P}_T$, and $\tilde{P}$ as in (3.1) and (3.2) above, but using $(q_1(\cdot), w_1(\cdot))$ in lieu of $(v(\cdot), w(\cdot))$, and define

$$\tilde{w}_1(t) = w_1(t) - \int_0^t q_1(s)ds.$$

Then, under $\tilde{P}$, the process $x(\cdot)$ satisfies

[4] The development is identical if $b(\cdot)$ depends on the control, or in the delay case, or if there is a reflection term.

$$\begin{aligned} dx_1(t) &= b_1(x(t))dt + \tilde{b}_1(x(t))dt + \sigma_1(x(t))d\tilde{w}_1(t), \\ dx_2(t) &= b_2(x(t))dt + \sigma_2(x(t))dw_2(t), \end{aligned} \tag{3.5}$$

and $(\tilde{w}_1(\cdot), w_2(\cdot))$ is a standard $\mathcal{F}_t$-Wiener process.

The process $\tilde{b}_1(x(\cdot))$ can be replaced by the more general form $b_c(x(t), u(t))$, where $b_c(\cdot)$ is bounded and measurable, $u(t)$ is U-valued, and $(u(\cdot), w(\cdot))$ is an admissible pair. The solution to (3.5) is weak-sense unique (under the new measure $\tilde{P}$) [34, 42, 52], as the original solution to (3.4) is weak-sense unique.

3.4 Cost Functions

In order not to overburden the text, the development of the numerical methods in the following chapters will concentrate on the discounted and ergodic cost functionals, with either boundary absorption or reflection. All of the standard cost functionals in [58] can be handled. In this section, we will describe the discounted cost functions for the models of Section 2, and results concerning the existence of optimal controls are given in the next section. The impulsive and singular control models are discussed in Section 6. The ergodic cost functional will be treated in Chapter 5.

The discounted cost function: Absorbing boundary. Consider the case where the boundary is absorbing. For a continuous function $\phi(\cdot)$, define the set $G = \{x : \phi(x) \le 0\}$ with nonempty interior $G^0 = \{x : \phi(x) < 0\}$, and suppose that G bounded and is the closure of G^0. The control process stops at the time $\tau_G = \inf_t\{x(t) \notin G^0\} = \inf_t\{x(t) \in \partial G\}$, the first time that the boundary ∂G of G is reached. If $x(t) \in G^0$ for all $t < \infty$, then define $\tau_G = \infty$. We will use the following assumptions.

A4.1. *The real-valued function* $g_0(\cdot)$ *in* (4.1) *is bounded and continuous. Whatever the case, the real-valued function* $k(\cdot)$ *satisfies the conditions on* $b(\cdot)$.

A4.2. *For each control, and w.p.1,* $\inf_t\{x(t) \notin G^0\} = \inf_t\{x(t) \notin G\}$.

Assumption (A4.2) is critical for the convergence of the numerical procedures when the process is to stop on first reaching a boundary. It ensures that (w.p.1) the paths are not tangent to the boundary at the first time that they reach it. Because all numerical procedures involve, either explicitly or implicitly, approximations of the original model, we need to ensure that the first hitting times of the approximations converge to that of the original model, and (A4.2) ensures this. The condition is discussed further in [58, Section 10.2], where it is shown how to avoid it by a simple and unrestrictive approximation of "smoothing" the absorption near the boundary.

Let $\beta > 0$ and let $E_{\hat{x}}^r$ denote the expectation under the initial condition $\hat{x} = \bar{x}(0)$, when the relaxed control $r(\cdot)$ is used on $[0, \infty)$. Then when the path only is delayed the discounted cost function is

$$W(\hat{x}, r) = E_{\hat{x}}^{r} \int_0^{\tau_G} \int_U e^{-\beta t} k(\bar{x}(t), \alpha) r'(d\alpha, t) dt + E_{\hat{x}}^{r} e^{-\beta \tau_G} g_0(x(\tau_G)). \quad (4.1)$$

Define[5]

$$V(\hat{x}) = \inf_r W(\hat{x}, r),$$

where the inf is over the admissible relaxed controls.

For the model (1.7), in which both the path and control are delayed, we need to specify the segment of the control on $[-\bar{\theta}, 0]$. This might be given in either ordinary or relaxed control form. In either case, let $\hat{r}$ denote the relaxed control representation of the initial control segment. In our terminology, $\hat{r} = \bar{r}(0)$. The discounted cost function corresponding with the model (1.7) is

$$\begin{aligned} W(\hat{x}, \hat{r}, r) = E_{\hat{x},\hat{r}}^{r} \int_0^{\tau_G} \int_{-\bar{\theta}}^{0} \int_U e^{-\beta t} \left[k(\bar{x}(t), \alpha, \theta) r'(d\alpha, t+\theta) \mu_c(d\theta) dt \right] \\ + E_{\hat{x},\hat{r}}^{r} e^{-\beta \tau_G} g_0(x(\tau_G)), \end{aligned} \quad (4.2)$$

where $E_{\hat{x},\hat{r}}^{r}$ is the expectation given initial conditions $\hat{x} = \bar{x}(0), \hat{r} = \bar{r}(0)$, and the use of relaxed control $r(\cdot)$ on $[0, \infty)$. Define

$$V(\hat{x}, \hat{r}) = \inf_r W(\hat{x}, \hat{r}, r),$$

where the infimum is over all relaxed controls with initial segments $\bar{r}(0) = \hat{r}$. Recall that, in our notation, for $\theta < 0$, $r'(d\alpha, t+\theta)dt = r(d\alpha, dt+\theta)$.

Optimal stopping. Suppose that we have the option of stopping the process at a time τ that can be no later than the first time τ_G that the process hits the boundary ∂G. Then τ is another control variable and the cost is either (4.1) or (4.2) with τ_G replaced by $\min\{\tau_G, \tau\}$ and, for (4.2), $V(\hat{x}, \hat{r}) = \inf_{r,\tau} W(\hat{x}, \hat{r}, r, \tau)$. In optimal stopping problems it is frequently the case that there is no control $u(\cdot)$.

Discounted cost function: Reflecting boundary. For the reflecting boundary model of Section 2, where (A2.1) and (A2.2) are used, the process $x(\cdot)$ never stops and we set $\tau_G = \infty$ in the cost function and put a weight

[5] Because we are working with weak-sense solutions, the Wiener process might not be fixed. For example, if Girsanov measure transformation methods are used, then the Wiener process will depend on the control. Then the inf in (4.1) should be over all admissible pairs $(r(\cdot), w(\cdot))$, with the given initial data. But to simplify the notation, we write simply $\inf_m$. This is essentially a theoretical issue. The numerical procedures give feedback controls and all that we need to know is that there is an optimal value function to which the approximating values will converge.

on the reflection terms. For a vector $q = \{q_i, i = 1, \ldots\}$, the discounted cost function analog of (4.1) is

$$W(\hat{x}, r) = E_{\hat{x}}^r \int_0^\infty \int_U e^{-\beta t} \left[k(\bar{x}(t), \alpha) r'(d\alpha, t) dt + q' dy(t) \right]. \tag{4.3}$$

The term $q'dy(t)$ can be used to penalize behavior on the boundary. For example, in models of communications systems, where a boundary denotes a buffer capacity, the reflection from that boundary denotes a buffer overflow that we might wish to penalize. The boundary penalty is also useful for the modeling of a "numerical" boundary. Suppose that the reflecting boundary is added simply to bound the state space for numerical purposes. One then tries to select the reflection direction so that the process is concentrated in the region of greatest importance. Adding an appropriate boundary penalty can be helpful in this regard.

The analog of (4.2) is

$$\begin{aligned} &W(\hat{x}, \hat{r}, r) \\ &= E_{\hat{x},\hat{r}}^r \int_0^\infty \int_{-\bar{\theta}}^0 \int_U e^{-\beta t} \left[k(\bar{x}(t), \alpha, \theta) r'(d\alpha, t+\theta) \mu_c(d\theta) dt + q' dy(t) \right]. \end{aligned} \tag{4.4}$$

For the model (2.12), replace (4.4) by

$$\begin{aligned} &W(\hat{x}, \hat{r}, r) \\ &= E_{\hat{x},\hat{r}}^r \int_0^\infty \int_{-\bar{\theta}}^0 \int_U e^{-\beta t} \left[k(x(t+\theta), \alpha, \theta) r'(d\alpha, t+\theta) d\mu_a(\theta) dt + q' dy(t) \right]. \end{aligned} \tag{4.5}$$

If the reflection term is delayed, then the cost function will also depend on the initial segment $\hat{z}$ of the reflection term. But we generally suppose that $z(t) = 0, t \leq 0$.

The $y_i(\cdot)$ in the representation $z(t) = \sum_i d_i y_i(t)$ are not necessarily defined uniquely if the process $x(\cdot)$ hovers around an edge or corner of G. This is not a problem if the noise does not allow such hovering, for example if

$$\inf_{\hat{x}} n_i' \sigma(\hat{x}) \sigma'(\hat{x}) n_i > 0 \tag{4.6}$$

for all faces i with $q_i \neq 0$, where n_i is the interior normal to the ith face [56, Theorem 4.2.1]. To handle the general problem, we use the following assumption, which will also ensure that the limit of the sequence of cost functionals for the numerical approximations is the cost for the limit processes.

A4.3. *Either the set of directions d_i for the faces meeting at any edge or corner are linearly independent or, if not, then the coefficients q_i are identical for the components of any linearly dependent (at any edge or corner) subset.* [In the latter case, it is only the sum of the relevant components $y_i(\cdot)$ that matters in the cost function.]

3.5 Existence of an Optimal Control

The proofs of convergence of the numerical procedures are facilitated by knowing that there is an optimal control. We will discuss the proof for the models of Section 1 and Subsections 2.1 and 2.2. The proof will illustrate one way of using the martingale method introduced at the end of Section 2.2 for identifying the limit of a sequence of processes as a diffusion and will serve as a template for the convergence proofs for the numerical methods in Chapters 7–9.

3.5.1 Reflecting or Absorbing Boundary: Discounted Cost

Theorem 5.1 establishes the existence of an optimal relaxed control for the discounted cost function and a reflecting or absorbing boundary. The proof of existence closely follows the standard procedure for the no-delay problem, say that of [58, Theorem 10.2.1]. We will outline the procedure and concentrate on the differences. See also [31, Theorem 2.1]. Theorem 5.2 says that the cost functionals are continuous in the initial conditions, uniformly in the control, and the proof is nearly identical to that of Theorem 5.1. Theorem 5.3 asserts that the use of relaxed controls does not affect the infimum of the costs. The no-delay form is [58, Theorem 10.1.2] and the proof is omitted because the adjustments for the current case are readily made, given the comments in the proof of Theorem 5.1. There is currently no proof that the optimal value function satisfies a Bellman equation. But this is not needed for the convergence of the numerical approximations. Theorem 5.4 will be of technical assistance in proving the convergence of the numerical algorithms in Chapters 7–9. The described controls are not for practical use. In the proof of Theorem 5.1, it is supposed that the sequence of reflection processes is tight. The proof of this fact is outlined in Theorem 5.5, and the method will be very useful when dealing with the singular control model in Section 6.

Theorem 5.1 *Use the model* (2.4) *with fixed initial condition* $\hat{x}, \hat{r}$*, where* $\hat{x}$ *is continuous on* $[-\bar{\theta}, 0]$ *and use the cost functional* (4.4). *Assume* (A1.2), (A1.3), (A2.1)–(A2.3), and (A4.3). *Then there is an optimal control. That is, there is a probability space on which is defined a set of processes* $(x(\cdot), w(\cdot), r(\cdot), z(\cdot))$ *solving* (2.4), *where* $(x(\cdot), r(\cdot), z(\cdot))$ *is nonanticipative with respect to the Wiener process* $w(\cdot)$, $\bar{r}(0) = \hat{r}$, $\bar{x}(0) = \hat{x}$, *and* $W(\hat{x}, \hat{r}, r) = V(\hat{x}, \hat{r})$. *If the control is not delayed so that the model is* (2.1), *then with* (A1.1) *and* (A1.4) *replacing* (A1.3) *and* (A2.3), *the conclusions continue to hold.*

Now assume the model (2.11) *with associated conditions* (A2.4)–(A2.6). *Then the conclusions continue to hold. Suppose that control stops when the boundary is reached. Use the cost function* (4.2), *drop* (A2.1), (A2.2), *and* (A4.3), *and assume* (A4.1), (A4.2). *Then the conclusions continue to hold.*

Proof. The proof will be given for (2.4) only. The proof for (2.11) is similar. Proofs of existence of optimal controls start by taking a minimizing sequence of controls, and showing that, for some subsequence, there is a limit control that is optimal. Because we are using weak-sense solutions, the probability space and driving Wiener process might depend on the control. This is only a technical matter, and the dependence will be suppressed in the discussion and notation.

Let $r^n(\cdot)$ be a minimizing sequence[6] of relaxed controls, with associated standard Wiener processes $w^n(\cdot)$, solutions $x^n(\cdot)$, reflection processes $z^n(\cdot)$, with initial conditions $\bar{x}^n(0) = \hat{x}$ and $\bar{r}^n(0) = \hat{r}$. That is, $x^n(\cdot)$ satisfies

$$x^n(t) = \hat{x}(0) + \int_{-\bar{\theta}}^{0} \left[\int_0^t \int_U b(\bar{x}^n(s), \alpha, \theta) r^n(d\alpha, ds + \theta) \right] \mu_c(d\theta) + \int_0^t \sigma(\bar{x}^n(s)) dw^n(s) + z^n(t). \tag{5.1}$$

There are nonanticipative, continuous, and nondecreasing processes $y_i^n(\cdot)$ such that $z^n(\cdot) = \sum_i d_i y^n(\cdot)$, so that we can work with either $z^n(\cdot)$ or $y^n(\cdot)$.

The $w^n(\cdot)$ are all standard Wiener processes, hence the set is tight, and any weak-sense limit is a standard Wiener process. The sequence of relaxed controls $r^n(\cdot)$ is always tight and any weak-sense limit has continuous paths, w.p.1. The processes defined by the ordinary and stochastic integral terms of (5.1) are also tight, and all weak-sense limits are continuous. The tightness and asymptotic continuity of the sequence $z^n(\cdot)$ and the continuity of any weak-sense limit follows from Theorem 5.5.[7]

Now, take a weakly convergent subsequence of $(x^n(\cdot), r^n(\cdot), z^n(\cdot), w^n(\cdot))$ with limit denoted by $(x(\cdot), r(\cdot), z(\cdot), w(\cdot))$, and index it by n also. Use the Skorohod representation Theorem 2.1.6 (see also [23, page 102]), so that we can assume that the convergence is w.p.1 in the topologies of the spaces of concern. By the weak convergence, we must have $x(t) \in G$, and that $z(\cdot)$ can change only at t where $x(t)$ is on ∂G. By the weak convergence and Skorokhod representation, we can suppose that $\sup_{s \le t} |x^n(s) - x(s)| \to 0$ for each $t > 0$. Then (A1.3) implies that, for all $\theta \in [-\bar{\theta}, 0]$ and $t \geq 0$, w.p.1,

$$\sup_{s \le t,\, \alpha} |b(\bar{x}^n(s), \alpha, \theta) - b(\bar{x}(s), \alpha, \theta)| \to 0,$$

and also for $\sigma(\cdot)$ and $k(\cdot)$ replacing $b(\cdot)$. The last sentence and the continuity and boundedness assumptions (A1.3) yield, w.p.1,

[6] As noted above and in Section 4, it is the pair $(r^n(\cdot), w^n(\cdot))$ of (control, Wiener process) that determines the cost. Thus a cost function $W(\hat{x}, \hat{r}, r)$ depends on the distribution of the pair $(r(\cdot), w(\cdot))$ where $w(\cdot)$ is the Wiener process that is used, together with the relaxed control $r(\cdot)$, to get the solution. For notational simplicity, we omit the $w(\cdot)$ argument because what is being defined is clear.

[7] The methods that are used for proving the tightness and asymptotic continuity of the reflection terms will be of use when dealing with the singular control problem as well, so we prefer to separate them out.

$$\int_0^t \int_U b(\bar{x}^n(s), \alpha, \theta) r^n(d\alpha, ds + \theta) \to \int_0^t \int_U b(\bar{x}(s), \alpha, \theta) r(d\alpha, ds + \theta)$$

for all $t \geq 0$ and $\theta \in [-\bar{\theta}, 0]$, w.p.1. From this and the dominated convergence theorem it follows that the first integral in (5.1) converges to the process obtained when the superscript n is dropped. The same argument shows that the first integral in the cost function (4.4), with $x^n(\cdot)$ and $r^n(\cdot)$ used, converges to the value for the limit.

Nonanticipativity of the limit processes with respect to the limit Wiener process $w(\cdot)$ is shown as follows, again using a standard method. Let $\phi_j(\cdot), j \leq J$, be continuous functions with compact support and write

$$\langle r, \phi_j \rangle (t) = \int_0^t \int_U \phi_j(\alpha, s) r(d\alpha\, ds).$$

For arbitrary $t > 0$ and integer $I > 0$, let $s_i \leq t$ for $i \leq I$, and let $h(\cdot)$ be an arbitrary bounded and continuous function. By the nonanticipativity, for each n and $T > 0$,

$$\begin{aligned} &Eh\left(x^n(s_i), z^n(s_i), w^n(s_i), \langle r^n, \phi_j \rangle (s_i), i \leq I, j \leq J\right) \\ &\quad \times \left(w^n(t+T) - w^n(t)\right) = 0. \end{aligned} \tag{5.2}$$

By the weak convergence, the continuity of the limit processes, and the uniform integrability of $\{w^n(t+T) - w^n(t)\}$, (5.2) holds with the superscript n dropped. Now, by Theorem 2.2.1, the arbitrariness of $h(\cdot), I, J, s_i, t,$ and $\phi_j(\cdot)$, implies that $w(\cdot)$ is a martingale with respect to the σ-algebra generated by $(x(\cdot), w(\cdot), z(\cdot), r(\cdot))$. Hence the limit processes are nonanticipative with respect to $w(\cdot)$.

The convergence of the stochastic integral is obtained by an approximation argument. For a measurable function $f(\cdot)$ and $\kappa > 0$, let $f_\kappa(\cdot)$ be the approximation that takes the value $f(n\kappa)$ on $[n\kappa, n\kappa + \kappa)$. Then, by the weak convergence and the continuity properties of $\sigma(\cdot)$ in (A1.2), for each κ and as $n \to \infty$,

$$\int_0^t \sigma(\bar{x}_\kappa^n(s)) dw^n(s) \to \int_0^t \sigma(\bar{x}_\kappa(s)) dw(s). \tag{5.3}$$

To get the limit of the stochastic integrals $\int_0^t \sigma(\bar{x}^n(s)) dw^n(s)$, we first bound the mean square value $E\int_0^t \left[\sigma(\bar{x}_\kappa^n(s)) - \sigma(\bar{x}^n(s))\right]^2 ds$ by a constant times

$$\begin{aligned} E\int_0^t \left[\sigma(\bar{x}_\kappa^n(s)) - \sigma(\bar{x}_\kappa(s))\right]^2 ds + E\int_0^t \left[\sigma(\bar{x}_\kappa(s)) - \sigma(\bar{x}(s))\right]^2 ds \\ + E\int_0^t \left[\sigma(\bar{x}^n(s)) - \sigma(\bar{x}(s))\right]^2 ds. \end{aligned} \tag{5.4}$$

Because $x(\cdot)$ is continuous w.p.1 and $x^n(\cdot) \to x(\cdot)$ uniformly on any bounded time interval w.p.1, the continuity properties of $\sigma(\cdot)$ imply that the first and third integrals go to zero as $n \to \infty$. The continuity properties of $\sigma(\cdot)$ and $x(\cdot)$ imply that the second integral goes to zero as $\kappa \to 0$. Hence the left-hand term of (5.3) approximates $\int_0^t \sigma(\bar{x}^n(s))dw^n(s)$ uniformly in n and in t on any bounded interval, for small κ. The convergence of the second integral of (5.4) and the nonanticipativity of $x(\cdot)$ imply the convergence (uniformly, w.p.1 on any bounded time interval) of the right-hand term in (5.3) to $\int_0^t \sigma(\bar{x}(s))dw(s)$. It follows from Theorem 5.5 that $z(\cdot)$ is the reflection process for $x(\cdot)$.

Factor the limit process $z(\cdot)$ as $z(\cdot) = \sum d_i y_i(\cdot)$. By Lemma 2.1, $\{y^n(t+T) - y^n(t); t < \infty, n < \infty\}$ is uniformly integrable for any $T < \infty$ and so are the increments of $z(\cdot)$ and $y(\cdot)$. This, the weak convergence, and the condition (A4.3), imply that component of the cost due to the reflection terms also converges to the value for the limit process. Finally, by the minimizing property of the sequence $r^n(\cdot)$, $W(\hat{x}, \hat{r}, r^n) \to W(\hat{x}, \hat{r}, r) = V(\hat{x}, \hat{r})$, the infimum of the costs.

Now turn to the case where the boundary is absorbing. The only new concern is the convergence of

$$W(\hat{x}^n, \hat{r}^n, r^n) = E^{r^n}_{\hat{x}^n,\hat{r}^n} \int_{-\bar{\theta}}^0 \left[\int_0^{\tau_G^n} \int_U e^{-\beta t} k(\bar{x}^n(t), \alpha, \theta) r^{n,\prime}(d\alpha, t+\theta) dt \right] \mu_c(d\theta)$$
$$+ E^{r^n}_{\hat{x}^n,\hat{r}^n} e^{-\beta \tau_G^n} g_0(x^n(\tau_G)),$$

where τ_G^n is the first escape time of $x^n(\cdot)$ from G^0. The terminal costs converge due to the weak convergence and the continuity of $g_0(\cdot)$, and the integral converges due to weak convergence and the condition (A4.2). ■

The next theorem is implied by the proof of Theorem 5.1.

Theorem 5.2. *Assume the conditions of Theorem 1.1, but let $(\bar{x}^n(0), \bar{r}^n(0))$ converge weakly to $(\hat{x}, \hat{r})$, where $\hat{x}(\cdot)$ is continuous. Then $V(\bar{x}^n(0), \bar{r}^n(0)) \to V(\hat{x}, \hat{r})$.*

The next theorem asserts that the use of relaxed controls does not change the minimal values. See [58, Theorem 10.1.2] for the no-delay case. The proof depends on the fact that for any relaxed control $r(\cdot)$ one can find a sequence of ordinary controls $u^n(\cdot)$, each taking a finite number of values in U, such that $(\hat{x}, r^n(\cdot), w(\cdot))$ converges weakly to $(\hat{x}, r(\cdot), w(\cdot))$ where $r^n(\cdot)$ is the relaxed control representation of $u^n(\cdot)$.

Theorem 5.3. *Assume the conditions of Theorem 1.2. Then, with fixed initial conditions,*

$$\inf_u W(\hat{x}, \hat{r}, u) = \inf_r W(\hat{x}, \hat{r}, r),$$

where the $\inf_u$ *(resp.,* $\inf_r$*) is over all controls (resp., relaxed controls).*

An approximately optimal control. The next result, which is [58, 11.1.6] except for the more general initial condition, will be useful in the proof of convergence of the numerical approximations to the optimal value. It constructs an ϵ-optimal control of a special form for the original model. Its purpose is simply to facilitate the proof, and it has no value in applications. The theorem is essentially a consequence of the weak-sense uniqueness of the solution for each control and initial condition.

Theorem 5.4. *Assume the conditions of Theorem* 5.1. *For* $\epsilon_0 > 0$, *let* $(x(\cdot), z(\cdot), r(\cdot), w(\cdot))$ *be an* ϵ_0*-optimal process. Let* $\epsilon > 0$, $\delta > 0$, *and let* U_ϵ *be a finite set in* U. *For small enough* ϵ_0, δ, *and fine enough approximation* U_ϵ *to* U, *there is a probability space with an* ϵ*-optimal solution* $(x^\epsilon(\cdot), z^\epsilon(\cdot), u^\epsilon(\cdot), w^\epsilon(\cdot))$ *of the following form. The control* $u^\epsilon(t)$ *takes values in* U_ϵ *and is constant on intervals* $[n\delta, n\delta + \delta), n \geq 0$. *There is* $\rho > 0$ *such that the probability law of* $u^\epsilon(n\delta)$, *conditioned on the data* $\{x^\epsilon(s), z^\epsilon(s), w^\epsilon(s), s \leq n\delta; u^\epsilon(l\delta), l < n\}$ *depends only on the initial condition and the samples* $\{w^\epsilon(l\rho), l\rho \leq n\delta; u^\epsilon(l\delta), l < n\}$. *It is continuous in the* $w^\epsilon(\cdot)$ *arguments and in the initial condition* $\hat{x} \in C(G; [-\bar{\theta}, 0])$ *for each value of* $\{u^\epsilon(l\delta), l < n\}$.

Tightness and asymptotic continuity of the reflection terms. It is difficult to prove directly that $\{z^n(\cdot)\}$ is tight and has continuous limits. In [58, Chapter 11] and [56, Sections 2.6 and 3.6], a time-stretching argument is used. An alternative proof in these references is by contradiction. One supposes that for some $\epsilon > 0$ there is an asymptotic discontinuity of magnitude $\geq \epsilon > 0$ in $z^n(\cdot)$ and then shows that (A2.1) and (A2.2) imply that such a discontinuity implies that there is $\epsilon_1 > 0$ such that there is an asymptotic jump of $x^n(\cdot)$ into the interior of G at a distance $\geq \epsilon_1 > 0$ from the boundary. This contradicts the fact that a reflection can occur only on the boundary. The result will have other uses in the book, and it is useful to phrase it in a way that is more general than what is needed for the proof of Theorem 5.1. We will give the proof of [56, Section 3.6] which is based on a time-stretching argument.

Assume the system

$$x^n(t) = x^n(0) + F^n(t) + z^n(t), \tag{5.5}$$

where $z^n(\cdot)$ is a "reflection" process satisfying (A5.1) below. Suppose that the right-continuous process $F^n(\cdot)$ has a discontinuity at t such that $q^n(t) = x^n(t-) + (F^n(t) - F^n(t-)) \notin G$. Then the reflection process increment $(z^n(t) - z^n(t-))$ at t acts in one of the reflection directions at $q^n(t)$ in the sense of the following condition.

A5.1. $z^n(\cdot)$ *is a reflection process satisfying* (A2.1) *and* (A2.2) *except possibly at the times* t *where* $q^n(t) \notin G$. *The discontinuities of* $F^n(\cdot)$ *are less than* $\epsilon_n \to 0$ *in magnitude. Thus* $z^n(\cdot)$ *can change only at* t *where* $x^n(t)$ *is within a distance of* ϵ_n *from the boundary* ∂G. *Let* $d_\epsilon(x)$ *denote the set of reflection directions at all points on* ∂G *within a distance of* ϵ *from* $x \notin G$. *Let* $z^n(0) = 0$ *and suppose that there are* $\mu_n \to 0$ *such that for* $q^n(t) \notin G$,

$$z^n(t) - z^n(t-) \in cone(d_{\mu_n}(q^n(t))), \tag{5.6a}$$

where cone(A) is the positive cone generated by the convex hull of A, and for each $T < \infty$

$$\sup_{t \le T} |z^n(t) - z^n(t-)| \to 0 \text{ in probability as } n \to \infty. \tag{5.6b}$$

Note that $x^n(\cdot) - q^n(\cdot) \to 0$ and that the process $z^n(\cdot)$ might actually send the path slightly interior to G. By (A2.1), (A2.2), and (A5.1), $z^n(\cdot)$ can be written as

$$z^n(t) = \sum_i y_i^n(t) d_i, \tag{5.7}$$

where $y_i^n(0) = 0$ and the $y_i^n(\cdot)$ are nondecreasing and can increase only at t where $x^n(t)$ is within a distance ϵ_n of ∂G_i.

Theorem 5.5. *Assume the form* (5.5), *and that $F^n(\cdot)$ satisfies the following condition: for each $\nu > 0$ and $T > 0$,*

$$\lim_{\delta \to 0} \limsup_n P\left\{ \sup_{t \le T} \sup_{s \le \delta} |F^n(t+s) - F^n(t)| \ge \nu \right\} = 0. \tag{5.8}$$

Assume (A2.1), (A2.2), *that $\{x^n(0)\}$ is tight, and that $z^n(\cdot)$ satisfies* (A5.1). *Then $\{x^n(\cdot), z^n(\cdot)\}$ is tight, the limit $(x(\cdot), z(\cdot))$ of any weakly convergent subsequence is continuous with probability one, $x(t) \in G$, $z(\cdot)$ can change only at t where $x(t) \in \partial G$, and*

$$z(t+s) - z(t) \in \text{cone}\left\{ d(x(u)) : t \le u \le t+s \right\}. \tag{5.9}$$

Proof. Because the sequence $F^n(\cdot)$ is tight and asymptotically continuous, using the Skorokhod representation of Theorem 2.1.6, without loss of generality for purposes of the proof we can suppose that $F^n(\cdot)$ converges uniformly to a continuous function. Indeed, by working with sample functions, we can suppose that the $F^n(\cdot)$ are not random. If the asymptotic continuity assertion is false, then without loss of generality (taking a subsequence if necessary) we can suppose that there is an asymptotic jump (say, of size $\ge \nu_0 > 0$) in some component $y_i^n(\cdot)$ somewhere on an interval that we denote by $[0, T]$. By shifting the time origin, we can suppose that there are $\nu_0 > 0$ and $\delta_n \to 0$ such that $y_i^n(\delta_n) \ge \nu_0$ for large n and some nonempty set of indices i.

Now, to prove a contradiction, we need only consider the functions on the interval $[0, \delta_n]$. Because $\delta_n \to 0$, we can suppose that $F^n(t)$ converges uniformly to zero. The condition (A5.1) implies that the asserted jump in $y^n(\cdot)$ cannot take the path $x^n(\cdot)$ interior to the state space G, say more than some $\nu_n \to 0$ from the boundary, at least for large n. So the problem reduces to a consideration of the possible asymptotic behavior of $y^n(\cdot)$ on or arbitrarily close to the boundary. It will be seen that (A2.1) and (A2.2) prohibit such jumps on the boundary.

By taking a subsequence if necessary, we can suppose (without loss of generality) that $x^n(0)$ converges to a point $x(0) \in \partial G$. Now the timescale transformation method that will be introduced will show that there is a solution to the equation $x(t) = x(0) + z(t)$, where $x(0) \in \partial G$, and $z(t) = \sum_i y_i(t) d_i$, where the $y_i(\cdot)$ are the components of the process $z(\cdot)$, and at least one of the $y_i(t)$ is positive for $t > 0$.

To prove the assertion, start by defining

$$T^n(t) = t + \sum_i y_i^n(t) \tag{5.10}$$

and

$$\tilde{T}^n(t) = \inf\{s : T^n(s) > t\}. \tag{5.11}$$

Define the time-stretched processes $\tilde{y}^n(t) = y^n(\tilde{T}^n(t))$, and so forth. The time-transformed set satisfies

$$\tilde{x}^n(t) = x(0) + \tilde{F}^n(t) + \sum_i \tilde{y}_i^n(t) d_i,$$

where $\tilde{F}^n(\cdot)$ can be assumed to go to zero uniformly in t as $n \to \infty$, and $\tilde{y}_i^n(\cdot)$ can increase only when the distance between $\tilde{x}^n(t)$ and ∂G_i is no greater than ϵ_n. The time transformation stretches out any real-valued component of the processes so that the distance between any discontinuities is at least equal to the size of the jump, and between jumps the processes are all Lipschitz continuous with Lipschitz constant no larger than unity.

Take a convergent subsequence with limits satisfying

$$\tilde{x}(t) = x(0) + \sum_i \tilde{y}_i(t) d_i. \tag{5.12}$$

By the assumption concerning the existence of an asymptotic discontinuity, for some nonempty set of indices i we must have $\tilde{y}_i(t) > 0$ for $t > 0$. Suppose that $x(0)$ is interior to the bounding face ∂G_i and $\tilde{y}_i(t) > 0$ for $t > 0$. For the other components we must have $\tilde{y}_j(t) = 0, j \neq i$. The condition (A2.2) implies that the vector d_i points inward, so that $\tilde{x}(t) \in G^0$ for small $t > 0$, which contradicts the impossibility of there being a jump to the interior of G.

Next, suppose that $x(0)$ lies on the edge or corner that is the intersection of ∂G_1 and ∂G_2. Then one or both of $\tilde{y}_1(\cdot), \tilde{y}_2(\cdot)$ are nonzero, with $\tilde{y}_i(\cdot) = 0$, $i \neq 1, 2$. If only one of the $\tilde{y}_i(\cdot)$ is nonzero, say $\tilde{y}_1(t) > 0$, then d_1 points either inward, outward, or into face 2. It cannot point outward, and if it points inward or into face 2, we are back to the case of the previous paragraph. Now, let both $\tilde{y}_1(\cdot)$ and $\tilde{y}_2(\cdot)$ be nonzero. The condition (A2.2) implies that no positive linear combination of d_1 and d_2 can be zero. Hence we cannot have $\tilde{x}(t) = x(0)$. Also, we cannot move to the interior or exterior. Condition

(A2.2) implies that $\langle \sum_i \tilde{y}_i(t) d_i, n_j \rangle > 0$ for j being either one of or both 1,2, which implies that we cannot stay on the edge or corner, which belongs to both faces 1 and 2. Hence $\tilde{x}(\cdot)$ must move away from the edge onto one of the adjoining faces. But then we are back to the case of the first paragraph. The proof when $x(0)$ lies on the intersection of more than two faces is done in the same way.

The facts that $z(\cdot)$ can change only at t where $x(t) \in \partial G$ and that (5.9) holds follows from (A5.1) and the convergence and asymptotic continuity. ■

3.6 Singular and Impulsive Controls

3.6.1 Singular Controls

The model. Singular controls [64, 41, 88] appear in many areas of control theory. Some applications to controlled routing and admission in communications and queueing systems are in [56]. Communication delays are often a crucial part of the physics and can play a key role in the performance.

Define $\lambda(\cdot) = \{\lambda_i(\cdot), i = 1, \ldots\}$, where each $\lambda_i(\cdot)$ is right-continuous, nondecreasing, bounded w.p.1 on each bounded interval, adapted to the standard Wiener process $w(\cdot)$ and with $\lambda_i(0) = 0$. If a function $g(\cdot)$ has a left-hand (resp., right-hand) limit at t, then $g(t-)$ (resp., $g(t+)$) denotes the left-hand (resp., right-hand) limit. A simple form of the singular control problem with boundary reflection and path delays takes the form [8]

$$dx(t) = b(\bar{x}(t))dt + q_1(\bar{x}(t-))d\lambda(t) + \sigma(\bar{x}(t))dw(t) + dz(t). \tag{6.1}$$

If there is a discontinuity in a component $\lambda_i(\cdot)$ at t, then $d\lambda_i(t) = \lambda_i(t) - \lambda_i(t-)$. The $\lambda_i(\cdot)$ are called singular controls because they are not necessarily representable as $d\lambda_i(t) = u_i(t)dt$ for some ordinary control $u_i(\cdot)$; i.e., $\lambda_i(\cdot)$, when considered as a measure on the real line, might be singular with respect to Lebesgue measure. If the singular control as well as the path is delayed, then one possible analog of (2.11) and (2.12) is

$$\begin{aligned} dx(t) = dt \int_U c(x(t)) + dt \int_{-\bar{\theta}}^{0} b(x(t+\theta),\theta) d\mu_a(\theta) + \sigma(x(t))dw(t) \\ + q_0(x(t-))d\lambda(t) + dt \int_{\theta=-\bar{\theta}}^{0} q_2(x((t+\theta)-),\theta) d_\theta \lambda(t+\theta) + dz(t). \end{aligned} \tag{6.2}$$

Equation (6.2) can be modified in various ways. For example, we could use only discrete delay values in the delayed singular control part of (6.2). The

[8] The values $b(\bar{x}(t-))$ and $\sigma(\bar{x}(t-))$ can be used as well, with no change in the solution.

terms in (2.1), (2.4), (2.11), or (2.12) and in the cost functions (4.1) and (4.2) that contain ordinary or relaxed controls can be added to the above models. Also, delays in the reflection process can be included if desired. We use the forms (6.1) and (6.2) in order to concentrate on the basic features associated with the singular control problem.

It is a common occurrence in applications that the optimal singular control acts to keep the state from leaving some (*a priori* unknown) region. In this case finding the singular control is equivalent to finding the boundaries of that region and the direction of action of the control there. With such controls, if the initial condition $x(0)$ is not in that region, there will be an initial jump in the control, whose purpose is to return the path to that region, and after that $\lambda(\cdot)$ would be continuous, but not necessarily absolutely continuous with respect to Lebesgue measure. The lack of absolute continuity is due to the need to counteract the role of the stochastic integral in pushing the path out of the region.

One possible discounted cost function has the form, where $\lambda(s) = 0, s \leq 0$,

$$W(\hat{x}, \lambda) = E_{\hat{x}}^{\lambda} \int_0^{\infty} \int_U e^{-\beta t} \left[k(\bar{x}(t))dt + q'dy(t) + q'_{\lambda} d\lambda(t) \right]. \tag{6.3}$$

We will use the following assumption.

A6.1. *The functions $b(\cdot), c(\cdot), q_0(\cdot)$, $q_2(\cdot)$, and $\sigma(\cdot)$ in (6.2) are continuous. The functions $b(\cdot), q_1(\cdot)$ and $\sigma(\cdot)$ in (6.1) and $k(\cdot)$ in (6.3) that might depend on the path memory segment have the form $b(\hat{x}(-\theta_i), i \leq L)$, where $-\bar{\theta} \leq \theta_i \leq 0, L < \infty$, and $b(\cdot)$ is continuous, and analogously for $k(\cdot), q_1(\cdot)$ and $\sigma(\cdot)$. The components $q_{\lambda,i}$ of the vector q_λ in (6.3) are positive. When (6.2) is used in Chapter 9, the $k(\bar{x}(t))$ in (6.3) is replaced by $k(x(t))$.*

Suppose that, for some small $\delta > 0$, the $q_2(\cdot)$ in (6.2) has the form $q_2(x, \theta) = q_2(x)$ for $\theta \in [-\bar{\theta}, -\bar{\theta} + \delta]$ and equals zero otherwise, and $\lambda(s) = 0$ for $s \leq 0$. Then, if $\lambda(\cdot)$ is continuous on $[t - \bar{\theta}, t - \bar{\theta} + \delta)$, the delayed singular control term in (6.2) is approximated by

$$q_2(x((t - \bar{\theta})-))[\lambda(t - \bar{\theta} + \delta) - \lambda(t - \bar{\theta})]dt.$$

In this way one can approximate point delays in the singular control.

3.6.2 Definition of and Existence of Solutions

Example of a potential problem. We need the assurance that there exists a well-defined solution for each initial condition and singular control and that there is an optimal control. When the $q_a(\cdot), a = 0, 1, 2$, depend on the path there is a particular problem that requires a careful definition of the solution. Consider, for example, the nondelay form $dx^n(t) = x^n(t-)d\lambda^n(t)$, where $\lambda^n(\cdot)$

is a piecewise-constant function with $\lambda^n(s) = 0, s \le t_0$, and that increases by $1/\sqrt{n}$ at times $t_0 + l/n, l = 1, 2, \ldots$, until the value J_1 is reached (at $J_1\sqrt{n}$ steps), after which it is constant. To simplify the discussion, until further notice suppose that the singular control does not try to take the path out of the set G, so that it does not cause any reflection terms. Then $x^n(t_0-) = x^n(0)$ and $x^n(t_0 + J_1/\sqrt{n}) = x(0)(1 + 1/\sqrt{n})^{J_1\sqrt{n}} \to x(0)\exp(J_1)$, as $n \to \infty$. On the other hand, if the entire set of jumps in $\lambda(\cdot)$ was concentrated at $t = t_0$ in that $\lambda(\cdot)$ had a jump of J_1 at $t = t_0$ and was constant thereafter, then we would have $x(t_0) = x(0) + J_1 x(0) < x(0)\exp(J_1)$.

When some $q_a(\cdot)$ depends on the path, the possible cumulative effect of many small jumps that is illustrated by the example makes it difficult to prove the existence of optimal controls by standard methods, where one takes limits of minimizing sequences, and the minimizing sequence of controls (which is not *a priori* known) could conceivably have the form in the example. A very useful approach to resolving this complication involves a time change argument, where the asymptotic discontinuity phenomena that was illustrated above is "smoothed out, " effectively by replacing it by a control that is a Lipschitz continuous function. The method is analogous to the "time-stretching" argument that was used in Theorem 5.5, and it works as follows.

Working with (6.1) and given $\lambda(\cdot)$, start by defining

$$\begin{aligned} T^\lambda(t) &= t + \sum_i \lambda_i(t) + \sum_i y_i(t), \\ \tilde{T}^\lambda(t) &= \inf\{s : T^\lambda(s) > t\}. \end{aligned} \tag{6.4}$$

Then we have the inverse relationship

$$T^\lambda(t) = \inf\{s : \tilde{T}^\lambda(s) > t\}. \tag{6.5}$$

Define the time-stretched "tilde" processes as in $\tilde{x}^\lambda(t) = x(\tilde{T}^\lambda(t))$. Then

$$\begin{aligned} \tilde{x}(t) = x(0) &+ \int_0^{\tilde{T}^\lambda(t)} b(\bar{x}(s))ds \\ &+ \int_0^{\tilde{T}^\lambda(t)} q_1(\bar{x}(s-))d\lambda(s) + \int_0^{\tilde{T}^\lambda(t)} \sigma(\bar{x}(s))dw(s) + z(\tilde{T}^\lambda(t)), \end{aligned}$$

where we suppose that $d\lambda(0) = 0$. With a change of variable, we can write

$$\begin{aligned} \tilde{x}^\lambda(t) = x(0) &+ \int_0^t b(\tilde{\bar{x}}^\lambda(s))d\tilde{T}^\lambda(s) + \int_0^t q_1(\tilde{\bar{x}}^\lambda(s-))d\tilde{\lambda}(s) \\ &+ \int_0^t \sigma(\tilde{\bar{x}}^\lambda(s-))d\tilde{w}^\lambda(s) + \tilde{z}^\lambda(t). \end{aligned} \tag{6.6}$$

With the new timescale the cost can be written as

$$W(\hat{x}, \lambda) = E_{\hat{x}}^\lambda \int_0^\infty \int_U e^{-\beta\tilde{T}^\lambda(t)} \left[k(\tilde{\bar{x}}^\lambda(t))d\tilde{T}^\lambda(t) + q' d\tilde{y}^\lambda(t) + q'_\lambda d\tilde{\lambda}(t)\right]. \tag{6.7}$$

As defined, the transformed process $\tilde{\lambda}(\cdot)$ is the sum of a Lipschitz continuous process (Lipschitz constant unity) and a pure jump process, where the time between two successive jumps is at least the magnitude of the first jump. The process $\tilde{T}^\lambda(\cdot)$ is Lipschitz continuous, with constant unity. If $\tilde{\lambda}(\cdot)$ and $\tilde{T}^\lambda(\cdot)$ are defined by (6.4) and (6.5), and there is a well-defined solution to (6.1), then the inverse-transformed set of processes yield that solution.

The "stretched out" form (6.6) will be the fundamental equation, in lieu of (6.1). Then to get the original model, use the inverse-transformed processes $x(t) = \tilde{x}^\lambda(T^\lambda(t))$ as the definition of the solution to (6.1), and analogously for (6.2).

The time-transformation procedure will be illustrated via a simple example. Consider the real-valued process $x(\cdot)$ on the state space $G = [0, 1]$:

$$dx(t) = x(t-)d\lambda(t) + \sigma dw(t) + dz(t).$$

Refer to Figure 6.1 for an illustration. In this motivational discussion, we suppose that $z(\cdot) = 0$. Then by (6.4), $\lambda(\cdot)$ determines $T^\lambda(\cdot)$ and $\tilde{T}^\lambda(\cdot)$ uniquely. Figure 6.1a plots an example of $\lambda(\cdot)$. There is a jump of size J_1 at time t_0 and then it increases linearly with slope a_1. Figure 6.1b plots the associated $T^\lambda(\cdot)$, and Figures 6.1c and 6.1d plot the stretched out timescale $\tilde{T}^\lambda(\cdot)$ and the stretched-out process $\tilde{\lambda}(\cdot)$, resp.

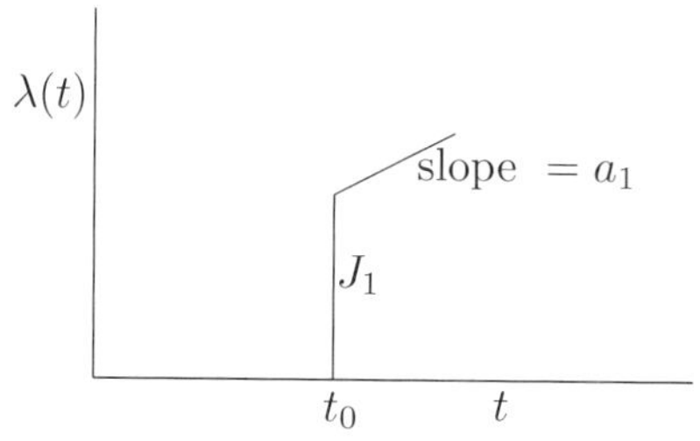

Figure 6.1a. An example of a singular control $\lambda(\cdot)$.

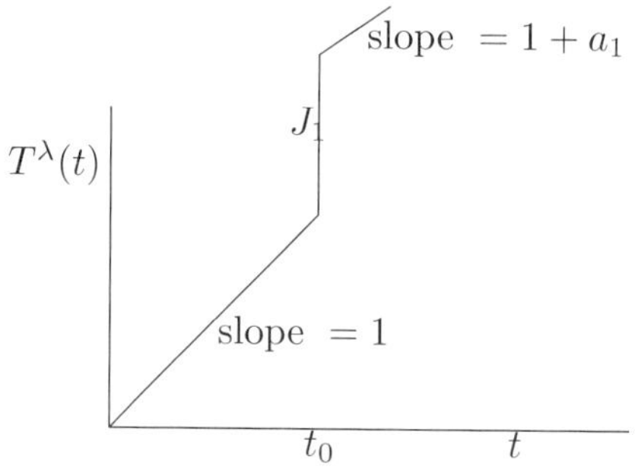

Figure 6.1b. The corresponding function $T^\lambda(\cdot)$.

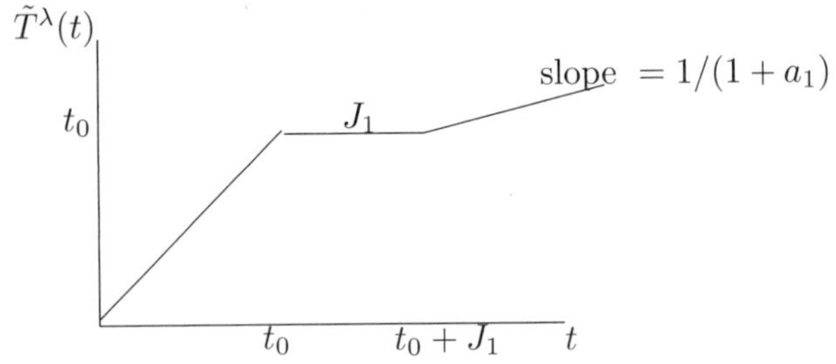

Figure 6.1c. The stretched-out timescale corresponding with $\lambda(\cdot)$.

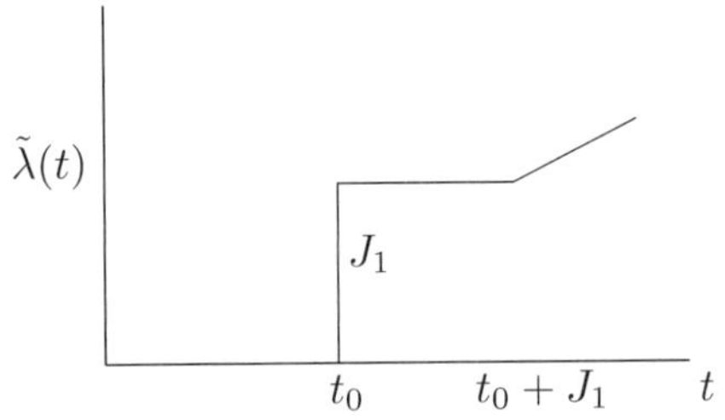

Figure 6.1d. The stretched-out process $\tilde{\lambda}(\cdot)$.

Now consider the limit $(\lambda(\cdot), \tilde{T}(\cdot), T(\cdot))$ of the sequence $(\lambda^n(\cdot), \tilde{T}^{\lambda^n}(\cdot), T^{\lambda^n}(\cdot))$ that is defined by the example at the beginning of the subsection. The limit $\tilde{\lambda}(\cdot)$ of $\tilde{\lambda}^n(\cdot)$ is plotted in Figure 6.2. The limit $\tilde{T}(\cdot)$ has the form in Figure 6.1c and the inverse process $T(\cdot)$ has the form in Figure 6.1b. Thus the graphs of $T^{\lambda}(\cdot)$ and $\tilde{T}^{\lambda}(\cdot)$ are the same as those for $T(\cdot)$ and $\tilde{T}(\cdot)$, resp., and the limit of the the singular controls $\lambda^n(\cdot)$ has a jump of J_1 at $t = t_0$. The graphs of $\tilde{\lambda}(\cdot)$ in Figure 6.1d (which is based on the example in Figure 6.1a) and that of $\tilde{\lambda}(\cdot)$ in Figure 6.2 are different. The first case entails an instantaneous jump of magnitude J_1, and in the second case $\tilde{\lambda}(\cdot)$ increases linearly with slope unity until the value J_1 is attained. Continue to suppose that there are no reflection terms. For the first case, $\tilde{x}(t_0) = \tilde{x}(t_0-) + J_1$, and in the second case, in the limit, we have the form of (6.6):

$$\tilde{x}(t_0 + J_1) = \tilde{x}(t_0-) + \int_{t_0}^{(t_0+J_1)} \tilde{x}(s-) d\tilde{\lambda}(s),$$

where $d\tilde{\lambda}(s) = ds$ in the range of concern for this example. This implies a jump of $x(t_0-)\exp(J_1)$ at t_0 in the limit, as in the example. If $x^n(t-)d\lambda^n(t)$ were replaced by $d\lambda^n(t)$, then the result would be the same in both cases.

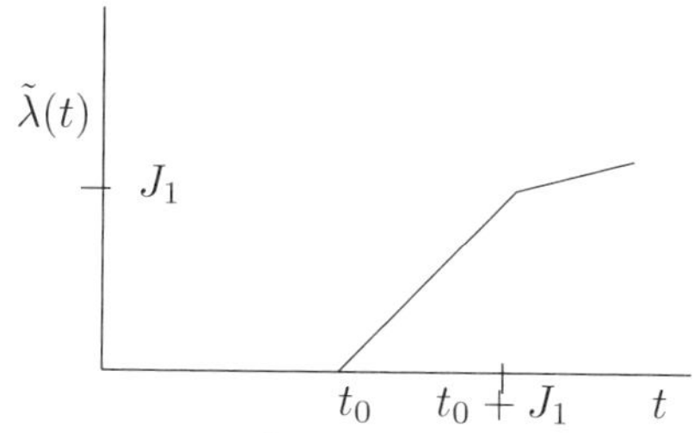

Figure 6.2. The limit $\tilde{\lambda}(\cdot)$ of the stretched-out processes.

The definition of a solution. Note that, in the example, when the limit $\tilde{T}(\cdot)$ or $\tilde{T}^\lambda(\cdot)$ is constant on an interval $[t_0, t_0 + J_1]$, and where the singular control has only one component, the limit $\lambda(\cdot)$ of the singular control jumps by J_1 at t_0. This holds whether the jump of $\lambda(\cdot)$ at t_0 is of value J_1 in just one step or whether it is the limit of the sum of a sequence of small jumps. In the second case, in the dynamical equation each of the small jumps was multiplied by the current value of the state.

Now drop the assumption that $z(\cdot) = 0$, but continue to suppose that $\lambda(\cdot)$ has only one component, and suppose that $\tilde{T}^\lambda(\cdot)$ is constant on an interval $[t_1, t_1 + J]$. If $\tilde{\lambda}(\cdot)$ does not try to take the path out of G on this interval, then we are back to the previous case. If $\tilde{\lambda}(\cdot)$ does try to take the path out of G on that interval, then the path is pushed back by the reflection term and the jump in the singular control will be $J_1 \le J$, with the increase in $\sum_i \tilde{y}_i(\cdot)$ on $[t_1, t_1 + J]$ being $J - J_1$. Note that the constancy of $\tilde{T}^\lambda(\cdot)$ on $[t_1, t_1 + J]$ implies a jump of $\lambda(\cdot)$ (of value J_1 in this one-dimensional example) at time t_0 defined by $\tilde{T}^\lambda(t_1) = t_0$.

Now consider the (real-valued) form

$$dx^n(t) = q_1(\bar{x}^n(t-))d\lambda^n(t) + \sigma dw(t) + dz^n(t),$$

where $\lambda^n(\cdot)$ is of the type introduced at the beginning of the subsection, with jumps $1/\sqrt{n}$, and write the formal limit as

$$dx(t) = q_1(\bar{x}(t-))d\lambda(t) + \sigma dw(t) + dz(t).$$

Suppose that the path stays in G so that there is no reflection term and that the limit $\tilde{T}^\lambda(\cdot)$ is constant on $[t_1, t_1 + J]$. Then, with t_0 defined by $\tilde{T}^\lambda(t_1) = t_0$, the limit jump in the path $x(\cdot)$ would occur at t_0 and be defined by $\tilde{x}(t_1 + J) - \tilde{x}(t_1-)$, where $\tilde{x}(t_1-) = x(t_0-)$ and

$$\tilde{x}(t_1 + J) = \tilde{x}(t_1-) + \int_{t_1}^{t_1+J} q_1(\tilde{\bar{x}}(s-))d\tilde{\lambda}(s) = \tilde{x}(t_1-) + \int_{t_1}^{t_1+J} q_1(\tilde{\bar{x}}(s-))ds. \tag{6.8}$$

The inverse transformation defined by (6.5) gives the correct value $x(t)$ at points t where $\lambda(\cdot)$ is continuous, and for all t if $q_1(\cdot)$ is constant. If the singular control has more than one component, the method is analogous, except for a minor change when several components jump simultaneously. In lieu of making specific assumptions that ensure that there is a weak-sense unique solution, we use the following condition.

A6.2. *There is a unique weak-sense solution to* (6.6) *and its analog for* (6.2) *for each singular control* $\lambda(\cdot)$ *that is bounded on each bounded time interval and each initial condition. That is, for each admissible pair* $(\lambda(\cdot), w(\cdot))$ *and initial condition, there is a probability space on which there are equivalent processes and a nonanticipative solution to* (6.6), *and the distribution of the* (*control, Wiener process, solution, reflection process*) *does not depend on the probability space.*

3.6.3 Existence of an Optimal Control

Theorem 6.1. *Assume* (A6.1), (A6.2), *system* (6.1) *or* (6.2), *cost function* (6.3), *and* (A2.1), (A2.2), *and* (A4.3). *Then there is an optimal control. If we use the cost function*

$$W(\hat{x}, \lambda) = E_{\hat{x}}^{\lambda} \int_0^{\tau_G} \int_U e^{-\beta t} \left[k(\bar{x}(t))dt + q'dy(t) + q_\lambda' d\lambda(t)\right] + E_{\hat{x}}^{\lambda} e^{-\beta\tau_G} g_0(x(\tau_G))$$

and replace the boundary conditions by (A4.1) *and* (A4.2), *then there is an optimal control.*

Proof. Recall the definition of the solution via the time-transformed model that was discussed in the previous subsection. We will confine the development to (6.1) with the reflecting boundary. The proof for the analog of (6.2) is nearly identical. The new issues that are due to the use of singular controls are the same in the other cases.

Let $(x^n(\cdot), \lambda^n(\cdot), w^n(\cdot), z^n(\cdot))$ be a minimizing sequence. Theorem 5.5 used a time-transformation method to show that the sequence of reflection processes was asymptotically continuous for the ordinary or relaxed control case. In the current case, there is no *a priori* assurance that $\{\lambda^n(\cdot)\}$ is tight, but the time-transformation method can be used to get the desired result. The proof follows the lines of the development in [56, Section 2.6]. See also [58, Section 11.1] and [60, Section 4], using the type of time transformation introduced in Theorem 5.5 and in the previous subsection. The positivity of the components of the weight vector q_λ implies that, for the minimizing sequence, the tails $\int_T^\infty$ of the costs goes to zero as $T \to \infty$ and that the costs are bounded, uniformly in n.

Analogously to what was done in Theorem 5.5 and in the previous subsection, define

$$T^n(t) = t + \sum_i y_i^n(t) + \sum_i \lambda_i^n(t), \tag{6.9}$$

and

$$\tilde{T}^n(t) = \inf\{s : T^n(s) > t\}. \tag{6.10}$$

Define the time-stretched processes $\tilde{x}^n(t) = x^n(\tilde{T}^n(t))$, and so forth. Then

$$\begin{aligned} \tilde{x}^n(t) = x(0) &+ \int_0^t b(\tilde{\bar{x}}^n(s))d\tilde{T}^n(s) \\ &+ \int_0^t q_1(\tilde{\bar{x}}^n(s-))d\tilde{\lambda}^n(s) + \int_0^t \sigma(\tilde{\bar{x}}^n(s))d\tilde{w}^n(s) + \tilde{z}^n(t), \end{aligned} \tag{6.11}$$

$$W(\hat{x}, \lambda^n) = E_{\hat{x}}^{\lambda^n} \int_0^\infty \int_U e^{-\beta \tilde{T}^n(t)} \left[k(\tilde{\bar{x}}^n(t))d\tilde{T}^n(t) + q' d\tilde{y}^n(t) + q'_\lambda d\tilde{\lambda}^n(t) \right]. \tag{6.12}$$

The set of time-transformed processes $(\tilde{x}^n(\cdot), \tilde{\lambda}^n(\cdot), \tilde{w}^n(\cdot), \tilde{y}^n(\cdot), \tilde{T}^n(\cdot))$ is tight, with $\tilde{T}^n(\cdot)$ being Lipschitz continuous, with constant unity. Each of the processes $\tilde{y}^n(\cdot), \tilde{\lambda}^n(\cdot)$ is the sum of a Lipschitz continuous process (Lipschitz constant unity) and a pure jump process, where the time interval between any two jumps is at least the magnitude of the first jump.

Extract a weakly convergent subsequence, indexed also by n, and with limit denoted by $(\tilde{x}(\cdot), \tilde{\lambda}(\cdot), \tilde{w}(\cdot), \tilde{y}(\cdot), \tilde{T}(\cdot))$. Then $\tilde{w}(\cdot)$ is a continuous martingale (with respect to the filtration induced by the full set of limit processes) and has quadratic variation process $I\tilde{T}(\cdot)$, where I is the identity matrix. Also, the limit set satisfies (6.6) in that

$$\begin{aligned} \tilde{x}(t) = x(0) &+ \int_0^t b(\tilde{\bar{x}}(s))d\tilde{T}(s) \\ &+ \int_0^t q_1(\tilde{\bar{x}}(s-))d\tilde{\lambda}(s) + \int_0^t \sigma(\tilde{\bar{x}}(s))d\tilde{w}(s) + \tilde{z}(t). \end{aligned} \tag{6.13}$$

The proof of nonanticipativity and of the form of the limit of the stochastic integrals is similar to that in Theorem 5.1.

Using Fatou's Lemma and the fact that $\{\lambda^n(\cdot)\}$ is a minimizing sequence, we have

$$\begin{aligned} &\lim_n W(\hat{x}, \lambda^n) = V(\hat{x}) \\ &\geq \tilde{W} = E_{\hat{x}}^{\tilde{\lambda}} \int_0^\infty \int_U e^{-\beta \tilde{T}(t)} \left[k(\tilde{\bar{x}}(t))d\tilde{T}(t) + q' d\tilde{y}(t) + q'_\lambda d\tilde{\lambda}(t) \right]. \end{aligned} \tag{6.14}$$

Hence there is an optimal control for (6.6), (6.7).

To relate (6.13) to (6.1), define the inverse transformation

$$T(t) = \inf\{s : \tilde{T}(s) > t\}. \tag{6.15}$$

Because $\sup_n E\lambda^n(t) < \infty$ for each t, the inverse is finite for each t and goes to infinity as $t \to \infty$ (all w.p.1). Now define the inverse processes $x(t) = \tilde{x}(T(t))$, and so forth. Then $w(\cdot)$ is a standard Wiener process with respect to the filtration induced by the set of inverse transformed processes.

Suppose that $\lambda(\cdot) = \tilde{\lambda}(T(\cdot))$ is continuous. Then so are the $y_i(\cdot)$, and the set $(x(\cdot), y(\cdot), \lambda(\cdot), w(\cdot))$ satisfies (6.1) with $z(t) = \sum_i d_i y_i(t)$. Also $\tilde{W} = W(\hat{x}, \lambda)$, which must equal $V(\hat{x})$.

Now suppose that the process $T(\cdot)$ has a jump of magnitude J at t_0. This corresponds with the process $\tilde{T}(\cdot)$ being constant on the interval $[t_1, t_1 + J]$ where $T(t_0-) = t_1$. Thus one or more components of $\lambda(\cdot)$ jump at t_0, and the sum of the jumps is $\leq J$. Suppose, for specificity, that $\lambda_i(\cdot)$ jumps J_i, $i = 1, 2$, $J \geq J_1 + J_2$. The value $x(t_0)$ is the solution at $t = t_1 + J$ of the differential equation

$$\tilde{x}(t) - \tilde{x}(t_1-) = \int_{t_1}^{t} q_1(\tilde{x}(s-))d\tilde{\lambda}(s) + \tilde{z}(t) - \tilde{z}(t_1-), \tag{6.16}$$

where $\tilde{z}(\cdot)$ is the boundary reflection process. ■

3.6.4 Impulsive Controls

By an impulsive control we mean a control that has the effect of a Dirac delta function: It forces an instantaneous change in the system [13, 41, 70]. Such controls occur as limits of systems operating in heavy traffic environments, as the traffic intensity goes to unity. See, for example, [58, Example 5, Section 8.1.5]. They occur if the control action or external interruptions shut down or alter part of the system for a period of time. They are also used as approximations to problems where a large "force" can be applied over a relatively short interval. The impulsive control problem differs from the singular control problem in that there is a positive "setup" cost as well as a cost associated with the magnitude of the control. When there is a setup cost, one is likely to have few control moments, but with larger forces used.

Let the sequence of random variables $\{\nu_i, \tau_i\}$ denote the values and the times of the impulses, and define

$$F(t) = \sum_{i:\tau_i \leq t} \nu_i.$$

It is always assumed that $F(\cdot)$ is nonanticipative with respect to $w(\cdot)$ and that $\tau_n \leq \tau_{n+1} \to \infty$ with probability one. But some τ_n might be infinite with a positive probability. The τ_i are $\mathcal{F}_t$-stopping times and ν_i is $\mathcal{F}_{\tau_i}$-measurable. For a matrix D, a simple "impulsively controlled" analog of (6.1) is

$$dx(t) = b(\bar{x}(t))dt + DdF(t) + \sigma(\bar{x}(t))dw(t) + dz(t). \tag{6.17}$$

One can add delays in the control, as well, with analogs of either (2.3) or (2.11) used.

Let $g(x, \nu)$ be a bounded and continuous function that is nondecreasing in ν for each x and satisfying $g(x, \nu) \geq c_0 > 0$ for all x, ν such that $\nu \neq 0$, and $g(x,0) = 0$. This positivity for $\nu \neq 0$ corresponds with the setup cost. A discounted cost function is

$$W(\hat{x}, F) = E_{\hat{x}}^{F} \left[\int_0^\infty e^{-\beta t} k(\bar{x}(t)) dt + \sum_i e^{-\beta \tau_i} g(x(\tau_i -), \nu_i) \right]. \qquad (6.18)$$

4

Approximations to the Dynamical Models

4.0 Outline of the Chapter

Because the state space for the delay equation is infinite-dimensional, the numerical problem can be quite difficult. A major part of the difficulty is due to the size of the state space that is required for the numerical approximation. Because of this, it is important to simplify the original dynamical model as much as possible without sacrificing the essential aspects of the results. No matter what the problem, the ability to get good numerical solutions depends in no small part on the relative insensitivity of the model to small variations, as the numerical algorithm itself is an approximation to the original model. This chapter is devoted to a set of model simplifications that have considerable promise when the path or path and/or control are delayed. The associated numerical algorithms are readily implementable, and the paths of the approximating models are often close to those of the original model. Many models with delays are very sensitive to variations in the delay or dynamics. Because of this, it is essential to use any approximation with care.

Section 1 is concerned with generic approximations to the drift function $b(\cdot)$ in the dynamics and $k(\cdot)$ in the cost rate. It is shown that the costs are insensitive to small variations in these functions, and in the case where the delays are continuously distributed, it allows us to discretize them. The results justify working with simpler control-value spaces, say those with finitely many points, and allow us to approximate the path memory segments in various ways.

Section 2 is concerned with approximations when only the path values (and not the control) are delayed. In the original model, the path memory segment $\bar{x}(t)$ evolves continuously and, even if the delays are discretized to be integral multiples of some basic value, one must still keep track of the infinite-dimensional quantity $\bar{x}(t)$. In order to get a finite-dimensional approximation, one needs to alter the way that the memory segment evolves in time. In the "periodic approximation," the delays vary periodically in time, oscillating about the true values with small variations. This resulting path

H.J. Kushner, *Numerical Methods for Controlled Stochastic Delay Systems*,
doi: 10.1007/978-0-8176-4621-9_4,

memory segment is finite-dimensional. The changes in the cost functions are minimal for small periods. This approximation can be used as a first step in developing numerical procedures.

For such periodic approximations, one must keep track of the time that has elapsed since the beginning of the current period. Depending on the numerical algorithm, discretizing this elapsed might or might not be convenient. In many cases, the number of points that are required for such a discretization will be very large. To handle this, Approximations 4 and 5 introduce models where the time advances randomly and that require only a finite (and usually not too large) number of points to keep track of its evolution. The numerical algorithms are readily adapted to such approximations. One can prove that the costs converge (uniformly in the controls) to those of the original model as the approximations are refined. The Bellman equation for the approximations involves the random renewal process that determines the evolution of time only via averages, and for a fine enough approximation it is essentially equivalent to that when the true elapsed time is used.

The simulations described in Section 3 illustrate the basic approximations. Paths for the original process are compared with those for the approximations, and it is seen that they can be quiet close, even with moderate levels of approximation. Evaluations of various cost functions support this conclusion. Although the approximations were good for the tested problems, as noted above one must always exercise care owing to the sensitivity to model variations that is common in systems with delays.

The model approximation problem is harder when the control or both the path and control are delayed. Because the paths are continuous functions, there are many ways of sampling and approximating them to get finite-dimensional results. But it is not usually known *a priori* whether the control processes have any regularity properties at all. Because of this there is an advantage in approximating the relaxed control representation. Some possibilities are discussed in Section 4. An alternative approach to systems with delays in the control and path is discussed in Chapter 9. Section 5 is concerned with system approximations when there is a delayed singular control. The methods used in Sections 4 and 5 should be taken as being well motivated and natural, but still preliminary. Much more work is needed on system approximation methods when the control is delayed. In the numerical algorithms (see Chapters 7 and 8), the path values are discretized and represented in various ways to minimize the memory requirement.

Delay equations might have rapidly time-varying terms, even rapidly varying delays. This complicates the numerical problem. But, under suitable conditions, there are limit and approximation theorems that allow us to replace the system by a simpler averaged one. Some such results are developed in Section 6.

4.1 Approximations of the Dynamical Systems

4.1.1 A Basic Approximation

A first step in simplifying the numerical problem is to simplify the original model. The idea is to work with a simpler model, but for which the values of the cost functions and the essential features are close to those for the original model. Indeed such approximations are an important part of the development of useful algorithms. The following theorems show that the cost function does not change much, uniformly in the control, if the dynamics are changed slightly, and they allow us to make the type of approximations just suggested, as well as more useful ones. Eventually the simplifications will lead to finite-dimensional approximations whose behavior is close to that of the original model and for which numerical algorithms can be conveniently developed.

We concentrate on the model (3.2.3), and cost function (3.4.4), and the analogs of the approximation results also hold for the other models discussed in Chapter 3. The next two theorems show that varying the memory structure slightly changes the costs only slightly, uniformly in the control. The theorems are intended to be suggestive of the possibilities, and their proofs use the methods of Theorem 3.5.1. In the system of concern in Theorem 1.1, the dynamical terms $b(\cdot), \sigma(\cdot), \mu_c(\cdot)$, and the cost rate $k(\cdot)$, are approximated by $b^n(\cdot), \sigma^n(\cdot), \mu_c^n(\cdot)$, and $k^n(\cdot)$, resp. In (A1.1), $v(\cdot)$ and $v^n(\cdot)$ are used for canonical paths in $C(G; [-\bar{\theta}, \infty))$, and the canonical path memory segments associated with them are denoted by $\bar{v}(t)$ and $\bar{v}^n(t)$, resp.

A1.1. *$b^n(\cdot)$ and $k^n(\cdot)$ are measurable $I\!R^r$-valued functions on $D(G; [-\bar{\theta}, 0]) \times U \times [-\bar{\theta}, 0]$. Suppose that $v^n(\cdot) \to v(\cdot)$ in $C(G; [-\bar{\theta}, \infty))$. Let $\bar{v}^n(t)$ (resp., $\bar{v}(t)$) denote the path memory segments at t associated with the paths $v^n(\cdot)$ and $v(\cdot)$, resp. Then $b^n(\bar{v}^n(t), \alpha, \theta) \to b(\bar{v}(t), \alpha, \theta)$ uniformly in α, θ on any bounded time interval. The real-valued function $k^n(\cdot)$ and the matrix-valued function $\sigma^n(\cdot)$ are measurable and satisfy the analogous property. Also, $\mu_c^n(\cdot)$ converges weakly to $\mu_c(\cdot)$.*

Theorem 1.1. *Use the model* (3.2.3) *with the cost function* (3.4.4). *Assume* (A3.1.2), (A3.1.3), (A3.2.1)–(A3.2.3), (A3.4.3) *and* (A1.1). *Let $W^n(\hat{x}, \hat{r}, r)$ denote the cost under $(b^n(\cdot), \sigma^n(\cdot), k^n(\cdot), \mu_c^n(\cdot))$. Then $W^n(\hat{x}, \hat{r}, r) \to W(\hat{x}, \hat{r}, r)$ uniformly in $\hat{x}, \hat{r}$ (where $\hat{x}$ is confined to some compact set in $C(G; [-\bar{\theta}, 0])$) and in $r(\cdot)$ as $n \to \infty$.*

Now drop the conditions (A3.1.2) *and* (A3.1.3), *and assume the model* (3.2.11) *and* (A3.2.4)–(A3.2.6). *Then the conclusions continue to hold. Suppose now that control stops when the boundary is reached, where the cost function is* (3.4.2). *Drop* (A3.2.1) *and* (A3.2.2) *and assume* (A3.4.1) *and* (A3.4.2). *Then the conclusions continue to hold.*

Comment on the notation. Consider a solution to (3.2.3), with relaxed control $r(\cdot)$ and Wiener process $w(\cdot)$. Because we are working with weak-sense solutions, when we abuse the notation by saying that we apply the same pair $(r(\cdot), w(\cdot))$ to the system that uses $(b^n(\cdot), \sigma^n(\cdot), k^n(\cdot), \mu_c^n(\cdot))$ and to the system with $(b(\cdot), \sigma(\cdot), k(\cdot), \mu_c(\cdot))$, we mean that there is a probability space on which are defined $(r^n(\cdot), w^n(\cdot))$ and a solution $x^n(\cdot)$, with reflection term $z^n(\cdot)$, where $(r^n(\cdot), w^n(\cdot))$ has the probability law of $(r(\cdot), w(\cdot))$.

Proof. The proof is very similar to that for the existence of an optimal control in Section 3.5, and we will only discuss the proof of the uniformity. Suppose that the uniformity (in the controls and initial condition) of the approximation does not hold as $n \to \infty$. Then there is a sequence of initial conditions $\hat{x}^n$ (confined to some compact set) and $\hat{r}^n$, admissible pairs $(r^n(\cdot), w^n(\cdot))$, and corresponding solutions $x^n(\cdot)$ with reflection terms $z^n(\cdot)$ such that

$$\begin{aligned} x^n(t) = \hat{x}^n(0) &+ \int_0^t ds \int_{-\bar{\theta}}^0 \int_U b^n(\bar{x}^n(s), \alpha, \theta) r^{n,\prime}(d\alpha, s+\theta)\mu_c^n(d\theta) \\ &+ \int_0^t \sigma^n(\bar{x}^n(s)) dw^n(s) + z^n(t), \end{aligned} \tag{1.1}$$

and, for some $\epsilon > 0$ and as $n \to \infty$,

$$|W^n(\hat{x}^n, \hat{r}^n, r^n) - W(\hat{x}^n, \hat{r}^n, r^n)| \geq \epsilon. \tag{1.2}$$

Let $(r_0^n(\cdot), w_0^n(\cdot))$ have the distribution of $(r^n(\cdot), w^n(\cdot))$, and let $x_0^n(\cdot), z_0^n(\cdot)$ be the associated solution and reflection term, resp., to (3.2.3) under the initial conditions $\hat{x}^n, \hat{r}^n$; i.e.,

$$\begin{aligned} x_0^n(t) = \hat{x}^n(0) &+ \int_0^t ds \int_{-\bar{\theta}}^0 \int_U b(\bar{x}_0^n(s), \alpha, \theta) r_0^{n,\prime}(d\alpha, s+\theta)\mu_c(d\theta) \\ &+ \int_0^t \sigma(\bar{x}_0^n(s)) dw_0^n(s) + z_0^n(t). \end{aligned} \tag{1.3}$$

Choose a weakly convergent subsequence of $(x^n(\cdot), y^n(\cdot),\ r^n(\cdot), w^n(\cdot), \hat{x}^n, \hat{r}^n)$ and $(x_0^n(\cdot), y_0^n(\cdot), r_0^n(\cdot), w_0^n(\cdot),\ \hat{x}^n,\ \hat{r}^n)$ with limits $(x(\cdot), y(\cdot), r(\cdot), w(\cdot), \hat{x}, \hat{r})$ and $(x_0(\cdot), y_0(\cdot), r_0(\cdot), w_0(\cdot), \hat{x}, \hat{r})$, resp. Let n index the subsequence. Use the Skorokhod representation so that the convergence can be supposed to be w.p.1 in the topology of the path spaces. By (A1.1), the continuity of the limit processes, and the fact that $r^n(d\alpha, ds+v)\mu_c^n(dv) \Rightarrow r(d\alpha, ds+v)\mu_c(dv)$ if $r^n(\cdot) \Rightarrow r(\cdot)$, the first integral in (1.1) converges to

$$\int_0^t ds \int_{-\bar{\theta}}^0 \int_U b(\bar{x}(s), \alpha, \theta) r'(d\alpha, s+\theta)\mu_c(d\theta).$$

The stochastic integrals are treated by the approximation procedure of Theorem 3.5.1. Thus, the limit of (1.1) satisfies

$$x(t) = \hat{x}(0) + \int_0^t ds \int_{-\bar{\theta}}^0 \int_U b(\bar{x}(s), \alpha, \theta) r'(d\alpha, s + \theta) \mu_c(d\theta) + \int_0^t \sigma(\bar{x}(s)) dw(s) + z(t).$$

Analogously, the limit of (1.3) satisfies

$$x_0(t) = \hat{x}(0) + \int_0^t ds \int_{-\bar{\theta}}^0 \int_U b(\bar{x}_0(s), \alpha, \theta) r_0'(d\alpha, s + \theta) \mu_c(d\theta) + \int_0^t \sigma(\bar{x}_0(s)) dw_0(s) + z_0(t).$$

The probability laws of $(r(\cdot), w(\cdot))$ and $(r_0(\cdot), w_0(\cdot))$ are identical and the initial conditions are the same for both cases. Hence the weak-sense uniqueness (A3.2.3) of the solution to (3.2.3) implies that $(x_0(\cdot), y_0(\cdot), r_0(\cdot), w_0(\cdot))$ and $(x(\cdot), y(\cdot), r(\cdot), w(\cdot))$ have the same probability laws. The proof that $W^n(\hat{x}^n, \hat{r}^n, r^n) \to W(\hat{x}, \hat{r}, r)$ and $W(\hat{x}^n, \hat{r}^n, r^n) \to W(\hat{x}, \hat{r}, r)$ (which contradicts (1.2)) is similar to the proof of the analogous result in Theorem 3.5.1 and the details are omitted. ■

Theorem 1.2. *Assume the conditions of Theorem* 1.1. *Let* $(r^{n,\epsilon}(\cdot), w^{n,\epsilon}(\cdot))$ *be an* ϵ*-optimal (control, Wiener process) pair, under* $(b^n(\cdot), \sigma^n(\cdot), k^n(\cdot), \mu_c^n(\cdot))$ *and initial conditions* $\hat{x}, \hat{r}$. *Then any weak-sense limit* $(r^\epsilon(\cdot), w^\epsilon(\cdot))$ *of* $(r^{n,\epsilon}(\cdot), w^{n,\epsilon}(\cdot))$ *is an* ϵ*-optimal pair under* $(b(\cdot), \sigma(\cdot), k(\cdot), \mu_c(\cdot))$.

Comment on the proof. Let $(x^{n,\epsilon}(\cdot), z^{n,\epsilon}(\cdot))$ denote the solution under $(r^{n,\epsilon}(\cdot), w^{n,\epsilon}(\cdot))$. Choose a weakly convergent subsequence (indexed by n) of $(x^{n,\epsilon}(\cdot), z^{n,\epsilon}(\cdot), r^{n,\epsilon}(\cdot), w^{n,\epsilon}(\cdot))$ with limit $(x(\cdot), z(\cdot), r^\epsilon(\cdot), w^\epsilon(\cdot))$. By the ϵ-optimality of $r^{n,\epsilon}(\cdot)$,

$$W^n(\hat{x}, \hat{r}, r^{n,\epsilon}) \le \inf_r W^n(\hat{x}, \hat{r}, r) + \epsilon. \tag{1.4}$$

We will use this and the weak convergence to show that

$$W(\hat{x}, \hat{r}, r^\epsilon) \le \inf_r W(\hat{x}, \hat{r}, r) + \epsilon. \tag{1.5}$$

Suppose that (1.5) is false. Then there is a $\delta > 0$ such that

$$\inf_r W(\hat{x}, \hat{r}, r) \le W(\hat{x}, \hat{r}, r^\epsilon) - \epsilon - \delta,$$

and a pair $(r_\delta(\cdot), w_\delta(\cdot))$ such that

$$W(\hat{x}, \hat{r}, r_\delta) \le W(\hat{x}, \hat{r}, r^\epsilon) - \epsilon - \delta/2. \tag{1.6}$$

Now apply the pair $(r_\delta(\cdot), w_\delta(\cdot))$ to (1.1) with initial conditions $(\hat{x}, \hat{r})$. That is, there is a probability space with processes $(r_\delta^n(\cdot), w_\delta^n(\cdot), x_\delta^n(\cdot), z_\delta^n(\cdot))$ defined on it such that

$$
\begin{aligned}
x_\delta^n(t) = \hat{x}(0) &+ \int_0^t ds \int_{-\bar{\theta}}^0 \int_U b^n(\bar{x}_\delta^n(s), \alpha, \theta) r_\delta^{n,\prime}(d\alpha, s+\theta) \mu_c^n(d\theta) \\
&+ \int_0^t \sigma^n(\bar{x}_\delta^n(s)) dw_\delta^n(s) + z_\delta^n(t),
\end{aligned}
\tag{1.7}
$$

and where the distribution of $(r_\delta^n(\cdot), w_\delta^n(\cdot))$ is that of $(r_\delta(\cdot), w_\delta(\cdot))$. By (1.6) and the uniqueness of the solution to (3.1.3), the cost corresponding to the limit of any weakly convergent subsequence of the processes in (1.7) must be $\leq W(\hat{x}, \hat{r}, r^\epsilon) - \epsilon - \delta/2$, which contradicts the ϵ-optimality (1.4) of $(r^{n,\epsilon}(\cdot), w^{n,\epsilon}(\cdot))$ for large n. Hence (1.5) holds. ■

The next theorem is a slight extension of Theorem 1.1. It will be used for the approximation of the memory in the next section. It allows us to use an approximation to the memory segment at time t that depends on the solution over an interval $[t - \bar{\theta}^n(t), t]$ in lieu of the interval $[t - \bar{\theta}, t]$ where, for each $T < \infty$, $\sup_{t \leq T} \left|\bar{\theta}^n(t) - \bar{\theta}\right| \to 0$ in probability as $n \to \infty$. It is motivated by approximations such as Approximations 2, 3, and 4 of the next section, which in turn are motivated by approximations that reduce the memory requirements of the numerical algorithms, such as those in Chapters 7–9. In Approximation 3, defined by (2.6) and called the periodic approximation, the maximum delay varies periodically between $\bar{\theta} - \delta/2$ and $\bar{\theta} + \delta/2$, and it is of interest to know that as $\delta \to 0$, the costs converge to those for the original model. In Approximation 4, defined by (2.7), the delays vary randomly and are determined by a renewal process, whose intervals are exponentially distributed with mean δ. The memory is the path from the present time back to $\bar{\theta}/\delta$ renewal times ago, mapped into a function on $[-\bar{\theta}, 0]$, and we wish to know that as $\delta \to 0$, the costs will converge to that for the original model.

In Approximation 5, defined by (2.8), the exponentially distributed intervals are replaced by Erlang distributed intervals of order δ/δ_0, of total mean δ. This will be called the periodic-Erlang approximation. We wish to know that, as $\delta_0 \to 0$, the costs converge to those for the periodic delay approximation. The motivation for this approximation, as a means of reducing the required memory for the numerical algorithms, is given in Chapter 8. Simulations that illustrate these approximations are in Section 3 where it is seen that they can be quite good. The limit of the memory segments at time t for the approximations mentioned in the previous paragraph is just the original memory segment $\bar{x}(t) = \{x(t+\theta), \theta \in [-\bar{\theta}, 0]\}$. But for the periodic-Erlang approximation noted in this paragraph, the limit (as $\delta_0 \to 0$, with δ fixed) has the form of the periodic memory segment that, for each t, is piecewise-constant and not continuous. This motivates Assumption A1.3 and Theorem 1.4. and was the motivation for the more general part of (A3.1.1) and (A3.1.3). The importance of these approximations for the numerical problem will be seen in Chapters 7–9.

In (A1.2), we use a sequence of approximations $\bar{x}_a^n(t), n < \infty$, to the path memory segment $\bar{x}(t)$. The subscript "a" will later be used to indicate the type of approximation.

A1.2. *Let the nonanticipative memory segment process* $\bar{x}_a^n(t) \in D(G; [-\bar{\theta}, 0])$ *be used in lieu of* $\bar{x}(t)$*, where* $\bar{x}_a^n(t)$ *might depend on the path* $x(t+\theta)$ *for a range* $\theta \in [-\bar{\theta}^n(t), 0]$ *in lieu of the range* $[-\bar{\theta}, 0]$ *(but mapped into a function on* $[-\bar{\theta}, 0]$ *), for some nonanticipative process* $\bar{\theta}^n(\cdot)$*. Let* $x^n(\cdot)$ *denote the solution under the memory segment* $\bar{x}_a^n(t)$*, assumed to be weak-sense unique for each initial condition and control. Suppose that, for each* $T < \infty$*, and as* $n \to \infty$*,*

$$\sup_{t \le T} \sup_{-\bar{\theta} \le \theta \le 0} |\bar{x}_a^n(t, \theta) - x^n(t+\theta)| \to 0 \tag{1.8}$$

in probability, uniformly in the control and initial condition.

In Theorem 1.3, $W_a^n(\hat{x}, \hat{r}, r)$ is the cost under the path memory segment $\bar{x}_a^n(\cdot)$.

Theorem 1.3. *Assume the conditions of Theorem* 3.5.1 *and* (A1.2)*. Use* $b(\bar{x}_a^n(t), \alpha, \theta)$ *instead of* $b^n(\bar{x}(t), \alpha, \theta)$*, and use* $\sigma(\bar{x}_a^n(t))$ *and* $k(\bar{x}_a^n(t), \alpha, \theta)$*. Then* $W_a^n(\hat{x}, \hat{r}, r) \to W(\hat{x}, \hat{r}, r)$ *uniformly in* $\hat{x}$ *in any compact set in* $C(G; [-\bar{\theta}, 0])$*, and in* $\hat{r}, r(\cdot)$*, as* $n \to \infty$*.*

Comment on the proof. The proof follows that of Theorem 1.1. Consider (3.2.3) under the current substitutions for the path memory segments:

$$\begin{aligned} x^n(t) = x(0) &+ \int_0^t ds \int_{-\bar{\theta}}^0 \int_U b(\bar{x}_a^n(s), \alpha, \theta) r'(d\alpha, s+\theta) \mu_c(d\theta) \\ &+ \int_0^t \sigma(\bar{x}_a^n(s)) dw(s) + z^n(t), \end{aligned}$$

which we write as

$$\begin{aligned} x^n(t) = x(0) &+ \int_0^t ds \int_{-\bar{\theta}}^0 \int_U b(\bar{x}^n(s), \alpha, \theta) r'(d\alpha, s+\theta) \mu_c(d\theta) \\ &+ \int_0^t \sigma(\bar{x}^n(s)) dw(s) + z^n(t) + \epsilon^n(t), \end{aligned}$$

where

$$\begin{aligned} \epsilon^n(t) = &\int_0^t ds \int_{-\bar{\theta}}^0 \int_U \left[b(\bar{x}_a^n(s), \alpha, \theta) - b(\bar{x}^n(s), \alpha, \theta) \right] r'(d\alpha, s+\theta) \mu_c(d\theta) \\ &+ \int_0^t \left[\sigma(\bar{x}_a^n(s)) - \sigma(\bar{x}^n(s)) \right] dw(s). \end{aligned}$$

By (A1.2), as $n \to \infty$, the max of the error term $\epsilon^n(t)$ over any finite interval goes to zero in mean, uniformly in the control and initial condition. The rest of the details are as in Theorem 1.1.

An extension of Theorem 1.3. In Theorem 1.3, it was assumed that the model used the path memory segment $\bar{x}(t)$, and it was $\bar{x}(t)$ that was approximated. But the original model might use a path memory that is different than

$\bar{x}(t)$. For example, it might be a piecewise-constant approximation to $\bar{x}(t)$, or depend on the values of $x(\cdot)$ over a time interval that is different than that used for $\bar{x}(t)$. Now let $\bar{x}_a(t) \in D(G;[-\bar{\theta},0])$ denote the value of the nonanticipative memory segment process that is to be used in the original model at time t. It is assumed to be piecewise-constant with intervals that do not depend on t and can be represented as a measurable function $F_a(t,x(s),s \le t)$ of the path and time. It also is used for the argument in the cost rate $k(\cdot)$ and in $\sigma(\cdot)$. Thus the original system is now

$$\begin{aligned} x(t) = x(0) &+ \int_0^t ds \int_{-\bar{\theta}}^0 \int_U b(\bar{x}_a(s),\alpha,\theta) r'(d\alpha, s+\theta)\mu_c(d\theta) \\ &+ \int_0^t \sigma(\bar{x}_a(s))dw(s) + z(t). \end{aligned} \tag{1.9a}$$

If the control is not delayed, then the system reduces to

$$x(t) = x(0) + \int_0^t ds \int_U b(\bar{x}_a(s),\alpha) r'(d\alpha,s) + \int_0^t \sigma(\bar{x}_a(s))dw(s) + z(t). \tag{1.9b}$$

A1.3. *Let the solution to (1.9) be weak-sense unique for each initial condition and control. Let the nonanticipative path memory segment* $\bar{x}_b^n(t) \in D(G;[-\bar{\theta},0])$ *be used in lieu of* $\bar{x}_a(t)$ *in the cost rate and in (1.9), with corresponding solution process* $x^n(\cdot)$, *assumed to be weak-sense unique for each initial condition and control. That is,*

$$\begin{aligned} x^n(t) = x(0) &+ \int_0^t ds \int_{-\bar{\theta}}^0 \int_U b(\bar{x}_b^n(s),\alpha,\theta) r'(d\alpha, s+\theta)\mu_c(d\theta) \\ &+ \int_0^t \sigma(\bar{x}_b^n(s))dw(s) + z^n(t). \end{aligned}$$

Suppose that, for each $T < \infty$, *and as* $n \to \infty$,

$$\sup_{t \le T} \sup_{-\bar{\theta} \le \theta \le 0} |\bar{x}_b^n(t,\theta) - \bar{x}_a^n(t,\theta)| \to 0 \tag{1.10}$$

in probability, uniformly in the control and initial condition, where $\bar{x}_a^n(t) = F_a(t,x^n(s),s \le t)$. *Suppose further that if* $x^n(\cdot)$ *converges to a process* $v(\cdot)$ *then* $\bar{x}_a^n(t)$ *converges to the memory segment process* $\bar{v}_a(t) = F_a(t,v(s),s \le t)$ *associated with* $v(\cdot)$ *in the sense that, for each* $T < \infty$ *and as* $n \to \infty$,

$$\sup_{t \le T} \sup_{-\bar{\theta} \le \theta \le 0} |\bar{v}_a(t,\theta) - \bar{x}_a^n(t,\theta)| \to 0.$$

Theorem 1.4. *Assume the conditions of Theorem 1.3 with (A1.3) replacing (A1.2). Then the conclusions of of Theorem 1.3 continue to hold.*

4.2 Approximations by Time-Varying Delays: Only Path Delayed

In this section, the path memory segments that approximate $\bar{x}(t)$ will be denoted by either $\bar{x}_a^\delta(t)$ or $\bar{x}^{\delta_0}\delta_a(t)$, where δ or (δ, δ_0) are parameters of the approximation and the subscript "a" denotes the type of approximation. These are nonanticipative functions of the solution to the system equation.

The numerical problem for delay equations involves a lot of memory because of the necessity of approximating the memory segments. In order to reduce the memory requirements, it is necessary to simplify the model, without sacrificing too much accuracy, and some approaches to attaining that goal are discussed in this section for the case where only the path is delayed. We use the models (3.2.1) or (in relaxed control form) (3.2.2). The approximations (2.4)–(2.8) will be used in Chapters 7–9. Theorems 1.1–1.4 will be applied to the various approximations in this section.

4.2.1 Discretized Delays

Discretized delays. Let $\delta > 0$ be small and such that $Q_\delta = \bar{\theta}/\delta$ is an integer. A simple way to approximate the delays is to discretize them to the finite set of values $\{0, \delta, 2\delta, \ldots, Q_\delta \delta\}$. Suppose, for concreteness, that we move any delays on the interval $(i\delta, i\delta + \delta], i < Q_\delta$, to the point delay $i\delta + \delta$, thus possibly increasing delays slightly. The precise approximation is as follows. Let $\bar{x}_d^\delta(t)$ denote the function on $[-\bar{\theta}, 0]$ with values $\bar{x}_d^\delta(t, \theta)$ given by

$$\begin{aligned} \bar{x}_d^\delta(t, 0) &= x(t) \\ \bar{x}_d^\delta(t, \theta) &= \begin{cases} x(t-\delta), & \theta \in [-\delta, 0), \\ \vdots & \\ x(t-\bar{\theta}), & \theta \in [-\bar{\theta}, -\bar{\theta}+\delta). \end{cases} \end{aligned} \tag{2.1}$$

The system (3.2.2) is now replaced by

$$x(t) = x(0) + \int_0^t \int_U b(\bar{x}_d^\delta(s), \alpha) r(d\alpha\, ds) + \int_0^t \sigma(\bar{x}_d^\delta(s)) dw(s) + z(t). \tag{2.2}$$

The solution corresponding to the memory segment $\bar{x}_d^\delta(\cdot)$ will depend on δ, but until further notice we suppress that dependence for notational simplicity. Let $W^\delta(\hat{x}, r)$ denote the associated discounted cost and $V^\delta(\hat{x})$ the minimal value. This terminology will also be used for the other approximations that are to follow that depend on a single parameter δ. If the delays were uniformly distributed on $[-\bar{\theta}, 0]$, then this approximation increases the average delay by $\delta/2$.

Theorem 1.1 applies, so as $\delta \to 0$, the costs for the approximation converge to the cost for the original model. Although this approach will simplify the representation of the delays, it will not help the memory problem because in order to compute the values $x(t - i\delta), i\delta \le \bar{\theta}$, for all t, we still need to track the entire (infinite-dimensional) segments $\bar{x}(t)$.

The following approximations all yield processes with a finite-dimensional representation. With them the effective delays vary either periodically or are randomly varying in time. As seen in Section 3, they can be quite good. But, given the sensitivity of the behavior of many models with delays to slight variations in the dynamics, one must always exercise care in the use of any approximation. There are numerous variations of each of the suggested approximations. They vary in convenience of use, and the forms that are described should be taken as one choice among many alternatives

Approximation 1. A direct approximation of $\bar{x}(t)$. Let $l\delta \le t < l\delta + \delta$. Redefine $\bar{x}_d^\delta(t)$ to be the restriction to the interval $[-\bar{\theta}, 0]$ of a piecewise-constant interpolation of the values

$$\big(x(t), x(l\delta), \ldots, x(l\delta - \bar{\theta})\big) \tag{2.3}$$

instead of those in the set in (2.1). In particular, we use the restriction of the piecewise-constant interpolation of the following points:

$$\bar{x}_d^\delta(t, 0) = x(t),$$

$$\bar{x}_d^\delta(t, \theta) = \begin{cases} x(l\delta), & \theta = -(t - l\delta), \\ x(l\delta - \delta), & \theta = -(t - l\delta) - \delta, \\ \quad\vdots & \\ x(l\delta - Q_\delta\delta), & \theta = -(t - l\delta) - Q_\delta\delta. \end{cases} \tag{2.4}$$

The construction is illustrated by the solid line in Figure 2.1

The maximum delay is $\bar{\theta}$ when $t = l\delta$, and it increases to $\bar{\theta} + \delta$ as $t \to l\delta + \delta$, at which point it reverts back to $\bar{\theta}$. An approximation that yields better results in the simulations is the linear interpolation, given by the dotted line in the figure, although the advantage is slight for small values of δ, and it requires more computation. To the extent that the linear interpolation is accurate, the maximum effective delay with the linear interpolation used remains close to $\bar{\theta}$. Either approximation can be used as basis for the numerical algorithms in Chapters 7–9. There is a computational burden associated with the fact that the approximation shifts continuously in time, and this is the motivation for the Approximations 2 and 3. For the numerical approximations, the values of the state are discrete. To use either of the forms of Figure 2.1, one needs to keep track of the continuously varying quantity $\tau^\delta(t) = t - l\delta = t(\text{mod } \delta)$, for $l\delta \le t < l\delta + \delta$. This introduces a quantity that takes values in a continuum and that must be approximated (as in Approximations 4 or 5 below) in order to adapt the model for the numerical algorithms.

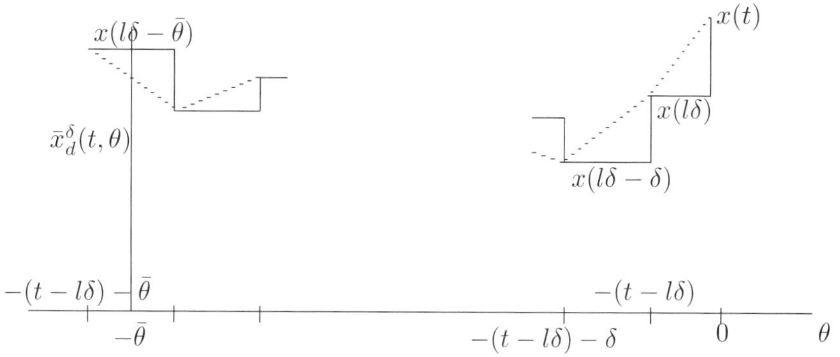

Figure 2.1. Illustration of (2.4).

4.2.2 Periodic Delays

Approximation 2. A periodic finite-memory approximation. The following approximation is an alternative interpolation of the values (2.3), but with the shift occurring at discrete times $i\delta, i = 1, 2, \ldots,$, rather than continuously. It requires less computation, although the effective delays vary periodically. Let $t \in [l\delta, l\delta+\delta)$. More precisely, for $l\delta \le t < l\delta+\delta$ let the memory segment be the function $\bar{x}_f^\delta(t)$ defined by:

$$\bar{x}_f^\delta(t,0) = x(t),$$
$$\bar{x}_f^\delta(t,\theta) = \begin{cases} x(l\delta), & \theta \in [-\delta, 0), \\ \vdots & \\ x(l\delta-\bar{\theta}+\delta), & \theta \in [-\bar{\theta}, -\bar{\theta}+\delta). \end{cases} \tag{2.5}$$

The construction is illustrated in Figure 2.2. If the delays are uniformly distributed on $[-\bar{\theta}, 0]$, then this procedure decreases the average delay by $\delta/2$. A piecewise-linear form can be used as well. Any approximation based on the set (2.3) and that converges in the sense of Theorems 1.1–1.3 can be used.

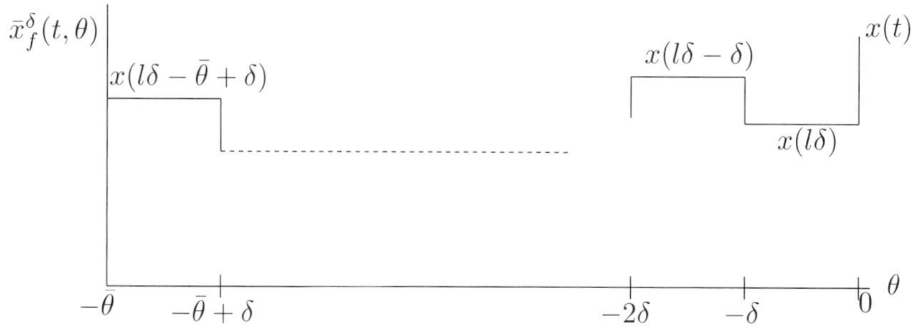

Figure 2.2. Path only delayed. Model (2.5), $l\delta \le t < l\delta+\delta$.

With the above model, we expect that the control at time $t \in [l\delta, l\delta + \delta)$ should depend only on $\bar{x}_f^\delta(t)$ and $t - l\delta$. For such controls, the process $X_n^\delta = (x(n\delta), x(n\delta - \delta), \ldots, x(n\delta - Q_\delta\delta + \delta))$ is Markov. Define $X^\delta(t) = (x(t), x(l\delta), \ldots, x(l\delta - Q_\delta\delta + \delta))$. The process defined by $(X^\delta(t), t - l\delta)$ for $t \in [l\delta, l\delta + \delta)$ is also Markov but with a periodic (period δ) transition function. The component $\tau^\delta(t) = t - l\delta = t(\text{mod } \delta)$ is necessary because (on each interval $[l\delta, l\delta + \delta)$) we need to keep track of the time that has elapsed since $l\delta$. For the numerical approximations in Chapters 7 and 8, the path values are discretized. In order to obtain finite-state approximations, it will be necessary to discretize the time as well. To prepare for this, we will introduce the Approximations 4 and 5 below. This point will be returned to in Chapter 7.

Approximation 3: Delays periodic in $\bar{\theta} \pm \delta/2$. In the model (2.5), the delays varied between $\bar{\theta} - \delta$ and $\bar{\theta}$. With the variation in the last paragraph, they varied between $\bar{\theta}$ and $\bar{\theta} + \delta$. Simulations (see the next section) show that the quality of the approximation is improved relative to those models if the maximum delay oscillates periodically between $\bar{\theta} - \delta/2$ and $\bar{\theta} + \delta/2$. To get such an approximation, define Q_δ^+ (assumed to be an integer) by $\bar{\theta} = (Q_\delta^+ + 1/2)\delta$ and let $l\delta \le t < l\delta + \delta$. Then define the memory segment $\bar{x}_p^\delta(t)$ as follows.

$$\bar{x}_p^\delta(t, 0) = x(t),$$

$$\bar{x}_p^\delta(t, \theta) = \begin{cases} x(l\delta), & \theta \in [-\delta/2, 0), \\ x(l\delta - \delta), & \theta \in [-\delta/2 - \delta, -\delta/2), \\ \vdots & \\ x(l\delta - Q_\delta^+\delta), & \theta \in [-\bar{\theta}, -\bar{\theta} + \delta). \end{cases} \tag{2.6}$$

The form is illustrated in Figure 2.3, and it is seen that the maximum delay varies periodically in the desired range. The value $\delta/2$ for the length of the rightmost interval was chosen because $t - l\delta$ varies from zero to δ as $t \to l\delta + \delta$, and the average value is $\delta/2$. This also explains the use of $\delta/2$ in (2.8) below. Other choices that maintain the correct mean values could be used as well. The use of a linear interpolation based on (2.6) gave slightly better results in simulations but involved more computation.

Discussion of the approximations and intervals in (2.6). Suppose that $b(\hat{x}, \alpha)$ has the form $b(x(t - \bar{\theta}), \alpha)$. With the approximation (2.6), as t varies from $l\delta$ and $l\delta + \delta$, the value $b(x(l\delta - Q_\delta^+\delta), \alpha)$ is used in the dynamical equation. Thus the delay varies between $\bar{\theta} - \delta/2$ to $\bar{\theta} + \delta/2$, with an average value of $\bar{\theta}$. Now consider the rightmost interval, whose length is $\delta/2$. For $t \in [l\delta, l\delta + \delta)$, we have $\bar{x}_p^\delta(t, -\delta/2) = x(l\delta)$. Thus the average delay of the approximation at $\theta = -\delta/2$ is just $\delta/2$. Such considerations motivate the choice of the intervals in (2.6). With the particular form for $b(\cdot)$ in this example, the values of the approximation on the interval $\theta \in [-\delta/2, 0]$ and,

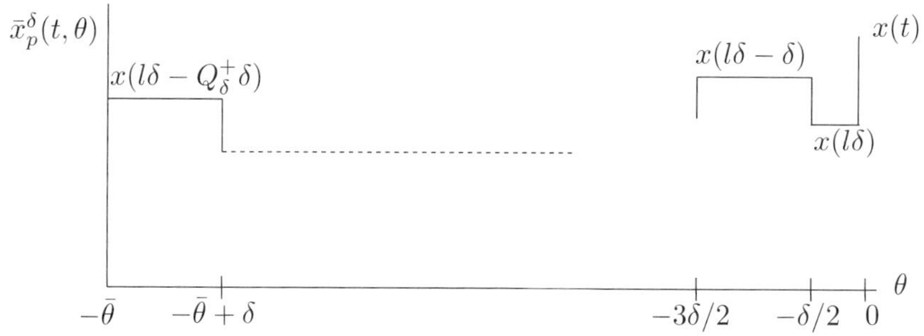

Figure 2.3. Illustration of $\bar{x}_p^\delta(t)$. Model (2.6), $l\delta \le t < l\delta + \delta$.

indeed, on $(-\bar{\theta}, 0)$ are irrelevant. But we need to keep track of the values $\tilde{X}^\delta(t) = \big(x(t), x(l\delta), x(l\delta - \delta), \ldots, x(l\delta - Q_\delta^+\delta)\big)$, no matter how they are interpolated.

Suppose now that the delay for the original problem is uniformly distributed on $[0, \bar{\theta}]$ in that $\mu_c(\cdot)$ is Lebesgue measure on $[-\bar{\theta}, 0]$. Let (2.6) (i.e., the form in Figure 2.3) be used for the approximation to the memory segment. Thus the distribution on $[-\bar{\theta}, 0]$ in Figure 2.3 is Lebesgue measure for each t. The approximation (2.6) concentrates the delay at the following values:

$$\left(Q_\delta^+\delta + (t - l\delta), \ldots, \delta + (t - l\delta), t - l\delta\right).$$

At time t, the interpolation effectively assigns a measure of value δ to each of the values, except the rightmost, to which the value $\delta/2$ is assigned. As t varies between $l\delta$ and $l\delta + \delta$, the range of the delays that are associated with the right-hand interval varies between $[0, 0]$ and $[0, \delta]$, with an average value $\delta/2$, which again helps justify the interpolation (2.6).

4.2.3 Randomly Varying Delays

For Approximation 3 and $l\delta \le t < l\delta + \delta$, one must keep track of the elapsed time $t - l\delta$ since the last shift. This can be a problem when the model is adapted for use with the numerical procedure. If the Markov chain approximation is such that the "interpolation interval" is constant (which is not usually the case) for each value of the discretization parameter, then we can get a complete finite-state approximation for numerical purposes. This is discussed in Chapter 7. Even then, when the interpolation interval is small, the number of points required to keep track of time will be very large.

Such considerations imply that we need to modify our approach to the measurement of the passage of time between shifts so that only a finite number of values are required. Such modifications are described by Approximations 4 and 5. In these approaches, time advances "randomly," and is determined by a renewal process with either exponentially or Erlang distributed intervals.

Approximation 4. Continue to suppose that only the path is delayed. The following model, where the delay varies randomly, is covered by Theorems 1.1

and 1.3. Let time be divided into mutually independent and nonanticipative intervals whose lengths are exponentially distributed and have mean length δ. In detail, let $\{v_n^\delta\}$ be i.i.d., random variables that are exponentially distributed with mean δ. They are the intervals between jumps of a Poisson process whose jump rate at time t, conditional on all data to t, is $1/\delta$. Thus they are independent of the Wiener process. Define $v_0^\delta = 0$ and $\sigma_n^\delta = \sum_{i=0}^{n-1} v_i^\delta$. The σ_n^δ are called the shift times or the δ-shift times for this approximation.

The memory will consist of the values of the path at the current time and at the most recent $Q_\delta = \bar{\theta}/\delta$ renewal or shift times. Thus the effective delay is randomly time-varying. In particular, define the path memory segment $\bar{x}_r^\delta(t)$ as follows. For $t \in [\sigma_l^\delta, \sigma_{l+1}^\delta)$,

$$\bar{x}_r^\delta(t,0) = x(t),$$
$$\bar{x}_r^\delta(t,\theta) = \begin{cases} x(\sigma_l^\delta), & -\delta \le \theta < 0, \\ \vdots & \\ x(\sigma_{l-Q_\delta+1}^\delta), & -\bar{\theta} \le \theta < \bar{\theta} + \delta. \end{cases} \tag{2.7}$$

See the illustration of $\bar{x}_r^\delta(t)$ in Figure 2.4.

If the subscript k in a term σ_k^δ in (2.7) is negative, then use the value of the initial condition at $\theta = -k\delta$. The mean value of $\sigma_l^\delta - \sigma_{l-Q_\delta+1}^\delta$ is $(Q_\delta - 1)\delta = \bar{\theta} - \delta$. The time between t and the previous renewal moment is exponentially distributed with mean δ [43, Chapter 5, Equation (6.5)]. Thus the average maximum delay is $\bar{\theta}$.

For $\sigma_l^\delta \le t < \sigma_{l+1}^\delta$ and (2.7) used, we expect that it is sufficient to have the control at time t dependent on $X_r^\delta(t) = (x(t), x(\sigma_l^\delta), x(\sigma_{l-1}^\delta), \ldots, x(\sigma_{l-Q_\delta+1}^\delta))$ only. Then $X_r^\delta(\cdot)$ is Markov and so is the discrete-parameter processes $X_{r,l}^\delta = (x(\sigma_l^\delta), x(\sigma_{l-1}^\delta), \ldots, x(\sigma_{l-Q_\delta+1}^\delta)), l = 0, 1, \ldots$. Because of the fact that the intervals are exponentially distributed and i.i.d., we need not keep track of the time since the last shift. Thus the state space is simpler than that for the previous case. This advantage is offset by the fact that the intervals are random.

Simulations indicate that there is too much randomness for such an approximation to be of general use, unless δ is small. The dimensionality of the path representation is inversely proportional to δ, so we prefer to approximate the passage of time by means other than by making δ very small. The following elaboration addresses this issue, by using a "periodic delay" approximation based on an Erlang distribution of high order.

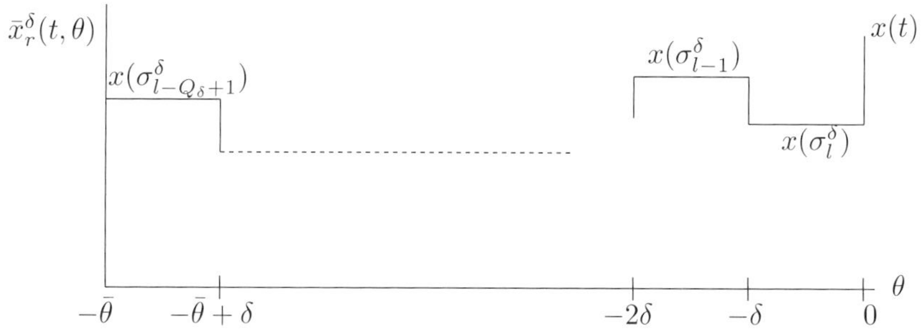

Figure 2.4. An illustration of $\bar{x}_r^\delta(t)$ defined by (2.7), for $t \in [\sigma_i^\delta, \sigma_{l+1}^\delta)$.

4.2.4 Periodic-Erlang Delays

Approximation 5. A compromise between (2.6) and (2.7): A "finer" random approximation. Equations (2.5) and (2.6) illustrated approximations that, for $t \in [l\delta, l\delta+\delta)$, required keeping track of the time $t - l\delta$. The approximation (2.7) did not require such a component of the memory, but the effective delay intervals were random. For small δ, this might not be a problem, but there are intermediate models between (2.5) (or (2.6)) and (2.7) that are well worth considering. One such model replaces the exponentially distributed intervals v_n^δ by Erlang distributed intervals, so that there is less randomness, although the memory requirement is increased.

Define the integer Q_δ^+ and the real number $\delta > 0$ by $\bar{\theta} = (Q_\delta^+ + 1/2)\delta$, as above (2.6). Let $\delta > \delta_0 > 0$ with $\bar{L}^{\delta_0,\delta} = \delta/\delta_0$, an integer. Each interval of length δ will be approximated by a sum of $\bar{L}^{\delta_0,\delta}$ random subintervals, with the length of each subinterval having an exponential distribution with mean δ_0. Thus the intervals between renewal times are sums of $\bar{L}^{\delta_0,\delta}$ of the subintervals and have an Erlang distribution of order $\bar{L}^{\delta_0,\delta}$ and total mean δ.

Now we give the formal definition. Let the random variables $\{v_n^{\delta_0}\}$ have the properties of the $\{v_n^\delta\}$ defined above (2.7), but with mean δ_0. Define $v_0^{\delta_0} = 0$ and $\sigma_n^{\delta_0} = \sum_{i=1}^{n-1} v_i^{\delta_0}$. We will divide $\{v_n^{\delta_0}\}$ into groups of $\bar{L}^{\delta_0,\delta}$ random variables as follows. Define $\bar{\sigma}_0^{\delta_0,\delta} = 0$ and

$$\bar{\sigma}_n^{\delta_0,\delta} = \sum_{i=1}^{n\bar{L}^{\delta_0,\delta}} v_i^{\delta_0}, \quad n = 1, 2, \ldots.$$

The intervals between renewal are $\bar{\sigma}_{l+1}^{\delta_0,\delta} - \bar{\sigma}_l^{\delta_0,\delta}, l = 0, 1 \ldots$. The intervals are mutually independent and are intended to approximate the constant interval of length δ. The renewal times $\bar{\sigma}_n^{\delta_0,\delta}$ are also called the shift times or δ-shift times. Define the path memory segment $\bar{x}_e^{\delta_0,\delta}(t)$ as follows. For $\bar{\sigma}_l^{\delta_0,\delta} \le t < \bar{\sigma}_{l+1}^{\delta_0,\delta}$,

$$\bar{x}_e^{\delta_0,\delta}(t,0) = x(t),$$

$$\bar{x}_e^{\delta_0,\delta}(t,\theta) = \begin{cases} x(\bar{\sigma}_l^{\delta_0,\delta}), & -\delta/2 \le \theta < 0, \\ \vdots \\ x(\bar{\sigma}_{l-Q_\delta^+}^{\delta_0,\delta}), & -\bar{\theta} \le \theta < \bar{\theta} + \delta. \end{cases} \tag{2.8}$$

The average value of $\bar{\sigma}_l^{\delta_0,\delta} - \bar{\sigma}_{l-Q_\delta^+}^{\delta_0,\delta}$ is $Q_\delta^+\delta = \bar{\theta} - \delta/2$. Because the intervals $\bar{\sigma}_{n+1}^{\delta_0,\delta} - \bar{\sigma}_n^{\delta_0,\delta}$ are Erlang distributed with mean δ and order $\bar{L}^{\delta_0,\delta}$, the approximation to a constant interval of length δ is better. The mean time between a random time and the last renewal time $\sigma_l^{\delta_0,\delta}$ before it is harder to calculate, but it is between $\delta/2$ and δ. As $\bar{L}^{\delta_0,\delta} \to \infty$, it converges to $\delta/2$. This argument provides additional support for our use of the interval $\delta/2$ in the top line of the brackets in (2.8).

Comment. When these approximations are adapted to the numerical problem in Chapters 7 and 8, a major concern is simplicity and the tracking of a state that can be embedded into a finite-state Markov chain and used in the dynamic programming equation. The representation of the path needs to be simple, and our choices satisfy that need. We have tried to motivate the particular choices that are made, but they are not exclusive and much more thought is required. We will use similar interpolations when the delayed control problem is discussed in Section 4. But there is no intrinsic requirement that the path and control have approximations of the same form.

The "Erlang" state. It will be helpful to formalize the process that marks the passage of time. The right-continuous Erlang state $L^{\delta_0,\delta}(t)$ is defined to be the number of δ_0-intervals that have passed since the last δ-shift, and it evolves as a cyclic Poisson process with $\bar{L}^{\delta_0,\delta}$ values, where the state moves $0 \to 1 \to 2 \cdots \to \bar{L}^{\delta_0,\delta} - 1 \to 0$, with the rate for each transition being $1/\delta_0$. Equivalently, the Erlang process changes values at the times $\sigma_n^{\delta_0}, n = 1, 2 \ldots$.

Let i and l be such that $\bar{\sigma}_l^{\delta_0,\delta} \le \sigma_i^{\delta_0} \le t < \sigma_{i+1}^{\delta_0} \le \bar{\sigma}_{l+1}^{\delta_0,\delta}$. If the control at time t depends only on $\left(x(t), x(\bar{\sigma}_l^{\delta_0,\delta}), \ldots, x(\bar{\sigma}_{l-Q_\delta^+}^{\delta_0,\delta})\right)$ and $L^{\delta_0,\delta}(t)$ (where, in this case, $L^{\delta_0,\delta}(t) = i - l\bar{L}^{\delta_0,\delta}$), then this process is Markov. We will then need to keep track of the $L^{\delta_0,\delta}(t)$, which takes only $\bar{L}^{\delta_0,\delta}$ values, and this is simpler than keeping track of the continuous variable t, modulo δ. The intervals have an Erlang distribution of high order, which is preferable to the exponential distribution. Thus we have a compromise between (2.5) (or (2.6)) and (2.7). The interpolations will be discussed further in the next section in connection with the simulations.

Note that while the passage of time is determined by the renewal process, the interpolations $\bar{x}_r^\delta(t)$ and $\bar{x}_e^{\delta_0,\delta}(\cdot)$ defined by (2.7) or (2.8) are piecewise-constant with interval δ (or $\delta/2$ for the rightmost interval). In principle, one could construct interpolations by replacing these constant interpolation intervals by the appropriate random renewal intervals, but this would entail the use of a random (and potentially unbounded) number of path values and of the associated values of the random intervals, all of which is to be avoided.

4.2.5 Convergence of Costs and Existence of Optimal Controls

The results of Section 3.5 concerning the existence of optimal controls for the various approximations and those of Section 1 concerning the continuity of the cost function with respect to the approximation parameters extend as follows, for the case where only the path memory segment is approximated. The proofs of existence of optimal controls are like that in Section 3.5. The proofs of the convergence of the various approximations as the approximation parameters go to their limits use Theorems 1.3 and 1.4. Theorem 1.4 is required for the assertions concerning Approximation 5, the periodic-Erlang model, where the memory segments and the path values at the shift times converge to those for the periodic model (Approximation 3) as $\delta_0 \to 0$. The details are omitted.

Theorem 2.1. *Use the system* (3.2.2) *and cost function* (3.4.3), *with* $\bar{x}_a^\delta(t)$ *replacing* $\bar{x}(t)$, *where* $\bar{x}_a^\delta(t)$ $(a = d, f, p, r)$ *is defined by any of the approximations given by* (2.1) *or* (2.4)–(2.7). *Let the cost and optimal cost be denoted by* $W^\delta(\hat{x}, r)$ *and* $V^\delta(\hat{x})$, *resp., for all cases. Assume* (A3.1.1), (A3.1.2), (A3.2.1)–(A3.2.3) *and* (A3.4.3). *Let* δ *be fixed. Then the conclusions of Theorems* 3.5.1–3.5.4 *hold as they pertain to these models. As* $\delta \to 0$, $W^\delta(\hat{x}, r) \to W(\hat{x}, r)$ *uniformly in* $r(\cdot)$ *and in* $\hat{x}$ *in any compact set in* $C(G; [-\bar{\theta}, 0])$. *Also,* $V^\delta(\hat{x}) \to V(\hat{x})$. *Now suppose that the controls are restricted to depend on one of these approximations to the path memory segment and consider* ϵ'*-optimal controls with this dependence. Given* $\epsilon > 0$, *there is* $\epsilon' > 0$ *that goes to zero as* $\epsilon \to 0$ *such that for small enough* δ *(which can depend on* ϵ*), an* ϵ'*-optimal control in any of the classes of approximations to the memory segments is* ϵ*-optimal for* (3.2.2).

Next suppose that control stops when the boundary is reached, as for (3.4.2). *Drop* (A3.2.1) *and* (A3.2.2) *and assume* (A3.4.1) *and* (A3.4.2). *Then the conclusions continue to hold.*

Now assume the model (3.2.11) *with conditions* (A3.1.1), (A3.1.2) *and* (A3.2.4)–(A3.2.6). *Let any of the approximations* (2.1) *or* (2.4)–(2.7) *be used to approximate the path component in the dynamical terms, and use* $\mu_c^n(\cdot)$ *such that* $\mu_c^n(\cdot) \Rightarrow \mu_c(\cdot)$. *Then the conclusions continue to hold.*

Now assume the approximation (2.8) *and let* $W^{\delta_0,\delta}(\hat{x}, r)$ *and* $V^{\delta_0,\delta}(\hat{x})$ *denote the cost function and the optimal cost, resp. Then, as* $\delta_0 \to 0$, $V^{\delta_0,\delta}(\hat{x}) \to V^\delta(\hat{x})$ *and* $W^{\delta_0,\delta}(\hat{x}, r) \to W^\delta(\hat{x}, r)$, *uniformly in the control and initial condition. The analog of the assertion concerning* ϵ, ϵ'*-optimal controls holds. Optimal controls exist in all cases.*

4.2.6 Differential Operator for the Periodic-Erlang Approximation

Let the path only be delayed and let $t \in [\bar{\sigma}^{\delta_0}\delta_l, \bar{\sigma}^{\delta_0}\delta_{l+1})$ and $L^{\delta_0,\delta}(t) < \bar{L}^{\delta_0,\delta} - 1$. Define $\tilde{X}_e^{\delta_0,\delta}(t) = \left(x(t), x(\bar{\sigma}_l^{\delta_0,\delta}), \ldots, x(\bar{\sigma}_{l-Q_\delta^+}^{\delta_0,\delta})\right)$, where $x(\cdot)$ is the solution process under the path memory segment $\bar{x}_e^{\delta_0\delta}(\cdot)$ defined in (2.8). In the time interval of concern, the only component of $\tilde{X}_e^{\delta_0,\delta}(t)$ that can change is the first, namely $x(t)$. Define $a(\cdot) = \sigma(\cdot)\sigma'(\cdot)$. With control $u(t)$ being used at time t, define the differential generator $\mathcal{L}^{u(t)}$, acting on twice continuously differential functions with compact support, as

$$\begin{aligned}\mathcal{L}^{u(t)} f(\tilde{X}_e^{\delta_0,\delta}(t)) &= f'_{x(t)}(\tilde{X}_e^{\delta_0,\delta}(t)) b(x_e^{\delta_0,\delta}(t), u(t)) \\ &+ \frac{1}{2}\sum_{i,j} f_{x_i(t)x_j(t)}(\tilde{X}_e^{\delta_0,\delta}(t)) a_{ij}(\bar{x}_e^{\delta_0,\delta}(t)).\end{aligned} \tag{2.9}$$

With the periodic-Erlang approximation, we need to keep track of the Erlang state $L^{\delta_0,\delta}(t)$ with values in $0, 1, \ldots, \bar{L}^{\delta_0,\delta}-1$. Now let $f(\cdot)$ be a function of $\tilde{X}_e^{\delta_0,\delta}(t)$ and $\delta_0 L^{\delta_0,\delta}(t)$, the elapsed time since the last shift, as it is counted by the Erlang state $L^{\delta_0,\delta}(t)$. The differential operator of the joint process $(\tilde{X}_e^{\delta,\delta_0}(t), \delta_0 L^{\delta_0,\delta}(t))$ is

$$\begin{aligned}A^{u(t)} f(\tilde{X}_e^{\delta_0,\delta}(t), \delta_0 L^{\delta_0,\delta}(t)) &= \mathcal{L}^{u(t)} f(\tilde{X}_e^{\delta_0,\delta}(t), \delta_0 L^{\delta_0,\delta}(t)) \\ &+ \frac{1}{\delta_0}\left[f(\tilde{X}_e^{\delta_0,\delta}(t), \delta_0 L^{\delta_0,\delta}(t) + \delta_0) - f(\tilde{X}_e^{\delta_0,\delta}(t), \delta_0 L^{\delta_0,\delta}(t))\right].\end{aligned} \tag{2.10}$$

If $f(X, \cdot)$ is continuously differentiable in the second variable, then, for small δ_0, the last term of (2.10) is close to the derivative of $f(\tilde{X}_e^{\delta_0,\delta}(t), \delta_0 L^{\delta_0,\delta}(t))$ in the second variable. This implies that in the interval of concern and for small δ_0, time effectively advances continuously, from the perspective of the differential operator. In computing cost functionals or in solving the Bellman equation, the actual exponentially distributed random variables do not occur. It is only the distribution function of the process that matters, and its closeness to that of the original model. The above observations give additional support to the use of the approximation.

4.3 Simulations Illustrating the Model Approximations

4.3.1 Simulations Based on the Periodic Approximation

As previously noted, the behavior of delay systems is often quite sensitive to variations in the model, and one always needs to experiment with the form of the approximation before embarking on the numerical computations, keeping in mind that the original model will often itself be only an approximation to the physical model. But for many classes of problems, the approximations in the previous sections are very acceptable. This will be demonstrated for one class, arising in the modeling of a "fluid approximation" to a form of TCP (transfer control protocol) regulation of Internet congestion in [89]. The system is ([89, Equation (4.2)])

$$\begin{aligned} dx(t) = & \frac{(1.3)x(t-\bar{\theta})(1-f(t-\bar{\theta}))}{\bar{\theta}^2 x(t)} dt - .5x(t)x(t-\bar{\theta})f(t-\bar{\theta})dt \\ & + \sigma \left[x(t-\bar{\theta})f(t-\bar{\theta}) \right]^{1/2} x(t)dw(t), \end{aligned} \tag{3.1}$$

where the real-valued $x(t)$ denotes a rate of transmission, $\bar{\theta}$ is the round-trip delay between the source and the buffer/router, the "transmission rate decrease" factor due to a lost packet is 1/2, and $f(t)$ is a measure of packets that are lost due to buffer overflows.

The origin of the model is a single-source system. A nonacknowledgment of receipt of a packet at a router causes the source to reduce its rate of transmission. The effect of the reduction is seen at the router after a delay of the round-trip time $\bar{\theta}$. In this model, packets are queued at the router in a buffer of size B_0, and lost packets are due to buffer overflow. The function $f(t)$, the fraction of packets lost at time t, depends on the transmission rate, system congestion due to other users, and the router service rate and buffer size. Modifying [89, page 67] slightly, we use the loss probability for a $M/M/1$ queue with a buffer size $B_0 = 5$, a service rate of $c = 10$, and exogenous (and uncontrolled) input rate unity, and $\bar{\theta} = 0.4$. Then, for $\rho(t) = x(t) + 1$,

$$f(t) = \frac{1 - \rho(t)/c}{1 - \left[\rho(t)/c\right]^{B_0+1}} \left[\frac{\rho(t)}{c}\right]^{B_0+1}.$$

Let us start with the deterministic form, as the presence of noise can confuse the comparison. So, first, let $\sigma = 0$. Figure 3.1 illustrates a typical simulation for the periodic approximation (2.6), for which the average delay is just $\bar{\theta}$. The outer curve is the original model and the other is for $Q_\delta^+ = 4$, $\delta = \bar{\theta}/(Q_\delta^+ + 1/2)$. The relative behavior is similar for all initial conditions, and we see that the path approximation is very good.

With the Approximation 2 of (2.5), the average delay is increased by $\delta/2$ and the performance can be seriously affected. See Figure 3.2, where the delay is increased by $\delta/2$ and has the value 0.444. The upper curve is for the original model, and the lower curve for the approximation, which is seen to be poor. The periodic approximation form (2.6), maintaining the same mean delay, is always superior.

Figure 3.3 is for the original model with $\sigma = .1$. We can see that even a very small noise value changes the paths considerably. With the periodic model (2.6) and $Q_\delta^+ = 4$, the paths and cost values are very close, with

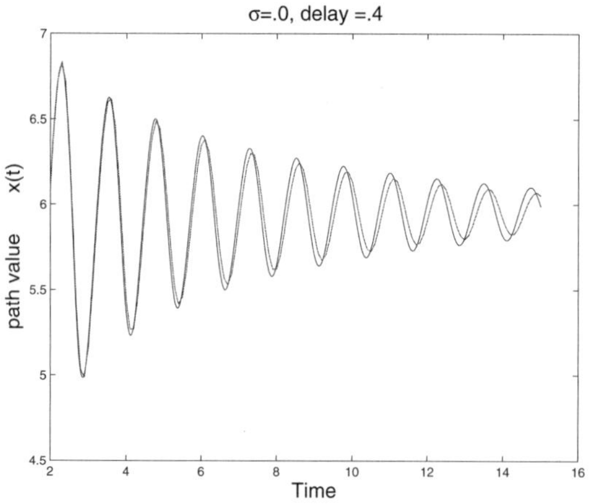

Fig. 3.1. Deterministic case: Approximation (2.6): original and periodic delay models, $Q_\delta^+ = 4$.

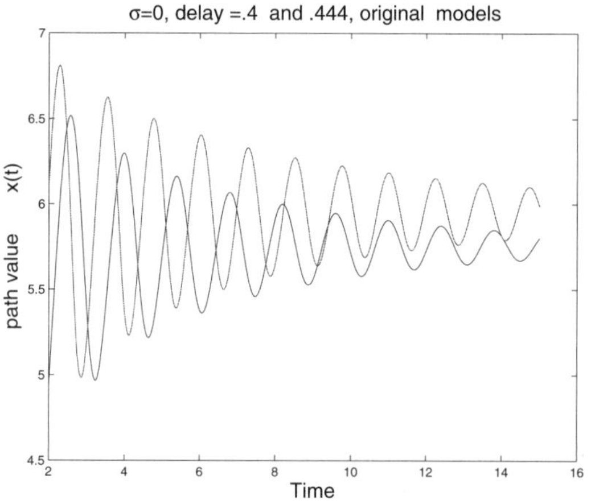

Fig. 3.2. Deterministic case: original model and model with delay increased by $\delta/2$.

sample differences in the means of the paths being 0.003 and in the variance being 0.005. Figure 3.4 plots a typical path of the original model on a larger time interval, to demonstrate the long-term behavior.

Figure 3.5 repeats the above for $\sigma = .3$. The sample mean and variance of the difference of the sample paths of the original model and those for the periodic model with $Q_\delta^+ = 4$ are .003 and .002, resp., with the values using $Q_\delta^+ = 3$ being only slightly higher. To facilitate comparisons, the same random numbers were used in the simulations of Figures 3.3–3.5, but the features of interest are typical. The values of the sample mean, variance, and of other simple cost functionals, are also very close. The use of the linear interpolation of Approximation 2 generally improves the approximation a little.

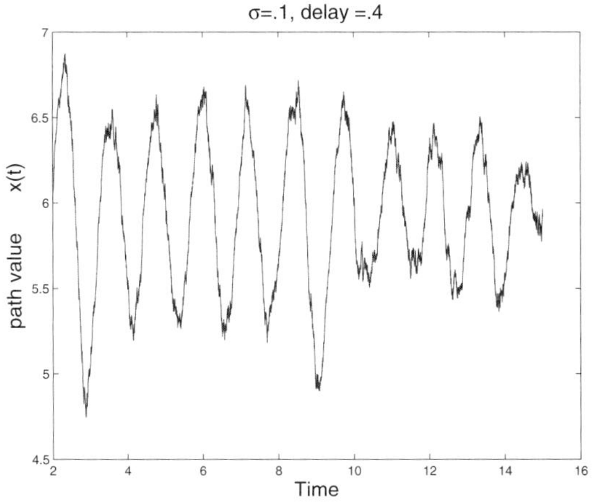

Fig. 3.3. Added noise, original model.

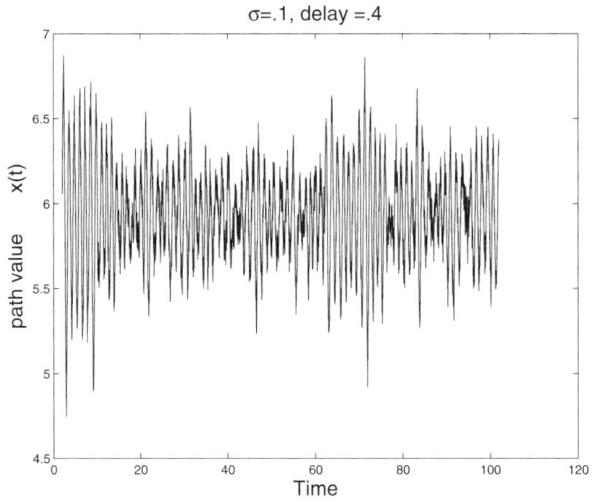

Fig. 3.4. Added noise, original model.

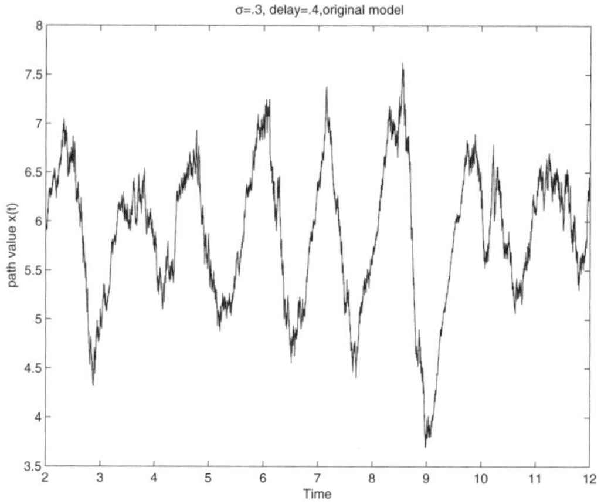

Fig. 3.5. Added noise.

4.3.2. Simulations Based on the Periodic-Erlang Approximation

The approximation based on (2.8) is illustrated in Figures 3.6 and 3.7.[1] Figure 3.6 compares the original model for $\sigma = 0$ with that for $\bar{L}^{\delta_0,\delta} = 20, Q_\delta^+ = 4$. The comparisons are similar for different initial conditions. With other sequences of random numbers used to get the "Erlang" intervals, the curves are still close with the slight shift of one over the other often reversed, so that on the average the approximation is very good. Keep in mind that when an approximation is used in the Bellman equation, there are no random numbers, only an average over the possibilities, which improves the quality of the approximation.

The initial condition that is used in these plots is "averaged" in that it uses a linear interpolation of the piecewise-constant approximation between the sample points. Generally, interpolating in this way gives a slight advantage. Figure 3.7 is for the same data, but with $\sigma = .1$. In the figure, the periodic-Erlang model has the sharpest and more extreme points in the first two cycles. This behavior is often reversed in plots based on other random numbers, so that the distribution of the paths for the original model and that for the approximation are close. With $Q_\delta^+ = 3$, the comparisons are similar but not as good.

[1] The random numbers used in Figure 3.7 were independent of those used in Figures 3.3–3.5

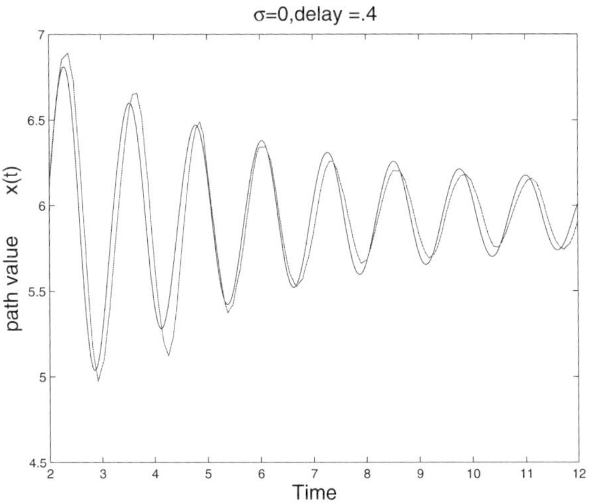

Fig. 3.6. Original model compared with periodic-Erlang delay approximation, $\bar{L}^{\delta_0,\delta} = 20, Q_\delta^+ = 4$.

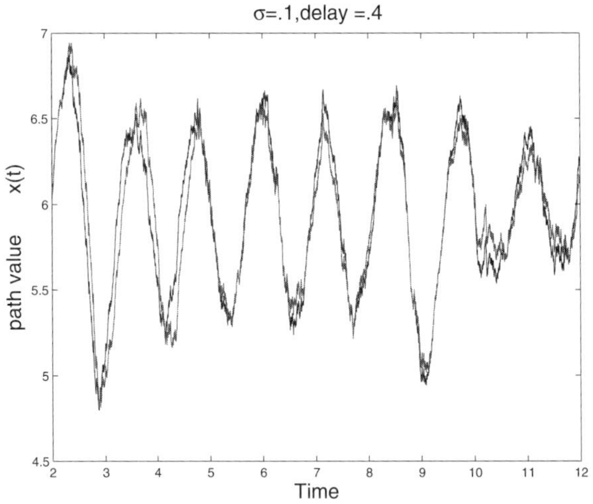

Fig. 3.7. Original model compared with periodic-Erlang delay approximation, $\bar{L}^{\delta_0,\delta} = 20, Q_\delta^+ = 4$.

4.4 Approximations: Path and Control Delayed

In this section, we will allow both the path and control to be delayed and discuss various approximations to the control memory segments. The approximation parameters will be δ for the periodic approximation and the pair δ_0, δ for the periodic-Erlang approximation, analogously to the usage for these forms in Section 2.

With ordinary controls used, the model with both the path and control delayed is (3.2.3). In relaxed control notation, the drift term is (3.1.6) and the full system is (3.2.4). If there are no boundary reflections, then delete the $z(\cdot)$-terms. The cost function is either (3.4.4) or (3.4.2). Unfortunately, in general, the control process $u(\cdot)$ has no *a priori* available useful regularity properties that can be exploited to get good approximations to the control memory segments. Because of this, in general, the control memory segment cannot be approximated in the manner that the path memory segment was, by using a piecewise-constant approximation to the process $u(\cdot)$ over the "memory interval." On the other hand, the fact that the relaxed control representation is continuous in time can be exploited and will be the basis of the approximation.

In the drift term (3.1.6), the control memory at time t can be represented as the derivative of the relaxed control over the time interval $[t-\bar{\theta}, t)$, namely $\{r'(d\alpha, t+\theta), \theta \in [-\bar{\theta}, 0)\}$. Recall the representation (3.1.8) where the derivative of the relaxed control was written as $\tilde{r}'(d\alpha, t, \theta)$. Because the approximation to the control memory segment at time t will depend on t, it is this form that will be approximated. Then for the periodic (parameter δ) approximations to the path and control memory segments, we write the approximating drift term as

$$\int_{-\bar{\theta}}^{0} \int_U b(\bar{x}_p^{\delta}(t), \alpha, \theta) \tilde{r}_p^{\delta,\prime}(d\alpha, t, \theta) \mu_c(d\theta). \tag{4.1}$$

The representation for the periodic-Erlang approximations will be written analogously.

The approximations that are to be used will yield a piecewise-constant control memory segment, whose values are obtained by averaging the controls over the intervals of constancy. This averaging procedure will yield the derivative of the relaxed control that is to be the approximation of the memory segment. There are many variations of the basic idea, and the particular forms that we describe are to be taken as suggestive of the possibilities. The described forms are successively less complicated from a numerical point of view. Analogs of Approximations 1 and 3 are described simply to illustrate how they might be adapted for use on the control memory segment. But they suffer from the same "memory" problems as when used for the path memory segment. For the path memory segment, the periodic-Erlang approximation resolved the main memory issues, but its adaptation to the control memory segment is still too complicated. This is partly resolved with the periodic-Erlang variant to be defined by Approximation 5a. All but Approximation 5a require tracking a quantity with values in a continuum and so would require additional approximations if they are to be adapted for numerical purposes.

An analog of Approximation 1. The following construction is an analog of what was done in Approximation 1 in Section 2 for the path memory segment. For $b > a$, define $r(d\alpha, [a, b]) = r(d\alpha, b) - r(d\alpha, a)$. For $\delta > 0$ and $l\delta \leq t < l\delta + \delta$, define the process $\tilde{r}_d^{\delta,\prime}(d\alpha, t, \theta)$ as the restriction to $\theta \in [-\bar{\theta}, 0]$ of the following function:

$$\tilde{r}_d^{\delta,\prime}(d\alpha, t, 0) = I_{\{u(t) \in d\alpha\}},$$

$$\tilde{r}_d^{\delta,\prime}(d\alpha, t, \theta) = \begin{cases} r(d\alpha, [l\delta, t])/(t - l\delta), & \theta \in [-(t - l\delta), 0), \\ r(d\alpha, [l\delta - \delta, l\delta])/\delta, & \theta \in [-(t - l\delta) - \delta, -(t - l\delta)), \\ \vdots & \\ r(d\alpha, [l\delta - \bar{\theta}, l\delta - \bar{\theta} + \delta])/\delta, & \\ \qquad \theta \in [-(t - l\delta) - \bar{\theta}, -(t - l\delta) - \bar{\theta} + \delta). & \end{cases} \tag{4.2}$$

A linear interpolation of (4.2) can be used as well. The process $r'(d\alpha, t)$ on the right side of (4.2) will generally depend on the parameter δ as the control that is actually used will generally depend on the approximation to the path memory segment that is used. But, to avoid excessive numbers of superscripts in the discussion of the various approximations, that dependence will be suppressed unless otherwise noted.

The problems with the use of (4.2) as a basis of the numerical approximation are similar to those raised for Approximation 1 in Section 2. One needs to keep track of the running time $t - l\delta$, and of the values of the relaxed control segment over this interval. The issues involved in the approximation of the control memory segment by a finite-state process are similar to those associated with the tracking of the time $t - l\delta$ in Approximations 2 and 3, and whose consideration led to the forms of Approximations 4 and 5, where time evolved "randomly." We next describe an analog of the periodic Approximation 3 and then analogs of the periodic-Erlang Approximation 5.

An analog of the periodic Approximation 3. Recall the definition of the integer $Q_\delta^+ : (Q_\delta^+ + 1/2)\delta = \bar{\theta}$. Let $l\delta \leq t < l\delta + \delta$. Then define $\tilde{r}_p^{\delta,\prime}(d\alpha, t, \theta)$ as follows.

$$\tilde{r}_p^{\delta,\prime}(d\alpha, t, 0) = I_{\{u(t) \in d\alpha\}},$$

$$\tilde{r}_p^{\delta,\prime}(d\alpha, t, \theta) = \begin{cases} r(d\alpha, [l\delta, t])/(t - l\delta), & \theta \in [-\delta/2, 0), \\ r(d\alpha, [l\delta - \delta, l\delta])/\delta, & \theta \in [-\delta/2 - \delta, -\delta/2), \\ \vdots & \\ r(d\alpha, [l\delta - Q_\delta^+\delta, l\delta - Q_\delta^+\delta + \delta])/\delta, & \theta \in [-\bar{\theta}, -\bar{\theta} + \delta). \end{cases} \tag{4.3}$$

An analog of the periodic-Erlang approximation for the control memory segment. Recall the notation of the periodic-Erlang Approximation 5 in Section 2. In this approximation, the passage of time was approximated by a cyclic Poisson process, taking the values δ_0 times $0, 1, \ldots, \bar{L}^{\delta_0,\delta} - 1, 0, 1, \ldots,$ in turn, with the value moving to the next step with rate $1/\delta_0$. Let $\bar{\sigma}_l^{\delta_0,\delta} \leq t < \bar{\sigma}_{l+1}^{\delta_0,\delta}$. An analogous approximation for the control memory segment is defined by $\tilde{r}_e^{\delta_0,\delta,\prime}(\cdot)$, as follows.

$$\tilde{r}_e^{\delta_0,\delta,\prime}(d\alpha, t, 0) = I_{\{u(t) \in d\alpha\}},$$

$$\tilde{r}_e^{\delta_0,\delta,\prime}(d\alpha, t, \theta) = \begin{cases} r(d\alpha, [\bar{\sigma}_l^{\delta_0,\delta}, t])/(t - \bar{\sigma}_l^{\delta_0 \delta}), & \theta \in [-\delta/2, 0), \\ r(d\alpha, [\bar{\sigma}_{l-1}^{\delta_0,\delta}, \bar{\sigma}_l^{\delta_0,\delta}])/\delta, & \theta \in [-\delta/2 - \delta, -\delta/2), \\ \vdots & \\ r(d\alpha, [\bar{\sigma}_{l-Q_\delta^+}^{\delta_0,\delta}, \bar{\sigma}_{l-Q_\delta^+ +1}^{\delta_0,\delta}])/\delta, & \theta \in [-\bar{\theta}, -\bar{\theta} + \delta). \end{cases} \tag{4.4}$$

Recall the discussion below Figure 2.3 concerning the motivation for the division of the interval $[-\bar{\theta}, 0]$ that is used in (4.4). As noted there, this is well motivated but is only one of many possibilities. Tracking the first line in the bracketed term is still a problem, as it involves tracking a continuous time variable. In the next approximation $t - \bar{\sigma}^{\delta_0}\delta_l$ will be approximated by $\delta_0 L^{\delta_0}\delta(t)$.

A numerically convenient form of the periodic-Erlang approximation for the control memory segment: Approximation 5a. The Approximation 5 of the path memory segment is useful for the numerical procedure. But the analog for the control memory segment is still too complicated. Henceforth suppose that U contains only a finite number of points $(\alpha_1, \ldots, \alpha_K)$. Then (4.4) requires tracking the time that control $u(t)$ takes the value $\alpha_i, i \leq K$, on each of the random time intervals $[\bar{\sigma}_l^{\delta_0,\delta}, t]$ or $[\bar{\sigma}_l^{\delta_0,\delta}, \bar{\sigma}_{l+1}^{\delta_0,\delta}]$. This time takes values in a continuum and will have to be discretized. Such issues will be addressed in Chapter 8, when the associated numerical problems are discussed. In preparation for that discussion, we will describe a form of the periodic-Erlang approximation for the discretization of the range of the values of $\tilde{r}_e^{\delta_0,\delta,\prime}(d\alpha, t, \theta)$.

Recall the definition of the Erlang state $L^{\delta_0}\delta(t)$ and its transitions, given below (2.8). We will now define a right-continuous process, $N_l^{\delta_0,\delta}(\alpha, \cdot)$, whose purpose is to count the number of times that the control takes the value α at the moments of change in the Erlang state in the lth cycle, where $\bar{\sigma}_l^{\delta_0,\delta} \leq t < \bar{\sigma}_{l+1}^{\delta_0,\delta}$. This quantity will be the basis of a feasible numerical approximation to (4.4). The precise definition is as follows. If $L^{\delta_0,\delta}(t)$ increases at time t, then define[2]

[2] In the definitions, we supposed that the controls are ordinary and not relaxed as that would be the case in the numerical approximations. If the controls are

$$N_l^{\delta_0,\delta}(\alpha,t) = N_l^{\delta_0,\delta}(\alpha,t-) + I_{\{u^{\delta_0,\delta}(t)=\alpha\}}.$$

If

$$L^{\delta_0,\delta}(t-) = \bar{L}^{\delta_0,\delta} - 1 \to L^{\delta_0,\delta}(t) = 0,$$

then $t = \bar{\sigma}_{l+1}^{\delta_0\delta}$, the lth cycle ends, the $(l+1)$th cycle begins, we define $\bar{N}^{\delta_0}\delta_l(\alpha) = N_l^{\delta_0,\delta}(\alpha,t-)$, and start the new cycle with the value

$$N_{l+1}^{\delta_0,\delta}(\alpha,t) = I_{\{u(t)=\alpha\}}.$$

The first term in the brackets in (4.4) is to be approximated by

$$N_l^{\delta_0,\delta}(\alpha,t)/[L^{\delta_0,\delta}(t)+1], \tag{4.5}$$

which is the fraction of times from the beginning of the lth cycle until the present at which the Erlang state has increased and the control took the value α simultaneously. The $r^{\delta_0}\delta_e(d\alpha,[\bar{\sigma}_{l-1}^{\delta_0,\delta},\bar{\sigma}_l^{\delta_0,\delta}])/\delta$ terms in (4.4) are to replaced by the terms

$$\frac{\bar{N}_{l-1}^{\delta_0,\delta}(\alpha)}{\bar{L}^{\delta_0,\delta}}, \tag{4.6}$$

which is the fraction of times in $(l-1)$st cycle that the control took the value unity when the Erlang state increased.

Summarizing, for $t \in [\bar{\sigma}_l^{\delta_0,\delta},\bar{\sigma}_{l+1}^{\delta_0,\delta})$, the approximation to the control memory segment that was just defined can be written as $\tilde{r}_{ee,l}^{\delta_0,\delta,\prime}(\cdot)$, where

$$\tilde{r}_{ee,l}^{\delta_0,\delta,\prime}(d\alpha,t,0) = I_{\{u(t)\}=\alpha\}},$$

$$\tilde{r}_{ee,l}^{\delta_0,\delta,\prime}(d\alpha,t,\theta) = \begin{cases} \dfrac{N_l^{\delta_0,\delta}(\alpha,t)}{L^{\delta_0,\delta}(t)+1}, & \theta \in [-\delta/2,0), \\ \dfrac{\bar{N}_{l-1}^{\delta_0,\delta}(\alpha)}{\bar{L}^{\delta_0,\delta}}, & \theta \in [-\delta/2-\delta,-\delta/2), \\ \vdots & \\ \dfrac{N_{l-Q_\delta^+}^{\delta_0,\delta}(\alpha)}{\bar{L}^{\delta_0,\delta}}, & \theta \in [-\bar{\theta},-\bar{\theta}+\delta). \end{cases} \tag{4.7}$$

Theorem 4.1. *Assume the conditions of Theorem 2.1, but with the control delayed as in this section, with $\mu_c(\cdot), b(\cdot)$, and $k(\cdot)$ satisfying* (A3.1.3). *For each of the approximations of this section, with the analogous forms used for the path memory segment, there is an optimal control. The cost and optimal cost functions for the periodic Approximation* 3 *converge to those for the original unapproximated model as $\delta \to 0$. If δ is fixed, then the cost and optimal cost*

---

relaxed but not ordinary controls, then in the various constructions, at time t use $r'(d\alpha,t)$ in (4.4). Theorem 4.1 will hold with this change.

functions for the periodic-Erlang Approximations 5 *and* 5a ((5a) *being used for the control memory segment) converge to those for the periodic Approximation* 3, *as* $\delta_0 \to 0$. *The assertions concerning* (ϵ, ϵ')*-optimal controls in Theorem* 2.1 *hold.*

Outline of the proof. We will outline the argument for the convergence of the periodic-Erlang approximations, starting with Approximation 5 used for both the path and control memory segments. Fix $\delta > 0$. Then $\bar{x}_e^{\delta_0\delta}(t)$ and $\tilde{r}_e^{\delta_0,\delta,\prime}(d\alpha, t, \theta)$ denote the memory segments corresponding to (2.8) and (4.4) at time t. In this proof, we will index the associated solution process and relaxed control by the pair (δ_0, δ). Thus $x^{\delta_0,\delta}(\cdot)$ denotes the solution process and $r^{\delta,\delta_0}(\cdot)$ the relaxed control representation of the control process from which the $\bar{x}_e^{\delta_0\delta}(t)$ and $\tilde{r}_e^{\delta_0,\delta,\prime}(d\alpha, t, \theta)$ are derived via (2.8) and (4.4).

For notational convenience, we suppose that the same Wiener process can be used for all δ_0. Then we can write

$$\begin{aligned} x^{\delta_0}\delta(t) = x(0) &+ \int_0^t ds \int_{-\bar{\theta}}^0 \int_U b(\bar{x}^{\delta_0}\delta_e(s), \alpha, \theta)\tilde{r}_e^{\delta_0,\delta,\prime}(d\alpha, s, \theta)\mu_c(d\theta) \\ &+ \int_0^t \sigma(\bar{x}^{\delta_0}\delta_e(s))dw(s) + z^{\delta_0}\delta(t). \end{aligned} \tag{4.8}$$

The set

$$\left(x^{\delta_0}\delta(\cdot), r^{\delta_0}\delta(\cdot), z^{\delta_0}\delta(\cdot), w(\cdot), \{\bar{\sigma}^{\delta_0}\delta_l, l < \infty\}\right) \tag{4.9}$$

is tight. Extract a weakly convergent subsequence (indexed by δ_0 for convenience) and with limit $(x^\delta(\cdot), r^\delta(\cdot), z^\delta(\cdot), w(\cdot), \{l\delta, l < \infty\})$. Use the Skorokhod representation so that we can assume w.p.1 convergence. By the convergence, the continuity of $x^\delta(\cdot)$ and the fact that $\bar{\sigma}^{\delta_0}\delta_l \to l\delta$, we have the convergence $\bar{x}^{\delta_0}\delta_e(t) \to \bar{x}_p^\delta(t)$ for all t, where $\bar{x}_p^\delta(\cdot)$ is defined by (2.6), with $x^\delta(\cdot)$ replacing $x(\cdot)$ on the right-hand side of (2.6).

By the above arguments and the continuity properties of $b(\cdot)$ in (A3.1.3), the difference between the drift term in (4.8) and

$$\int_0^t ds \int_{-\bar{\theta}}^0 \int_U b(\bar{x}_p^\delta(s), \alpha, \theta)\tilde{r}_e^{\delta_0,\delta,\prime}(d\alpha, s, \theta)\mu_c(d\theta) \tag{4.10}$$

goes to zero as $\delta_0 \to 0$. We would like to show that, asymptotically, we can replace the $\tilde{r}_e^{\delta_0,\delta,\prime}(d\alpha, s, \theta)$ in (4.10) by $\tilde{r}_p^{\delta,\prime}(d\alpha, s, \theta)$, where $\tilde{r}_p^{\delta,\prime}(\cdot)$ is obtained from $r^\delta(\cdot)$. To do this we need to show that

$$\int_0^t ds \int_{-\bar{\theta}}^0 \int_U b(\bar{x}_p^\delta(s), \alpha, \theta) \left[\tilde{r}_e^{\delta_0,\delta,\prime}(d\alpha, s, \theta) - \tilde{r}_p^{\delta,\prime}(d\alpha, s, \theta)\right] \mu_c(d\theta)$$

goes to zero as $\delta_0 \to 0$. But this follows from the fact that $b(\bar{x}_p^\delta(s), \alpha, \theta)$ is continuous in (s, α, θ) for $s \in (l\delta, l\delta + \delta), l = 0, 1, \ldots$, the fact that $\bar{\sigma}_l^{\delta_0,\delta} \to l\delta$, both w.p.1 and in mean, and the constructions (4.3) and (4.4). From this point on, the details of the proof for Approximation 5 follow the lines of Theorem 3.5.1, and are omitted.

Now we turn to the case where (2.8) is used for the path memory segment and Approximation 5a, defined by (4.7), is used for the control memory segment. Let the processes in (4.9) now denote those resulting from the use of these approximations (2.8) and (4.7). As above, the set (4.9) is tight and we use the same notation $(x^\delta(\cdot), \ldots)$ for the limits. Assume Skorokhod representation.

The numerators of the terms in the brackets in (4.7) are values of point processes, as is $L^{\delta_0,\delta}(t)$. We will decompose (δ_0 times) these terms in terms of compensators and martingales. For the zeroth cycle, where $t < \bar{\sigma}_1^{\delta),\delta}$, we can write the compensator-martingale decomposition as

$$\delta_0 L^{\delta_0,\delta}(t) = t + M^{\delta_0}\delta_L(t),$$

$$\delta_0 N^{\delta_0}\delta_0(\alpha, t) = \int_0^t I_{\{u^{\delta_0}\delta(s)=\alpha\}} ds + M^{\delta_0}\delta_N(t),$$

where the quadratic variation of the martingale $M^{\delta_0}\delta_L(\cdot)$ is $\delta_0 t$, and that of the martingale $M^{\delta_0}\delta_N(\cdot)$ is $\delta_0 \int_0^t I_{\{u^{\delta_0}\delta(s)=\alpha\}} ds$. There is an analogous decomposition on any interval $[\bar{\sigma}^{\delta_0}\delta_l, \bar{\sigma}^{\delta_0}\delta_{l+1})$. For $\bar{\sigma}_l^{\delta_0,\delta} \le t < \bar{\sigma}_{l+1}^{\delta_0,\delta}$, we have

$$\int_{\bar{\sigma}_l^{\delta_0,\delta}}^t I_{\{u^{\delta_0}\delta(s)=\alpha\}} ds = r^{\delta_0}\delta(d\alpha, [\bar{\sigma}_l^{\delta_0,\delta}, t]).$$

This, the weak convergence, and the fact that the quadratic variations of the martingales are proportional to δ_0, implies that for all $l = 0, 1, \ldots$, as $\delta_0 \to 0$ we have the convergence

$$\begin{aligned} \frac{N_l^{\delta_0,\delta}(\alpha, t)}{L^{\delta_0,\delta}(t) + 1} &\to \frac{r^\delta(d\alpha, [l\delta, t])}{t - l\delta}, \\ \frac{\bar{N}_l^{\delta_0,\delta}(\alpha)}{\bar{L}^{\delta_0,\delta}} &\to \frac{r^\delta(d\alpha, [l\delta, l\delta + \delta])}{\delta}, \end{aligned} \tag{4.11}$$

in mean, and this limit is the periodic approximation defined by (4.3).

Let $\tilde{r}_e^{\delta_0,\delta,\prime}(\cdot)$ denote the process that is computed from (4.4), but based on the relaxed control that is a consequence of the use of the memory segments defined by (2.8) and (4.7). Redefine $\bar{x}_p^\delta(\cdot)$ to be the process computed from (2.6), but based on the new limit $x^\delta(\cdot)$, and redefine $\bar{x}_e^{\delta_0,\delta}(\cdot)$ to be the process computed from (2.8), but based on the solution that is a consequence of the use of (4.7). As with (4.10), as $\delta_0 \to 0$ the difference between the drift term with the use of $\bar{x}_e^{\delta_0,\delta}(\cdot)$ and

$$\int_0^t ds \int_{-\bar{\theta}}^0 \int_U b(\bar{x}_p^\delta(s), \alpha, \theta) \tilde{r}_{ee}^{\delta_0,\delta,\prime}(d\alpha, s, \theta) \mu_c(d\theta) \tag{4.12}$$

goes to zero. The main issue concerns the relaxed control term in (4.12).

Using the above compensator-martingale decompositions and (4.11), we can write (4.12) as

$$\int_0^t ds \int_{-\bar{\theta}}^0 \int_U b(\bar{x}_p^\delta(s), \alpha, \theta) \tilde{r}_e^{\delta_0,\delta,\prime}(d\alpha, s, \theta)\mu_c(d\theta) + \epsilon^{\delta_0,\delta}(t), \tag{4.13}$$

where the error term is a consequence of the martingale error terms $M^{\delta_0}\delta_L(\cdot)$ and $M_N^{\delta_0}\delta(\cdot)$ in the above expansions. The sets of integrals (indexed by δ_0) in (4.12) and (4.13) are tight. Hence the set $\{\epsilon^{\delta_0}\delta(\cdot)\}$ indexed by δ_0 is tight. It follows from this and the fact that $E|\epsilon^{\delta_0}\delta(t)| \to 0$ as $\delta_0 \to 0$ that $\epsilon^{\delta_0,\delta}(\cdot)$ converges to zero. We now proceed as for the first case of the proof. ■

4.5 Singular Controls

Approximations analogous to those of the previous section can be written for the singular control models of Section 3.6. In the model (3.6.1), the singular control does not appear in delayed form, so only the path memory segment needs to be approximated, and this can be done with the methods of Section 2. We will briefly consider the model (3.6.2) where the singular control is delayed and at time t the control memory segment is $\bar{\lambda}(t) = \{\lambda(t) - \lambda(t+\theta), \theta \in [-\bar{\theta}, 0)\}$. Analogs of the periodic and periodic-Erlang approximations will be briefly commented on. The motivation for these approximations, as part of an effort to reduce the memory requirements, is the same as that for the case where the control is not singular.

A periodic approximation to the singular control memory segment. Let $\bar{\lambda}_p^\delta(t,\theta), \theta \in [-\bar{\theta}, 0)$ denote the periodic approximation to the control memory segment at time t, to be defined below. We extend the domain of the definition to the closed interval $[-\bar{\theta}, 0]$ by defining $d\bar{\lambda}_p^\delta(t,0) = d\lambda(t)$. For $l\delta \le t < l\delta + \delta$, a simple analog of the periodic approximation (4.3) to the control memory segment is defined by the piecewise-constant function whose differences at the change points are given by

$$d_\theta \bar{\lambda}_p^\delta(t,\theta) = \begin{cases} d\lambda(t), & \theta = 0, \\ \lambda([l\delta, t]), & \theta = -\delta/2, \\ \lambda([l\delta - \delta, l\delta]), & \theta = -\delta/2 - \delta, \\ \vdots & \\ \lambda([l\delta - Q_\delta^+, l\delta - Q_\delta^+ + \delta]), & \theta = -\bar{\theta}. \end{cases} \tag{5.1}$$

The associated solution is

$$
\begin{aligned}
dx(t) = dt \int_U c(x(t)) + dt \int_{-\bar{\theta}}^{0} b(x(t+\theta), \theta) d\mu_a(\theta) + \sigma(\bar{x}(t)) dw(t) \\
+ q_0(x(t-)) d\lambda(t) + dt \int_{\theta=-\bar{\theta}}^{0} q_2(x((t+\theta)-), \theta) d_\theta \bar{\lambda}_p^\delta(t, \theta) + dz(t).
\end{aligned}
\tag{5.2}
$$

Using the conditions and methods of Theorem 3.6.1, it can be proved that the solutions, controls, and costs converge to those for (3.6.2), as $\delta \to 0$. As for case of an ordinary control, one needs to keep track of the time that has elapsed since the start of the current cycle, and the next approximation is a first attempt to deal with this.

An analog of the periodic-Erlang Approximation 5. Recall the discussion concerning (4.4). Let $\bar{\lambda}_e^\delta(t, \theta), \theta \in [-\bar{\theta}, 0)$, denote the approximation to the control memory segment at time t. Extend the domain of definition to $[-\bar{\theta}, 0]$ by using the first line of (5.1). Let $\bar{\sigma}_l^{\delta_0 \delta} \le t < \bar{\sigma}_{l+1}^{\delta_0 \delta}$. Then an analog of (4.4) is

$$
d_\theta \bar{\lambda}_e^{\delta_0, \delta}(t, \theta) = \begin{cases} d\lambda(t), & \theta = 0, \\ \lambda([\bar{\sigma}_l^{\delta_0, \delta}, t]), & \theta = -\delta/2, \\ \lambda([\bar{\sigma}_{l-1}^{\delta_0, \delta}, \bar{\sigma}_l^{\delta_0, \delta}]), & \theta = -\delta/2 - \delta, \\ \vdots & \\ \lambda([\bar{\sigma}_{l-Q_\delta^+}^{\delta_0, \delta}, \bar{\sigma}_{l-Q_\delta^+ + 1}^{\delta_0, \delta}]), & \theta = -\bar{\theta}. \end{cases}
\tag{5.3}
$$

Using the conditions and methods of Theorem 3.6.1, it can be proved that the solutions, controls, and costs with the use of (5.3) converge to those for the periodic approximation (5.2), as $\delta_0 \to 0$. The next approximation completes the analog of (4.7), where both time and value advance discretely.

An analog of the periodic-Erlang Approximation 5a in (4.7). The approximation (5.3) needs to be simplified so that the memory has only a finite number of values. One major difference with the situation dealt with in the approximation (4.7) is that here there is no *a priori* bound on the increment in the singular control on any given time interval, so that an additional truncation will have to be done when the numerical approximation is dealt with. We also need to adjust the way that the memory increment is updated.

Let us consider a case where the singular control has only one component. Suppose that we only allow increments of size $k\delta_1, k = 0, 1, \ldots$. Then it can be shown that as $\delta_1 \to 0$, the costs will converge to those for the model without the discretization. With such a discretization, the procedure of Approximation 5a in (4.7) can be carried over. Drop the denominators in (4.7). Let $\bar{\sigma}_l^{\delta_0, \delta} \le t < \bar{\sigma}_{l+1}^{\delta_0, \delta}$. Let $N_l^{\delta_0, \delta}(\alpha, t)$ in (4.7) be replaced by the δ_1 times the number of times since the start of the current (lth) cycle that the singular control increased. For arbitrary l, replace $\bar{N}_l^{\delta_0,\delta}(\alpha)$ by δ_1 times the number of times in the lth cycle that the singular control has increased.

4.6 Rapidly Varying Delays

In this section, we will illustrate one way in which models with distributed delays arise. The delays in the physical model to be considered are time-varying and vary rapidly. It is shown that the model can be approximated by one with distributed delays, where the distribution is an asymptotic average of the delays in the physical models.

Rapidly time-varying delays in the drift function. Consider the model (3.2.4) with time varying delays. The rapidly varying delays in the path will be separated from the control and we use the system

$$\begin{aligned} dx^\epsilon(t) &= b^1(x^\epsilon(t+\theta_k^\epsilon(t)), k \le K)\, dt \\ &+ dt \int_{-\bar{\theta}}^0 \int_U b^2(\bar{x}^\epsilon(t), \alpha, v) r'(d\alpha, t+v)\mu_c(dv) + \sigma(\bar{x}^\epsilon(t))dw^\epsilon(t) + dz^\epsilon(t). \end{aligned} \tag{6.1}$$

The first term on the right models the time variations in the delays in the path. Rapidly varying delays in the control are much harder to deal with and are omitted. The cost function will be (3.4.4), but with cost rate

$$\begin{aligned} &k^1(x^\epsilon(t+\theta_k^\epsilon(t)), k \le K) \\ &\quad + \int_{-\bar{\theta}}^0 \int_U k^2(\bar{x}^\epsilon(t), \alpha, v) r'(d\alpha, t+v)\mu_c(dv). \end{aligned} \tag{6.2}$$

In addition to the cost rate (6.2), there is the cost rate $q'dy(t)$ associated with the reflection terms, if the boundary is reflecting. Let E_t^ϵ denote the expectation given the data to time t for all of the processes in (6.1).

A6.1. *The process $\theta^\epsilon(\cdot) = (\theta_i^\epsilon(\cdot), i = 1, \ldots, K)$ is nonanticipative with respect to the Wiener process $w^\epsilon(\cdot)$ and takes the form $\theta_i^\epsilon(s) = \theta_i(s/\epsilon)$, where $\theta_i(t) \in [-\bar{\theta}, 0]$. The process $\theta(\cdot)$ is asymptotically stationary with stationary measure $\nu(\cdot)$ in the sense that the following mixing condition holds: Uniformly in $g(\cdot)$ in any compact set in $C(G; [-\bar{\theta}, 0])$, and for any $\Delta > 0$, as $\epsilon \to 0$,*

$$\begin{aligned} \epsilon E_{n\Delta}^\epsilon \int_{n\Delta/\epsilon}^{(n\Delta+\Delta)/\epsilon} & b^1(g(\theta_j(s)), j \le K) ds \to \Delta \hat{b}^1(g(\cdot)) \\ &= \Delta \int_{-\bar{\theta}}^0 b^1(g(\theta_j), j \le K)\nu(d\theta) \end{aligned} \tag{6.3}$$

in mean, uniformly in n. The analogous condition holds for $k^1(\cdot)$, with the definition $\hat{k}^1(g(\cdot)) = \int_{-\bar{\theta}}^0 k^1(g(\theta_j), j \le K)\nu(d\theta)$.

A6.2. *$b^1(\cdot)$ is a continuous $I\!R^r$-valued function and $k^1(\cdot)$ is a continuous real-valued function, both on on $[I\!R^r]^K$.*

A6.3. *$b^2(\cdot)$ ($I\!R^r$-valued) and $k^2(\cdot)$ (real-valued) satisfy the continuity and boundedness condition on $b(\cdot)$ in (A3.1.3) and $\mu_c(\cdot)$ satisfies (A3.1.3).*

A6.4. *The system*

$$\begin{aligned} dx(t) &= dt \int \hat{b}^1(x(t-\theta_i), i \le K)\nu(d\theta) \\ &+ dt \int_{-\bar{\theta}}^{0} \int_U b^2(\bar{x}(t), \alpha, v) r'(d\alpha, t+v)\mu_c(dv) + \sigma(\bar{x}(s))dw(s) + dz(t) \end{aligned} \tag{6.4}$$

has a unique weak-sense solution for each initial condition and relaxed control.

Let $\hat{W}(\hat{x}, \hat{r}, r)$ denote the discounted cost with the model (6.4) and cost rate $\hat{k}^1(\cdot) + k^2(\cdot)$ with reflection term cost rate $q'dy(t)$. Let $W^\epsilon(\hat{x}, \hat{r}, r)$ denote the cost with the model (6.1) and cost rate (6.2) with reflection term cost rate $q'dy(t)$. The theorem is written for reflected diffusions. The analog for the boundary absorption case is similar.

Theorem 6.1. *Use the system model (6.1) and let the cost function be (3.4.2) with the rate given in the previous paragraph. Assume (A6.1)–(A6.4), (A3.2.1)–(A3.2.2) and (A3.4.3). Let $\sigma(\cdot)$ satisfy (A3.1.2). Then, as $\epsilon \to 0$, $W^\epsilon(\hat{x}, \hat{r}, r) \to \hat{W}(\hat{x}, \hat{r}, r)$ uniformly in $(\hat{x}, \hat{r})$, where $\hat{x}$ is confined to some compact set in $C(G; [-\bar{\theta}, 0])$.*

Now, suppose that $\sigma(\cdot)$ has the form $\sigma(x^\epsilon(t+\theta_i^\epsilon(t)), i \le K)$. Define $a(\cdot) = \sigma(\cdot)\sigma'(\cdot)$. Suppose that there is $\hat{a}(\cdot)$ such that, for any $\Delta > 0$ and continuous function $g(\cdot)$, as $\epsilon \to 0$,

$$\begin{aligned} \epsilon E_{n\Delta}^\epsilon \int_{n\Delta/\epsilon}^{(n\Delta+\Delta)/\epsilon} & a(g(\theta_i(s)), i \le K)ds \to \Delta \hat{a}(g(\cdot)) \\ &= \int \hat{a}(g(\theta_i), i \le K)\nu(d\theta) \end{aligned}$$

in mean, uniformly in n. Let $\hat{a}(\cdot)$ have a continuous square root $\hat{\sigma}(\cdot)$ and suppose that weak-sense uniqueness holds with $\hat{\sigma}(\cdot)$ used. Then the theorem holds with $\hat{\sigma}(\cdot)$ replacing $\sigma(\cdot)$ in (6.4).

Proof. When uniformity in $\hat{x}$ is referred to, it is always assumed that $\hat{x}$ lies in some compact set. We will use the martingale method to identify the limit process. The proof is based on that of Theorem 3.5.1. Let $\hat{x}^\epsilon, \hat{r}^\epsilon, r^\epsilon(\cdot)$ denote the initial conditions and relaxed control that are applied to the ϵ-system. We can suppose that there are $\hat{x}, \hat{r}$ such that $\hat{x}^\epsilon \to \hat{x}, \hat{r}^\epsilon \to \hat{r}$ in their respective topologies. Let $f(\cdot)$ be a real-valued function on $I\!R^r$ with compact support and whose first and second partial derivatives are continuous. For $h(\cdot), \phi(\cdot), s_i, t, T, I, J$, as in (3.5.2), Itô's Lemma applied to (6.1) implies that

$$\begin{aligned}
Eh\left(x^\epsilon(s_i), y^\epsilon(s_i), w^\epsilon(s_i), \langle \phi_j, r^\epsilon(s_i)\rangle, i \le I, j \le J\right) \Big[f(x^\epsilon(t+T)) - f(x^\epsilon(t)) \\
- \int_t^{t+T} f'_x(x^\epsilon(s)) b^1(x^\epsilon(s+\theta_i^\epsilon(s)), i \le K) ds \\
- \int_t^{t+T} \int_U \int_{-\bar{\theta}}^0 f'_x(x^\epsilon(s)) b^2(\bar{x}^\epsilon(s), \alpha, v) r^{\epsilon,\prime}(d\alpha, s+v) \mu_c^\epsilon(dv, s) ds \\
- \int_t^{t+T} f'_x(x^\epsilon(s)) dz^\epsilon(s) - \frac{1}{2} \int_t^{t+T} \sigma'(\bar{x}^\epsilon(s)) f_{xx}(x^\epsilon(s)) \sigma(\bar{x}^\epsilon(s)) ds \Big] = 0.
\end{aligned} \tag{6.5}$$

As in Theorem 3.5.1, the set $\{x^\epsilon(\cdot), y^\epsilon(\cdot), r^\epsilon(\cdot), w^\epsilon(\cdot), \epsilon > 0\}$ is tight and all weak-sense limit processes are continuous. Now take a weakly convergent subsequence, and abusing notation, index it by ϵ as well, and denote the limit by $(x(\cdot), y(\cdot), r(\cdot), w(\cdot))$. We will use the Skorokhod representation.

Next we derive the limit of the term involving $b^1(\cdot)$ in (6.5). Thus, consider the component of (6.5) defined by

$$\begin{aligned}
Eh\left(x^\epsilon(s_i), y^\epsilon(s_i), w^\epsilon(s_i), \langle \phi_j, r^\epsilon(s_i)\rangle, i \le I, j \le J\right) \\
\times \left[\int_t^{t+T} f'_x(x^\epsilon(s)) b^1(x^\epsilon(s+\theta_j^\epsilon(s)), j \le K) ds \right].
\end{aligned} \tag{6.6}$$

For $\Delta > 0$ (and assuming, w.l.o.g., that t and T are integral multiples of Δ), write the integral in (6.6) as

$$\sum_{i:\, i\Delta = t}^{t+T-\Delta} \int_{i\Delta}^{i\Delta+\Delta} f'_x(x^\epsilon(s)) b^1(x^\epsilon(s+\theta_j^\epsilon(s)), j \le K) ds.$$

By the weak convergence, given $\rho > 0$ there is $\Delta > 0$ such that for small $\epsilon > 0$ the sum can be approximated within ρ, in the mean, by

$$\sum_{i:\, i\Delta = t}^{t+T-\Delta} \int_{i\Delta}^{i\Delta+\Delta} f'_x(x^\epsilon(i\Delta)) b^1(x^\epsilon(i\Delta+\theta_j^\epsilon(s)), j \le K) ds. \tag{6.7}$$

For the purpose of evaluating the expectation (6.6), we can use the approximation (6.7) and evaluate the conditional expectations in the sum

$$E_t^\epsilon \sum_{i:\, i\Delta = t}^{t+T-\Delta} E_{i\Delta}^\epsilon \int_{i\Delta}^{i\Delta+\Delta} f'_x(x^\epsilon(i\Delta)) b^1(x^\epsilon(i\Delta+\theta_j^\epsilon(s)), j \le K) ds, \tag{6.8}$$

where the ith summand can be written as

$$\begin{aligned}
& f'_x(x^\epsilon(i\Delta)) E_{i\Delta}^\epsilon \int_{i\Delta}^{i\Delta+\Delta} b^1(x^\epsilon(i\Delta+\theta_j(s/\epsilon)), j \le K) ds \\
&= \epsilon f'_x(x^\epsilon(i\Delta)) E_{i\Delta}^\epsilon \int_{i\Delta/\epsilon}^{(i\Delta+\Delta)/\epsilon} b^1(x^\epsilon(i\Delta+\theta_j(s)), j \le K) ds.
\end{aligned} \tag{6.9}$$

The proof of tightness in Theorem 3.5.1 can also be used to show that the doubly indexed set $\{x^\epsilon(t+\cdot);\epsilon,t\}$ is tight in $D(G;[-\bar{\theta},\infty))$. Because the processes in this set are all continuous, the set is tight in $C(G;[-\bar{\theta},\infty))$. Then, for each $\gamma>0$, there is a compact set $K_\gamma\in C(G;[-\bar{\theta},0])$ such that

$$\inf_{t,\epsilon} P\left\{\left(x^\epsilon(t+\theta),-\bar{\theta}\le\theta\le 0\right)\in K_\gamma\right\}\ge 1-\gamma. \tag{6.10}$$

For fixed small $\gamma>0$ and $\delta>0$, choose a finite number of disjoint sets $S_l^{\gamma,\delta}\subset C(G;[-\bar{\theta},0])$, $l=1,2,\ldots,A_{\gamma,\delta}$, each of which has diameter less than δ and are such that $K_\gamma\subset\cup_l S_l^{\gamma,\delta}$. For each l, choose any $g_l(\cdot)\in S_l^{\gamma,\delta}$. We have

$$\begin{aligned}\sup_{-\bar{\theta}\le s\le 0}&\left|x^\epsilon(i\delta+s)-g_l(s)\right| I_{\{(x^\epsilon(i\Delta+s),s\in[-\bar{\theta},0])\in S_l^{\gamma,\delta}\}}\\ &\le \delta I_{\{(x^\epsilon(i\Delta+s),s\in[-\bar{\theta},0])\in S_l^{\gamma,\delta}\}}.\end{aligned} \tag{6.11}$$

Using the fact that the values of $(x^\epsilon(i\Delta+s), s\in[-\bar{\theta},0])$ are known at time $i\Delta$, and that the delays on the interval $[i\Delta,i\Delta+\Delta]$ depend on the "future" values $\theta^\epsilon(s), s\ge i\Delta$, for small enough γ and δ, (6.11), and the continuity of $b^1(\cdot)$ implies that we can approximate (6.9) uniformly in i by

$$\sum_l I_{\{(x^\epsilon(i\Delta+s),s\in[-\bar{\theta},0])\in S_l^{\gamma,\delta}\}} f_x'(x^\epsilon(i\Delta))\left[E_{i\Delta}^\epsilon\int_{i\Delta}^{i\Delta+\Delta} b^1(g_l(\theta_j^\epsilon(s)),j\le K)ds\right],$$

which equals

$$\begin{aligned}&\sum_l I_{\{(x^\epsilon(i\Delta+s),s\in[-\bar{\theta},0])\in S_l^{\gamma,\delta}\}}\times\\ &\quad f_x'(x^\epsilon(i\Delta))\left[\epsilon E_{i\Delta}^\epsilon\int_{i\Delta/\epsilon}^{(i\Delta+\Delta)/\epsilon} b^1(g_l(\theta_j(s)),j\le K)ds\right].\end{aligned} \tag{6.12}$$

Using the approximations (6.7), (6.8), (6.11) and (6.12), the arbitrariness of γ,Δ, and δ, condition (A6.1) (which implies that the convergence of the conditional expectation in (6.12) is uniform in l), and the weak convergence yields that, as $\epsilon\to 0$, (6.6) converges to

$$\begin{aligned}&Eh\left(x(s_i),y(s_i),w(s_i),\langle\phi_j,r(s_i)\rangle,i\le I,j\le J\right)\\ &\qquad\times\int_t^{t+T}\int f_x'(x(s))\hat{b}^1(x(s+\theta_i),i\le K)\nu(d\theta)ds.\end{aligned}$$

We omit the details concerning the convergence of the reflection term in (6.5) and of the approximation when $\sigma(\cdot)$ depends on rapidly varying delays. Finally, we have

$$
\begin{aligned}
&Eh\left(x(s_i), y(s_i), w(s_i), \langle \phi_j, r(s_i)\rangle, i \le I, j \le J\right) \Big[f(x(t+T)) - f(x(t)) \\
&\quad - \int_t^{t+T} \int f_x'(x(s))\hat{b}^1(x(s+\theta_i), i \le K)\nu(d\theta)ds \\
&\quad - \int_t^{t+T} \int_{-\bar{\theta}}^0 \int_U f_x'(x(s))b^2(\bar{x}(s), \alpha, v)r'(d\alpha, s+v)\mu_c(dv)ds \\
&\quad - \int_t^{t+T} f_x'(x(s))dz(s) - \frac{1}{2}\int_t^{t+T} \sigma'(\bar{x}(s))f_{xx}(x(s))\sigma(\bar{x}(s))ds \Big] = 0
\end{aligned}
$$

(with $\hat{\sigma}(\cdot)$ used in lieu of $\sigma(\cdot)$ if $\sigma(\cdot)$ has rapidly varying delays). By the remarks connected with (2.2.6), there is a Wiener process such that (6.4) holds. With the above convergence results available, the proof of the convergence of the cost functions is similar to that in Theorem 3.5.1, and the details are omitted. ■

5

The Ergodic Cost Problem

5.0 Outline of the Chapter

This chapter is concerned with the average cost per unit time (ergodic cost) problem. For such problems, one generally prefers to work with feedback controls, and we use a generalized form of such controls, called relaxed feedback controls, that are analogous to the extension of ordinary controls to relaxed controls in Section 3.1. We assume that the systems are nondegenerate in that the noise covariance matrix $\sigma(x)\sigma'(x)$ is positive definite. The Girsanov transformation method of Section 3.3 is used to define the process in terms of an uncontrolled model.

Because our aim is the preparation for the needs of the numerical algorithms, the issues of model complexity and simplification that were of concern in Chapter 4 are also of concern for the ergodic cost problem. This is complicated by the fact that there is not much relevant ergodic theory for nonlinear controlled delay equations. We will develop ergodic theorems for the process $\bar{x}(t)$, and for various finite-dimensional approximations that are, in fact, sufficient for the model approximations and the numerical problem. Only the case where the path is delayed is considered. Nothing seems to be known about the problem where the control is delayed as well.

Section 1 is devoted to the construction of the controlled process via the Girsanov transformation and to showing the mutual absolute continuity of the measures of $\bar{x}(t)$ for the controlled process, with respect to those for the uncontrolled process, and the irreducibility and aperiodicity of the process $\bar{x}(t)$. These results are used in Section 2 to show that a Doeblin condition holds, and hence that $\bar{x}(t)$ is geometrically ergodic, all uniformly in the control. Using this, it is then shown that the costs per unit time converge to the ergodic cost and that the normalized discounted costs converge as well, as the discount factor goes to zero. This latter fact justifies the use of the discounted cost function as a substitute for the ergodic cost function, when the discount factor is small. The proofs of all of these facts depend heavily on estimates

H.J. Kushner, *Numerical Methods for Controlled Stochastic Delay Systems*,
doi: 10.1007/978-0-8176-4621-9_5,

of the Radon–Nikodym derivatives in the measure transformations that are used in the Girsanov transformation.

Section 3 is concerned with general approximations of the models. It is shown that one can approximate the dynamics and control value space U, while sacrificing little in the value of the cost. In essence, continuity properties of the invariant measures and ergodic costs as functions of the dynamical terms are proved. Approximations such as dealt with in Chapter 4, where the delays vary either periodically or periodically with time intervals determined by a renewal process, are of great interest due to their role in reducing the memory requirements for the numerical problem. The relevant ergodic theorems and approximations to the cost functions for the periodic approximation are discussed in Section 4, and those for the periodic-Erlang approximations are discussed in Section 5.

It is shown that the optimal ergodic cost for the original model can be well approximated by the cost for any of these model approximations, which justifies their use in the numerical algorithms. It is also shown that nearly optimal controls for any of these approximations are nearly optimal for the original problem where the memory segment in the dynamics is not approximated.

The development in this chapter is for the case where there is no delay in the diffusion coefficient $\sigma(\cdot)$: At time t, it depends only on $x(t)$. How to handle the case where there is past dependence in this term is shown at the end of Section 4. But, for the most part, we stick to the simpler case to avoid overcomplicating the development.

Unfortunately, little is known at present concerning the general nonlinear control problem when there are delays, particularly for the ergodic cost problem. For example, it is not known whether there is a feedback optimal control. Because this is not the place to develop such results, we make various assumptions to cover our ignorance. The key assumption is that the use of relaxed feedback controls cannot be improved upon by the use of relaxed controls. This property holds if there are no delays, and it seems very likely that it holds quite generally in the delay case. The methods that were used in the proof of this result for the no-delay case do not seem to be able to be carried over.

5.1 The Basic Model

5.1.1 Relaxed Feedback Controls

The system of interest for the bulk of the chapter, when ordinary feedback controls are used, is the reflected diffusion model

$$dx(t) = b(\bar{x}(t), u(\bar{x}(t)))dt + \sigma(x(t))dw(t) + dz(t),$$

with cost function

$$\lim_{T\to\infty} \frac{1}{T} E_{\hat{x}}^{u} \left[\int_0^T k(\bar{x}(t), u(\bar{x}(t)))dt + q'y(T) \right].$$

where $\hat{x} = \bar{x}(0)$ is the initial condition. Since the ergodic cost problem is of concern over an infinite time interval, and there is no discounting of the future, it is important for the numerical problem that we can ensure boundedness of the path throughout. This is one motivation for the use of a reflected diffusion model. If the reflecting boundaries are added for numerical purposes, to bound the state space, then one must experiment with the boundaries to ensure that they have only a minimal effect on the results.

When working with average cost per unit time control problems over an infinite interval, it is common to suppose that the controls are feedback so that the process is Markov, and appropriate ergodic theorems can be applied to show the convergence of the average cost per unit time to the "stationary" or "ergodic" cost and to obtain optimality and approximation results. For the discounted or finite-time problem we saw that the use of relaxed controls facilitated the proofs of approximation and existence theorems. Analogously, for the ergodic cost problem, the use of relaxed feedback controls [9, 56] facilitates the proofs of such results.

Suppose that $m(\hat{x}, \cdot)$ is a probability measure on the Borel sets of U for each $\hat{x} \in D(G; [-\bar{\theta}, 0])$ and that $m(\cdot, A)$ is Borel measurable for each Borel set $A \subset U$. Then we say that $m(\cdot)$ is a *relaxed feedback control.* Any ordinary feedback control $u(\hat{x})$ has a representation as the relaxed feedback control $m(\cdot)$, where $m(\hat{x}, B) = I_{\{u(\hat{x}) \in B\}}$. In terms of the relaxed feedback control, the system is

$$dx(t) = \int_U b_m(\bar{x}(t))dt + \sigma(x(t))dw_m(t) + dz(t), \tag{1.1}$$

where $w_m(\cdot)$ is a standard vector-valued Wiener process, and we define

$$b_m(\bar{x}(t)) = \int_U b(\bar{x}(t), \alpha)m(\bar{x}(t), d\alpha). \tag{1.2}$$

The subscript m is used on the Wiener process in (1.1) as the solutions will be constructed by the Girsanov transformation method. The corresponding cost function is

$$\gamma(m) = \lim_{T\to\infty} \frac{1}{T} E_{\hat{x}}^{m} \left[\int_0^T \int_U k_m(\bar{x}(t))dt + q'y(T) \right], \tag{1.3}$$

where $k_m(\cdot)$ is defined analogously to $b_m(\cdot)$. Under the conditions to be imposed, the limit in (1.3) will exist uniformly in the initial condition and $m(\cdot)$.

The system (1.1) will be constructed using the Girsanov transformation, starting with the uncontrolled system

$$dx = \sigma(x)dw + dz, \tag{1.4}$$

where $w(\cdot)$ is a standard vector-valued Wiener process. Unless mentioned otherwise, throughout the chapter we always assume the conditions (A3.2.1) and (A3.2.2) on the set G, condition (A3.1.1) on the function $k(\cdot)$, and the following assumption on $b(\cdot), \sigma(\cdot)$ and the solution to (1.4). Additional assumptions will be stated as neeeded. Whenever the convergence or approximations of the cost functions are involved, we also assume the additional condition (A3.4.3) on the reflection directions $\{d_i\}$.

A1.1. *$\sigma(\cdot)$ is continuous on G, $\sigma^{-1}(\cdot)$ exists and is continuous, and $a(\cdot) = \sigma(\cdot)\sigma'(\cdot)$ is Hölder continuous. The system (1.4) is defined on a complete probability space $(\Omega, \mathcal{F}, P_x)$ with filtration $\{\mathcal{F}_t, t < \infty\}$, $\mathcal{F} = \lim_t \mathcal{F}_t$, and Ω is the set of continuous functions on $[0, \infty)$ with the local sup norm topology. The system (1.4) has a unique weak-sense solution for each initial condition $x \in G$. The initial condition $\hat{x} = \bar{x}(0)$ for the system with delays is continuous on $[-\bar{\theta}, 0]$, and $\hat{x}(\theta) \in G$ for $\theta \in [-\bar{\theta}, 0]$. The function $b(\cdot)$ is bounded and measurable.*

In (A1.1) and in the (uncontrolled and controlled, resp.) system equations (1.4) and (1.9), the function $\sigma(\cdot)$ does not depend on delayed arguments. See the comments at the end of Section 4 concerning the approach that is to be taken when $\sigma(\hat{x})$ is used in lieu of $\sigma(x)$.

Construction of the controlled system. Recall the discussion of the Girsanov transformation method in Section 3.3. The controlled system will be constructed as follows. We start with the uncontrolled model (1.4) and construct the desired solution via the Girsanov transformation. Let Ω_T denote the restriction of Ω to functions on $[0, T]$ and let $m(\cdot)$ be a relaxed feedback control. For $T > 0$ define

$$\zeta(T, m) = \int_0^T \left[\sigma^{-1}(x(s)) b_m(\bar{x}(s))\right]' dw(s) - \frac{1}{2} \int_0^T \left|\sigma^{-1}(x(s)) b_m(\bar{x}(s))\right|^2 ds, \tag{1.5}$$

and set

$$R(T, m) = e^{\zeta(T,m)}. \tag{1.6}$$

The process $x(\cdot)$ in (1.5) is the solution of (1.4), whose initial condition is simply $x(0) \in G$. There are no delays. But to coordinate notation with the delay case, we will sometimes write the initial condition as $\hat{x}$, with $\hat{x}(0) = x(0)$. There are two ways of writing the transition probability for (1.4) that will be used, depending on the application. The standard transition probability for the uncontrolled $x(\cdot)$ is denoted by $P_0(x, t, B) = P_x\{x(t) \in B\}$, for $B \in \mathcal{B}(G)$, the Borel sets in G. When we are concerned with segments of the uncontrolled path, for $A \in \mathcal{B}(C(G; [-\bar{\theta}, 0]))$ and $t \geq 0$, we will also have use for the form $P(\hat{x}, t, A) = P_{\hat{x}}\{\bar{x}(t) \in A\}$. For $t > \bar{\theta}$ this is $P_{x(0)}\{\bar{x}(t) \in A\}$. For $t > \bar{\theta}$, we will also use the definition $P(x, t, A) = P_x\{\bar{x}(t) \in A\}$ where $x \in G$. Let $P_{\hat{x},T}$ denote the measure induced by (1.4) on $\mathcal{B}(C(G; [-\bar{\theta}, T]))$.

For each $(\hat{x}, T, m(\cdot))$, define the probability measure $P^m_{\hat{x},T}$ on $(\Omega_T, \mathcal{F}_T)$ via the Radon–Nikodym derivative $R(T, m)$:

$$dP^m_{\hat{x},T} = R(T, m)dP_{\hat{x},T}. \tag{1.7}$$

For each $(\hat{x}, m(\cdot))$, the family $P^m_{\hat{x},T}$ of measures indexed by T can be extended uniquely to a measure $P^m_{\hat{x}}$ on $(\Omega, \mathcal{F})$ that is consistent with the $P^m_{\hat{x},T}$. When there is no control (i.e., where the system is (1.4)), we omit the superscript or subscript m. The process $w_m(\cdot)$ defined by

$$dw_m(t) = dw(t) - \left[\sigma^{-1}(x(s))b_m(\bar{x}(s))\right] dt \tag{1.8}$$

is an $\mathcal{F}_t$-standard Wiener process on $(\Omega, P^m_x, \mathcal{F})$ [42]. Now, as in Section 3.3, rewrite the uncontrolled model (1.4) as

$$dx(t) = b_m(\bar{x}(t))dt + \sigma(x(t))dw_m(t) + dz(t). \tag{1.9}$$

Under the measures $\{P^m_{\hat{x}}, \hat{x} \in C(G; [-\bar{\theta}, 0])\}$, the process $\bar{x}(\cdot)$ obtained from the solution $x(\cdot)$ of (1.9) is a Markov process and we use $P^m(\hat{x}, t, \cdot)$ for its transition function.

The reference [83] contains a wealth of material on ergodic problems for infinite-dimensional systems. Most of the work concerning particular systems of stochastic differential equations involve Hilbert space valued processes driven by an infinite-dimensional Wiener process (the cylindrical Wiener process). Ergodic theorems for nondegenerate finite-dimensional linear delay systems (nonreflected) are dealt with, but the material concerning nonlinear systems requires that the linear part dominate and that the nonlinear part satisfies a Lipschitz condition, which would exclude general relaxed feedback controls. See also [12] for similar work for Hilbert space valued processes.

The work that is closest to ours is [86], which also uses Girsanov transformation methods to construct the solutions and verify the ergodic properties. The development supposes that $\sigma(\cdot)$ is a constant and it does not deal with control problems, approximations, or with reflected diffusions (so that the diffusions are unbounded). But it does show the Doeblin condition and convergence to the invariant measure. From a numerical point of view, one needs to work in a bounded space, either directly or via an appropriate transformation. We follow the procedure that was used in [56, Chapter 4]. The paper [18] also uses Girsanov transformation techniques (for the nonreflected model) and shows, under its assumptions, that the transition probabilities converge to the invariant measure, and that the cost and invariant measure are continuous functions of the control. The conditions are more restrictive than what is needed here. We note that the first work to deal with the continuity of the invariant measure and stationary costs in the control for a diffusion process is [54]. See also [56, Section 4.5].

5.1.2 Density Properties and Preliminary Results

The proof of convergence of the measures $P^m(\hat{x}, t, \cdot)$ to an invariant measure as $t \to \infty$ depends on the fact that for any $t_1 > \bar{\theta}$, the measures defined by $P_{\hat{x}}^m\{x(t) \in B\}, B \subset G,\ t \geq t_1$, are absolutely continuous with respect to Lebesgue measure, uniformly in $m(\cdot), t$ and in $\hat{x}$ in any compact set. This fact will be proved in the next few theorems.

Theorem 1.1. *The process defined by (1.4) is a Feller process in that $E_x f(x(t))$ is continuous for each bounded and continuous function $f(\cdot)$. It is also a strong Markov process. For each relaxed feedback control, the process (1.9) is a strong Markov process and the solution is weak-sense unique for each initial condition $\hat{x} \in C(G; [-\bar{\theta}, 0])$.*

Comments on the proof. The Feller property of (1.4) is a consequence of its weak-sense uniqueness for each initial condition and the continuity of $\sigma(\cdot)$. The strong Markov property and the weak-sense uniqueness of the solution to (1.9) is a consequence of the strong Markov property and weak-sense uniqueness of the solution to (1.4) and the Girsanov transformation method of constructing (1.9).

For the uncontrolled system (1.4), the next result is Theorem 4.2.1 of [56]. The proof depended only on the properties ((A3.2.1), (A3.2.2)) of G, the boundedness of the drift terms, and the nondegeneracy of $\sigma(\cdot)\sigma'(\cdot)$. Hence it remains valid for the problem with delays. The result for the controlled system follows from the fact that for any set $A \in \mathcal{F}_T$ such that $P_{\hat{x}}\{A\} = 0$, we have $P_{\hat{x}}^m\{A\} = E_{\hat{x}} R(T, m) I_A = 0$.

Theorem 1.2. *Let $N_\delta(\partial G)$ denote a δ-neighborhood of the boundary ∂G. Then, for $0 < t_1 < t_2 < \infty$,*

$$\begin{aligned} &P_{\hat{x}}^m\{x(t) \in \partial G\} = 0 \text{ for } t > 0 \text{ and all } m(\cdot) \text{ and } \hat{x}, \\ &\lim_{\delta \to 0} \sup_{t_1 \leq t \leq t_2} \sup_{\hat{x}} \sup_{m} P_{\hat{x}}^m\{x(t) \in N_\delta(\partial G)\} = 0. \end{aligned} \tag{1.10}$$

The next result is Theorem 4.2.2 and Theorem 4.3.3 of [56]. The proof depends on the properties of the uncontrolled system (1.4) and the boundedness of the second moment of the Radon–Nikodym derivative $R(T, m)$ that is used in the Girsanov transformation. So the delays in the drift term play no role. The symbol $l(B)$ denotes the Lebesgue measure of the set B. .

Theorem 1.3. *For $t > 0$ the transition probability $P_{\hat{x}}^m\{x(t) \in \cdot\}$ is uniformly mutually absolutely continuous with respect to Lebesgue measure in the following sense. Let $0 < t_1 < t_2 < \infty$. Given $\delta > 0$ there is $\epsilon > 0$ such that, if $l(B) \leq \epsilon$ for $B \in \mathcal{B}(G)$, then*

$$\sup_{t_1 \leq t \leq t_2} \sup_{m} \sup_{\hat{x}} P_{\hat{x}}^m\{x(t) \in B\} \leq \delta. \tag{1.11}$$

Also, for any $\epsilon > 0$ there is $\delta > 0$ such that

$$\inf_{t_1 \le t \le t_2} \inf_{\hat{x}} \inf_{m} \inf_{\{B: l(B) \ge \epsilon\}} P_{\hat{x}}^m \{x(t) \in B\} \ge \delta. \tag{1.12}$$

The control $m(\cdot)$ can depend on time as well as on the path. By using the Markov property of the process $\bar{x}(t)$, we can replace $t \le t_2$ by $t < \infty$.

Comment on Theorem 1.3. The theorem implies that $P_0(x, t, \cdot)$ and $l(\cdot)$ are mutually absolutely continuous, uniformly in x and $t \ge t_1 > 0$. In particular, for each $\epsilon > 0$ and $t \ge t_1$, there are $\infty > \epsilon_2 > \epsilon_1 > 0$ such that for each initial condition x_0 there is a set $B_\epsilon(x_0, t)$ with $l(B_\epsilon(x_0, t)) \le \epsilon$ such that, for $x \notin B_\epsilon(x_0, t)$,

$$\epsilon_2 \ge P_0(x_0, t, dx)/dx \ge \epsilon_1.$$

This implies that for $x \notin B_\epsilon(x_i, t), i = 1, 2$,

$$\frac{\epsilon_2}{\epsilon_1} \ge \frac{P_0(x_0, t, dx)}{P_0(x_1, t, dx)} \ge \frac{\epsilon_1}{\epsilon_2}. \tag{1.13}$$

Theorem 1.4. *Let $0 = t_0 < t_1 < \cdots < t_k < \infty$, with $t_{i+1} - t_i \ge \delta_1 > 0$. Let $B_i \in \mathcal{B}(G)$. Then for $\delta > 0$ and $k < \infty$, there is $\epsilon > 0$ (that does not depend on the $\{t_i\}$) such that $l(B_i) \le \epsilon$ for any i implies that*

$$P_{\hat{x}}^m \{x(t_1) \in B_1, \ldots, x(t_k) \in B_k\} \le \delta, \text{ all } m(\cdot), \hat{x}. \tag{1.14}$$

Also, for any $\epsilon > 0$ there is $\delta > 0$ such that

$$\inf_{t_i, i \le k} \inf_{\hat{x}} \inf_{m} \inf_{\{B_i: l(B_i) \ge \epsilon, i \le k\}} P_{\hat{x}}^m \{x(t_1) \in B_1, \ldots, x(t_k) \in B_k\} \ge \delta. \tag{1.15}$$

The control $m(\cdot)$ can depend on time as well as on the path.

Comment. (1.14) and (1.15) are obtained by using the Markov property

$$\begin{aligned} &P_{\hat{x}}^m \{x(t_1) \in B_1, \ldots, x(t_k) \in B_k\} \\ &\quad = E_{\hat{x}}^m P^m \left\{x(t_1) \in B_1, \ldots, x(t_k) \in B_k \mid \mathcal{F}_{t_{k-1}}\right\} \\ &\quad = E_{\hat{x}}^m P^m \left\{x(t_1) \in B_1, \ldots, x(t_{k-1}) \in B_{k-1} \mid \mathcal{F}_{t_{k-1}}\right\} \\ &\qquad \times P^m \{x(t_k) \in B_k \mid x(t_{k-1})\}, \end{aligned}$$

then bounding the right-hand term above or below, as required, by Theorem 1.3, and iterating backwards.

Comment on an extension to random times. By the strong Markov property of (1.4), the times t and t_i in Theorems 1.2–1.4 can be replaced by random times. In particular, for each n, let $\phi_i^n, i = 1, \ldots, k$, be mutually independent and identically distributed random variables that are independent of the process $x(\cdot)$ of (1.4) and such that

$$\lim_{\epsilon \to 0} \sup_{i,n} P\{\phi_i^n \le \epsilon, \text{ or } \phi_i^n \ge 1/\epsilon\} = 0$$

Then the theorems continue to hold, uniformly in n, for ϕ_1^n replacing t in Theorem 1.1 and $\sum_{j=1}^{i} \phi_j^n$ replacing t_i otherwise.

Theorem 1.5. *Let $A \in \mathcal{B}(C(G;[-\bar{\theta},0]))$. For any $t > 0$ there are $\epsilon > 0, \epsilon_1 > 0$, and $x_0 \in G$, such that $P(x_0, t + \bar{\theta}, A) = P_{x_0}\{\bar{x}(t + \bar{\theta}) \in A\} \ge \epsilon_1$ implies that $P(x, t + \bar{\theta}, A) = P_x\{\bar{x}(t + \bar{\theta}) \in A\} \ge \epsilon$ for all $x \in G$.*

Proof. The proof follows from the comments after Theorem 1.3. Consider the set $B_0(A) = \{\psi(-\bar{\theta}) : \psi(\cdot) \in A\}$. If $P_{x_0}\{\bar{x}(t + \bar{\theta}) \in A\} \ge \epsilon_1$, then we must have $P_0(x_0, t, B_0(A)) \ge \epsilon_1$. Hence there is $\epsilon_2 > 0$ such that $l(B_0(A)) \ge \epsilon_2$ no matter what A. By the discussion after Theorem 1.3, for any $\infty > \epsilon_4 > \epsilon_3 > 0$, $\epsilon_4 \ge P_0(x_1, t, dx)/P_0(x_0, t, dx) \ge \epsilon_3$, except possibly on a set $B_{\epsilon_3,\epsilon_4}(x_0, x_1, t)$ whose Lebesgue measure goes to zero uniformly in x_0, x_1 as $\epsilon_3 \to 0$ and $\epsilon_4 \to \infty$. Since the distribution of the paths on $[t, t + \bar{\theta}]$ is determined by the initial condition $x(t)$, we can conclude that there is $\epsilon > 0$ such that $P_x\{\bar{x}(t + \bar{\theta}) \in A\} \ge \epsilon$ for all x. ■

The next theorem is a consequence of the irreducibility and aperiodicity of the process $\bar{x}(t)$ obtained from the uncontrolled form (1.4) and the absolute continuity of the measure $P_{\hat{x}}^m(\cdot)$ with respect to $P(\hat{x}, t, \cdot)$ for $t > \bar{\theta}$.

Theorem 1.6. *The process $\bar{x}(t)$ is irreducible and aperiodic for each relaxed feedback control $m(\cdot)$ and initial condition $\hat{x}$.*

Theorem 1.7. *Let $t > 2\bar{\theta}$. Then $P^m(\hat{x}, t, \cdot)$ is absolutely continuous with respect to $P(x, t, \cdot)$, for any $x \in G$ and $\hat{x} \in C(G;[-\bar{\theta},0])$, and conversely. The absolute continuity is uniform in $m(\cdot)$ and $\hat{x}$. [The measures are all on $\mathcal{B}(C(G;[-\bar{\theta},0]))$.] In particular, for each $\epsilon > 0$ such that the set A satisfies $P(x_0, t, A) \ge \epsilon$ for some x_0, there is $\epsilon_2 > 0$ such that $P^m(\hat{x}, t, A) \ge \epsilon_2$ for all $\hat{x}$ and $m(\cdot)$. Also, if $P^m(\hat{x}, t, A) \ge \epsilon$ for some $\hat{x}$ and $m(\cdot)$, then there is $\epsilon_2 > 0$ such that $P^{m_1}(\hat{x}_1, t, A) \ge \epsilon_2$ for all $\hat{x}_1$ and $m_1(\cdot)$. In both cases, the ϵ_2 will depend on ϵ and t, but it will be positive for all $t > 2\bar{\theta}$.*

Proof. We will only prove that if $P(x_0, t, A) \ge \epsilon$ for some x_0, then there is $\epsilon_2 > 0$ such that $P^m(\hat{x}, t, A) \ge \epsilon_2$ for all $\hat{x}$ and $m(\cdot)$. The other assertions are proved similarly. Recall that $P^m(\hat{x}, t, A) = E_{\hat{x}} R(t, m) I_{\{\bar{x}(t) \in A\}}$ and define the set $B_k^m = \{\zeta(t, m) \ge -k\}$. Then

$$\begin{aligned} E_{\hat{x}} R(t, m) I_{\{\bar{x}(t) \in A\}} &\ge e^{-k} E_{\hat{x}} I_{\{B_k^m\}} I_{\{\bar{x}(t) \in A\}} \\ &\ge e^{-k} \left[E_{\hat{x}(0)} I_{\{\bar{x}(t) \in A\}} - E_{\hat{x}} I_{\{\Omega - B_k^m\}} \right]. \end{aligned} \tag{1.16}$$

By the hypothesis and Theorem 1.5, we can suppose that there is $\epsilon_1 > 0$ depending only on ϵ and such that $E_{\hat{x}(0)} I_{\{\bar{x}(t) \in A\}} = P(\hat{x}(0), t, A) \ge \epsilon_1$ for all

$\hat{x}(0)$. Thus

$$P^m(\hat{x}, t, A) \geq e^{-k} \left[\epsilon_1 - E_{\hat{x}} I_{\{\Omega - B_k^m\}} \right].$$

Finally, choose k so that $E_{\hat{x}} I_{\{\Omega - B_k^m\}} \leq \epsilon_1/4$ for all $m(\cdot)$ and $\hat{x}$. ∎

Recurrence. Let $\phi(\cdot)$ be a probability measure on $\mathcal{B}(C(G; [-\bar{\theta}, 0]))$ We say that $\bar{x}(\cdot)$ is ϕ-recurrent under $m(\cdot)$ if for some $\delta > 0$ and $\phi(A) > 0$.

$$\sum_{n=1}^{\infty} P^m(\hat{x}, n\delta, A) = \infty, \qquad \text{for all } \hat{x}. \tag{1.17}$$

The next theorem is a direct consequence of Theorem 1.7 and the Markov property of $\bar{x}(t)$.

Theorem 1.8. *For each $m(\cdot)$, the process $\bar{x}(t)$ is ϕ-recurrent for $\phi(\cdot) = P(x_0, t, \cdot)$ for any $x_0 \in G$ and $t > \bar{\theta}$.*

The existence of an invariant measure for each $m(\cdot)$. In order to prove that the measures $P^m(\hat{x}, t, \cdot)$ converge to an invariant measure, we need to know that there is at least one invariant measure for each $m(\cdot)$.

Theorem 1.9. *For each $m(\cdot)$ there is at least one invariant measure for the process $\bar{x}(t)$.*

Proof. Let $f(\cdot)$ be a continuous and bounded real-valued function on the space $C(G; [-\bar{\theta}, 0])$. Then $E_{\hat{x}}^m f(\bar{x}(t))$ is continuous in $\hat{x}$ by the weak-sense uniqueness of the solution to (1.9). Hence $\bar{x}(t)$ is a Feller process for each $m(\cdot)$. By Lemma 3.2.2, for each $m(\cdot)$ and initial condition $\hat{x}$, the set of processes $\{x(t + \cdot); t < \infty\}$ is tight in $C(G; [0, \infty))$. Hence the set of random variables $\{\bar{x}(t); t < \infty\}$ is tight and there is a sequence $t_n \to \infty$ such that $\int_0^{t_n} P^m(\hat{x}, t, \cdot) dt/t_n$ converges weakly to some measure $\nu(\cdot)$. Then by [83, Theorem 3.1.1], $\nu(\cdot)$ is an invariant measure under $m(\cdot)$. ∎

5.2 The Doeblin Condition

Consider a discrete-parameter Markov chain $\{Y_n\}$ with values in a complete separable metric space. If there is a measure $\phi(\cdot)$ such that for some integer n_0 and $0 < \epsilon_0 < 1, \delta_0 > 0$,

$$\phi(A) > \epsilon_0 \Rightarrow P(Y, n_0, A) \geq \delta_0 \tag{2.1}$$

for all Y in the state space, then we say that the chain satisfies a Doeblin condition.

The Doeblin condition implies the following [71, Theorems 16.0.2 and 16.2.3], [85, Chapter 4].

Theorem 2.1. *Let the chain $\{Y_n\}$ be aperiodic and ϕ-irreducible and let there exist an invariant measure $\pi(\cdot)$. Then the chain is geometrically ergodic in the sense that there are $R > 0$ and $\rho < 1$ such that*

$$||P(Y, n, \cdot) - \pi(\cdot)|| \leq R\rho^n, \tag{2.2}$$

where $||\nu(\cdot)||$ is the variation norm. The measure $\pi(\cdot)$ is unique and is absolutely continuous with respect to $\phi(\cdot)$. The values of R and ρ depend only on $\epsilon_0, \delta_0, n_0$ and not on other properties of the process $\{Y_n\}$.

Theorem 2.2. *The Doeblin condition holds for the process $\bar{x}(t)$ for any $m(\cdot)$, where $\phi(\cdot) = P(x_0, t, \cdot)$ for any $x_0 \in G$ and $t > 2\bar{\theta}$. The invariant measure $\mu^m(\cdot)$ is unique for each $m(\cdot)$ and there are $C < \infty$ and $\rho \in (0, 1)$, not depending on $\hat{x}$ or $m(\cdot)$, such that*

$$||P^m(\hat{x}, t, \cdot) - \mu^m(\cdot)|| \leq C\rho^t. \tag{2.3}$$

The $\mu^m(\cdot)$ are mutually absolutely continuous.

Proof. The process is ϕ-irreducible and aperiodic by Theorem 1.6, and by Theorem 1.9, there is at least one invariant measure. By Theorem 1.7 and the definition (2.1), the constants C and ρ can be taken to be the same for all $m(\cdot)$ and $\hat{x}$. The mutual absolute continuity assertion follows from the mutual absolute continuity of the $P^m(\hat{x}, t, \cdot)$ for all $m(\cdot)$, $\hat{x}$, and any $t > 2\bar{\theta}$. ■

The cost function. Define

$$\gamma(\hat{x}, T, m) = E_{\hat{x}}^m \frac{1}{T}\left[\int_0^T \int_U k(\bar{x}(t), \alpha) m(\bar{x}(t), d\alpha) dt + q'y(T)\right] \tag{2.4}$$

and

$$\gamma(m) = \int \int_U k(\hat{v}, \alpha) m(\hat{v}, d\alpha) \mu^m(d\hat{v}) + \int E_{\hat{v}}^m \left[q'y(1)\right] \mu^m(d\hat{v}). \tag{2.5}$$

Theorem 2.3. $\gamma(\hat{x}, T, m) \to \gamma(m)$, *uniformly in $\hat{x}$ and $m(\cdot)$ as $T \to \infty$.*

Proof. By Theorem 2.2, $||P^m(\hat{x}, t, \cdot) - \mu^m(\cdot)|| \to 0$, uniformly in $\hat{x}$ and $m(\cdot)$. Then, for $f^n(\cdot)$ being measurable and uniformly bounded, $\int f^n(\hat{v})[P^m(\hat{x}, t, d\hat{v}) - \mu^m(\delta\hat{v})] \to 0$, uniformly in $\hat{x}$ and $m(\cdot)$, no matter what the $f^n(\cdot)$. For the component involving $k(\cdot)$, the theorem statement then follows from

$$E_{\hat{x}}^m \int_U k(\bar{x}(t),\alpha)m(\bar{x}(t),d\alpha) = \int\int_U k(\hat{v},\alpha)m(\hat{v},d\alpha)P^m(\hat{x},t,d\hat{v})$$
$$\to \int\int_U k(\hat{v},\alpha)m(\hat{v},d\alpha)\mu^m(d\hat{v}).$$

Write

$$E_{\hat{x}}^m q'[y(t+1)-y(t)] = \int P^m(\hat{x},t,d\hat{v})E_{\hat{v}}^m q'y(1),$$

where $E_{\hat{v}}^m q'y(t)$ is a measurable function of $\hat{v}$. By the moment bounds in Lemma 3.2.1, $\sup_{m,\hat{x}} E_{\hat{x}}^m |y(1)|^2 < \infty$. This uniform integrability and the uniform convergence $||P^m(\hat{x},t,\cdot) - \mu^m(\cdot)|| \to 0$ implies that the limit of the average reflection costs per unit time is the second term in (2.5). ■

Convergence of the discounted costs to the stationary cost. The proof of Theorem 2.3 yields the following result.

Theorem 2.4. *In addition to the assumptions of Section* 1*, assume the cost function* (3.4.3) *but with a relaxed feedback control and where we now denote the discounted cost by* $W_\beta(\hat{x},m)$*. Then as* $\beta \to 0$

$$\beta W_\beta(\hat{x},m) \to \gamma(m) \tag{2.6}$$

uniformly in $\hat{x}$ *and* $m(\cdot)$*.*

5.3 Approximations of the Models

For the discounted cost case in Chapter 4, various results were proved concerning the continuity of the costs in the model parameters, and some "finite-memory" approximations were developed, all for the purpose of facilitating the numerical approximations. Similar considerations are important to the ergodic cost problem. Some preliminary results will be presented in this section, where it is shown that the transition probabilities, invariant measures, and ergodic costs are continuous in the dynamical term $b(\cdot)$, uniformly in $m(\cdot)$ and $\hat{x}$. We will make use of the inequality

$$\left|e^a - e^b\right| \le \left|e^a + e^b\right| |a-b| . \tag{3.1}$$

Theorem 3.1. *Let* $b^n(\cdot)$ *be measurable functions of* $\hat{x}$ *and* α *for each* n*. Let* $\sup_n \sup_{\hat{x},\alpha} |b^n(\hat{x},\alpha)| < \infty$ *and* $b^n(\hat{x},\alpha) \to b(\hat{x},\alpha)$*, uniformly in* $\alpha \in U$ *for each continuous function* $\hat{x}$*. Let* $P_n^m(\hat{x},t,\cdot)$ *denote the measure of* $\bar{x}(t)$ *associated with* $b^n(\cdot)$ *replacing* $b(\cdot)$ *and with control* $m(\cdot)$ *and initial condition* $\hat{x}$*. Then, for each* $t > 2\bar{\theta}$*,*

$$||P_n^m(\hat{x},t,\cdot) - P^m(\hat{x},t,\cdot)|| \to 0 \tag{3.2}$$

for each continuous $\hat{x}$. The convergence is uniform in $m(\cdot)$. The convergence is also uniform in any compact $\hat{x}$-set of continuous initial conditions if the convergence of $b^n(\cdot) \to b(\cdot)$ is uniform on each such compact set.

Proof. Define $R^n(t,m) = \exp \zeta^n(t,m)$, where

$$\zeta^n(t,m) = \int_0^t \left[\sigma^{-1}(x(s))b_m^n(\bar{x}(s))\right]' dw(s) - \frac{1}{2}\int_0^t \left|\sigma^{-1}(x(s))b_m^n(\bar{x}(s))\right|^2 ds,$$

where $b_m^n(\hat{x}) = \int_U b^n(\hat{x},\alpha)m(\hat{x},d\alpha)$. We have

$$[P_n^m(\hat{x},t,A) - P^m(\hat{x},t,A)] = E_{\hat{x}}\left[R^n(t,m) - R(t,m)\right] I_{\{\bar{x}(t)\in A\}}.$$

Because $\sup_{m,n}\sup_{\hat{x}} E_{\hat{x}}[R_n^m(t,m)]^2 < \infty$, using (3.1), the boundedness of $b(\cdot)$ and $b^n(\cdot)$ and Schwarz's inequality yields

$$\begin{aligned} &||P_n^m(\hat{x},t,\cdot) - P^m(\hat{x},t,\cdot)|| \\ &\quad \le K_t E_{\hat{x}}^{1/2}\int_0^t ds \int_U \left|\left(b^n(\bar{x}(s),\alpha) - b(\bar{x}(s),\alpha)\right) m(\bar{x}(s),d\alpha)\right| \\ &\quad \le K_t E_{\hat{x}}^{1/2}\int_0^t \sup_\alpha \left|b^n(\bar{x}(s),\alpha) - b(\bar{x}(s),\alpha)\right| ds, \end{aligned} \tag{3.3}$$

for some constant K_t. The process $\bar{x}(t)$ in (3.3) is from the uncontrolled form (1.4). As $n \to \infty$, the expression (3.3) goes to zero for each continuous $\hat{x}$, as the path segments $\bar{x}(t)$ are continuous for all t w.p.1.

The assertion concerning uniformity of convergence on each compact set of initial conditions can be proved by a weak convergence argument. Let $\hat{x}_n \to \hat{x}$ uniformly on $[-\bar{\theta},0]$, and let $x_n(\cdot)$ denote the solution to (1.4) under initial condition $\hat{x}_n$. We need to show that

$$E_{\hat{x}_n}^{1/2}\int_0^t \sup_\alpha \left|b^n(\bar{x}_n(s),\alpha) - b(\bar{x}_n(s),\alpha)\right| ds \to 0. \tag{3.4}$$

The set $\{x_n(\cdot)\}$ is tight in $D(G;[-\bar{\theta},\infty))$ and because all the paths are continuous, it is tight in $C(G;[-\bar{\theta},\infty))$. Extract a weakly convergent subsequence, indexed also by n, and use the Skorokhod representation so that we can suppose that the convergence is w.p.1, uniformly on each bounded time interval. Then, using the fact that the set of segments $\{\bar{x}_n(t), 0 \le t < \infty\}$ is tight in $C(G;[-\bar{\theta},0])$ and the uniform convergence of $b^n(\cdot)$ on each compact set, (3.4) holds, which implies the uniformity assertion, as each subsequence has a further subsequence that is weakly convergent. ■

Theorem 3.2. *Let $b^n(\cdot)$ be as in Theorem 3.1. Define*

$$\gamma_n(\hat{x}, T, m) = E_{\hat{x}}^m \frac{1}{T} \left[\int_0^T \int_U k(\bar{x}^n(t), \alpha) m(\bar{x}^n(t), d\alpha) dt + q' y^n(T) \right], \quad (3.5)$$

where $x^n(\cdot)$ is the solution to (1.1) or, equivalently, to (1.9), under $b^n(\cdot)$ and control $m(\cdot)$, with associated reflection process components $y^n(\cdot)$ and invariant measure $\mu_n^m(\cdot)$. Then, as $n \to \infty$, $||\mu_n^m(\cdot) - \mu^m(\cdot)|| \to 0$. Also $||P_n^m(\hat{x}, t, \cdot) - \mu^m(\cdot)|| \to 0$ and

$$\gamma_n(\hat{x}, t, m) \to \gamma(m) \quad (3.6)$$

no matter how $n \to \infty$ and $t \to \infty$. The convergence is uniform in $m(\cdot)$. It is also uniform in $\hat{x}$, if the convergence of $b^n(\cdot)$ is uniform on each compact set.

Proof. By Theorem 2.2, $||P_n^m(\hat{x}, t, \cdot) - \mu_n^m(\cdot)|| \to 0$ uniformly in $n, \hat{x}$ and $m(\cdot)$, as $t \to \infty$. By Theorem 3.1, for each t and $\hat{x}$, $||P_n^m(\hat{x}, t, \cdot) - P^m(\hat{x}, t, \cdot)|| \to 0$ uniformly in $m(\cdot)$ as $n \to \infty$. We have

$$||\mu_n^m(\cdot) - \mu^m(\cdot)|| \le ||\mu_n^m(\cdot) - \mu^m(\cdot) \mp P_n^m(\hat{x}, t, \cdot) \pm P^m(\hat{x}, t, \cdot)||. \quad (3.7)$$

This is bounded by

$$||\mu_n^m(\cdot) - P_n^m(\hat{x}, t, \cdot)|| + ||\mu^m(\cdot) - P^m(\hat{x}, t, \cdot)|| + ||P_n^m(\hat{x}, t, \cdot) - P^m(\hat{x}, t, \cdot)||. \quad (3.8)$$

Given $\epsilon > 0$, choose t large enough so that the first two terms in (3.8) are less than $\epsilon/3$ for all $m(\cdot), n, \hat{x}$. Then, for that t, the third term goes to zero as $n \to \infty$, uniformly in $m(\cdot)$ by the computations in Theorem 3.1. If the convergence of $b^n(\cdot)$ is uniform on compact $\hat{x}$-sets, then so is the convergence of the third term, by Theorem 3.1. This compactness requirement on the initial condition can be dropped, by noting that the set $\{\bar{x}^n(t), n < \infty, t > \bar{\theta}\}$ is tight in $C(G; [-\bar{\theta}, 0])$ (see the last part of the proof of Theorem 3.1), and using $\bar{x}^n(t_0)$ in lieu of $\hat{x}$ in (3.7) and (3.8), where $t_0 > \bar{\theta}$.

These computations imply that $||\mu_n^m(\cdot) - \mu^m(\cdot)|| \to 0$. By Theorem 2.2, $||P_n^m(\hat{x}, t, \cdot) - \mu_n^m(\cdot)|| \to 0$, uniformly in $m(\cdot), n$, and $\hat{x}$, as $t \to \infty$. Thus,

$$||P_n^m(\hat{x}, t, \cdot) - \mu^m(\cdot)|| \to 0$$

as $t \to \infty$ and $n \to \infty$ in any way at all. The convergence of the cost component involving $k(\cdot)$ follows from this. If the convergence of $b^n(\cdot)$ is uniform on compact $\hat{x}$-sets, then so is the convergence of this cost component for each compact set of initial conditions.

Recall that if $||\nu^n(\cdot) - \nu(\cdot)|| \to 0$ for measures $\nu^n(\cdot)$ and $\nu(\cdot)$, then $\int [\nu^n(d\hat{x}) - \nu(d\hat{x})] f(\hat{x}) \to 0$, uniformly in $f(\cdot)$, if the functions $f(\cdot)$ are measurable and uniformly bounded. Thus, using the uniform integrability in Lemma 3.2.1,

$$\int [\mu^m(d\hat{v}) - P_n^m(\hat{x}, t, d\hat{v})] E_{n,\hat{v}}^m y^n(1) \to 0$$

as $n \to \infty$ and $t \to \infty$, where $E^m_{n,\hat{v}}$ denotes the expectation under initial condition $\hat{v}$ and the use of $b^n(\cdot)$ and control $m(\cdot)$. Since for each initial condition $\hat{v}$ (and uniformly on any compact $\hat{v}$-set), $E^m_{n,\hat{v}} y(1) \to E^m_{\hat{v}} y(1)$ as $n \to \infty$, the convergence of the reflection term component of the costs follows. ■

The theorem implies that we can approximate U by a finite subset. It also implies that we can approximate $\bar{x}(t)$ in both $b(\cdot)$ and $k(\cdot)$ by the piecewise-constant interpolation (intervals δ) of the δ-samples $x(t-\delta), \ldots, x(t-\bar{\theta})$, with $\bar{x}(t,0) = x(t)$ and $\bar{\theta}/\delta$ is an integer. The full Markov state of the process at time t is still $\bar{x}(t)$. Further approximations will be discussed in the next section.

Note on Theorem 3.2 and the order of taking the limits. The fact that t and n can go to their limits in any way at all is important in applications. If as n grew to infinity we required an ever larger value of t for the approximation of the invariant measure by the transition probability, or if the t, n required for a good approximation depended on $m(\cdot)$ or on $\hat{x}$, then the result would not be very useful. The result says that for large enough n, t, the approximation is good, uniformly in $m(\cdot)$ and $\hat{x}$.

5.4 Approximations with Periodic Delays

5.4.1 Limit and Approximation Results for Periodic Delays

The general assumptions of this section are those of Section 1 and that $b(\cdot)$ satisfies (A3.1.1). Recall the periodic Approximation 3 that was defined by (4.2.6). We will consider an analog for the ergodic cost problem. Let $l\delta \le t < l\delta + \delta$ and, as in Subsection 4.2.2, define $\tilde{X}^\delta(t) = \left(x(t), x(l\delta), x(l\delta - \delta), \ldots, x(l\delta - Q^+_\delta \delta)\right)$ and $\tau^\delta(t) = t - l\delta = t(\text{mod } \delta)$, where δ and the integer Q^+_δ satisfy $\bar{\theta} = (Q^+_\delta + 1/2)\delta$. Recall the definition of $\bar{x}^\delta_p(t)$ from (4.2.6) and that the full system state at time t is $(\bar{x}^\delta_p(t), \tau^\delta(t))$ or, equivalently, $(\tilde{X}^\delta(t), \tau^\delta(t))$. Define $\tilde{X}^\delta_l = \left(x(l\delta), x(l\delta - \delta), \ldots, x(l\delta - Q^+_\delta \delta)\right)$, the vector of terms in the brackets in (4.2.6).

Relaxed feedback controls $m(\cdot)$ of the form defined above (1.1) will be used, as well as extensions that are periodic in time (period δ) and depend on $(\tilde{X}^\delta(t), \tau^\delta(t))$. For a (nonperiodic) relaxed feedback control $m(\cdot)$ and memory segment $\bar{x}^\delta_p(t)$ used, the controlled process is (1.9) with control $m(\bar{x}^\delta_p(t), d\alpha)$, where $\bar{x}^\delta_p(t)$ is obtained from $x(\cdot)$, and is defined from (1.4) via the measure $P^{\delta,m}_{\hat{x}}$ obtained from the Girsanov transformation with Radon–Nikodym derivative $R^\delta(t,m) = \exp \zeta^\delta(t,m)$, where

$$
\begin{aligned}
\zeta^\delta(t,m) = & \int_0^t \left[\sigma^{-1}(x(s)) \int_U b(\bar{x}_p^\delta(s),\alpha)m(\bar{x}(s),d\alpha)\right]' dw(s) \\
& -\frac{1}{2}\int_0^t \left|\sigma^{-1}(x(s)) \int_U b(\bar{x}_p^\delta(s),\alpha)m(\bar{x}(s),d\alpha)\right|^2 ds.
\end{aligned} \tag{4.1}
$$

The system is

$$
dx(t) = dt \int_U b(\bar{x}_p^\delta(t),\alpha)m(\bar{x}(t),d\alpha) + \sigma(x(t))dw_m(t) + z(t). \tag{4.2a}
$$

With the periodic relaxed feedback control used, the system is defined as above, but with drift term

$$
\int_U b(\bar{x}_p^\delta(t),\alpha)m(\bar{x}_p^\delta(t),\tau^\delta(t),d\alpha),
$$

and it is written as

$$
dx(t) = dt \int_U b(\bar{x}_p^\delta(t),\alpha)m(\bar{x}_p^\delta(t),\tau^\delta(t),d\alpha) + \sigma(x(t))dw_m(t) + z(t). \tag{4.2b}
$$

That is, $m(\bar{x}^\delta(s),\tau^\delta(s),d\alpha)$ is used in lieu of $m(\bar{x}(s),d\alpha)$ in (4.1). In the analysis we will also need the system (1.9) with a periodic relaxed feedback control, namely,[1]

$$
dx(t) = dt \int_U b(\bar{x}(t),\alpha)m(\bar{x}_p^\delta(t),\tau^\delta(t),d\alpha) + \sigma(x(t))dw_m(t) + z(t). \quad \text{(4.2c)},
$$

For a control $m(\cdot)$ that is periodic or not, let $P^{\delta,m}(\hat{x},t,\cdot)$ denote the periodic transition probability for the process $\bar{x}(t)$ defined by either (4.2a) or (4.2b). It will be periodic because the memory segment is the periodic function $\bar{x}_p^\delta(t)$. For either a periodic or a nonperiodic control $m(\cdot)$ and models (4.2), define the finite time-average cost $\gamma^\delta(\hat{x},T,m)$ analogously to (2.4). Since the maximum delay varies from $\bar{\theta}-\delta/2$ to $\bar{\theta}+\delta/2$, w.l.o.g., we can suppose that the transition probability is defined for $\hat{x}$ being a function on an interval $[-\bar{\theta}-\bar{\delta},0]$, where $\bar{\delta}$ is an integral multiple of δ and is larger than any of the (small) values of $\delta/2$ that are of concern.

The next theorem follows from Theorem 2.2 and shows that the transition probabilities of the δ-sampled processes converge to the associated invariant measures.

Theorem 4.1. *The process $\bar{x}(n\delta), n = 1,2,\ldots,$ where $x(\cdot)$ is defined by either (4.2a) or (4.2b), satisfies the Doeblin condition, uniformly in the control $m(\cdot)$*

---

[1] For simplicity, we use the same symbol $w_m(\cdot)$ for the Wiener processes in (4.2a)–(4.2c). They are not the same, since the actual Wiener process would depend on the form of the controlled drift term. The analysis involves only the Wiener process $w(\cdot)$.

and in $\hat{x}$, whether the control is relaxed feedback or periodic relaxed feedback. For the reference measure, we can use $\phi(\cdot) = P(x_0, t_0, \cdot)$ for any $x_0 \in G$, as in Theorem 2.2, where we can suppose that t_0 is a multiple of δ. For each $\delta > 0$ and periodic or nonperiodic relaxed feedback control $m(\cdot)$, there is a unique invariant measure $\mu^{\delta,m}(\cdot)$ for $\bar{x}(n\delta)$ and

$$||P^{\delta,m}(\hat{x}, n\delta, \cdot) - \mu^{\delta,m}(\cdot)|| \leq C\rho^{n\delta}, \tag{4.3}$$

where $C < \infty$ and $\rho \in (0,1)$ do not depend on $m(\cdot), \hat{x}$, or δ.

The next theorem shows that the finite-time costs converge to the ergodic cost.

Theorem 4.2. *As $T \to \infty$, for both models* (4.2a) *and* (4.2b), *$\gamma^\delta(\hat{x}, T, m) \to \gamma^\delta(m)$ for each δ. The convergence is uniform in periodic and nonperiodic $m(\cdot)$ and in $\hat{x}$ and δ.*

Comment on the proof. Let δ_1 be an integral multiple of δ. It is bounded but might depend on δ. Consider (4.2b) and a periodic relaxed feedback control. The average cost on $[0, N\delta_1]$ that corresponds to the rate $k(\cdot)$ can be written as

$$\frac{1}{N\delta_1} \sum_{n=0}^{N-1} \int P^{\delta,m}(\hat{x}, n\delta_1, d\hat{y}) \int_0^{\delta_1} ds\, P^{\delta,m}(\hat{y}, s, d\hat{v}) \int_U k(\hat{v}_p^\delta, \alpha) m(\hat{v}_p^\delta, \tau^\delta(s), d\alpha), \tag{4.4}$$

where, at time s, $\hat{v}_p^\delta$ is the periodic memory segment derived from the current path segment $\hat{v}$ and the time $\tau^\delta(s)$ since the last shift. Then the proof for the component of the cost that contains $k(\cdot)$ follows from (4.3), which implies that $P^{\delta,m}(\hat{x}, n\delta, \cdot)$ converges to $\mu^{\delta,m}(\cdot)$ in variation norm as $n\delta \to \infty$, uniformly in $\delta, m(\cdot)$, and $\hat{x}$. The argument for the reflection term component of the cost is similar and is omitted. ■

The next theorem shows that if the dynamics and cost rate depend on the memory segment form $\bar{x}_p^\delta(t)$, but the control is a nonperiodic relaxed feedback control, then as $\delta \to 0$, the costs converge to those for the case where the dynamics and cost rate depend on the full memory segment $\bar{x}(t)$.

Theorem 4.3. *Consider the model* (4.2a). *For each nonperiodic relaxed feedback control $m(\cdot)$, and as $\delta \to 0$, $||\mu^{\delta,m}(\cdot) - \mu^m(\cdot)|| \to 0$ and $\gamma^\delta(m) \to \gamma(m)$, uniformly in $m(\cdot)$.*

Proof. The proof uses arguments like that of Theorems 3.1 and 3.2. We have

$$P^{\delta,m}(\hat{x}, t, A) - P^m(\hat{x}, t, A) = E_{\hat{x}} \left[R^\delta(t,m) - R(t,m)\right] I_{\{\bar{x}(t) \in A\}}. \tag{4.5}$$

As in Theorem 3.1, there are constants K_t such that

$$
\begin{aligned}
&||P^{\delta,m}(\hat{x},t,\cdot) - P^m(\hat{x},t,\cdot)|| \\
&\qquad \le K_t E_{\hat{x}}^{1/2} \int_0^t ds \int_U \left| \left(b(\bar{x}_p^\delta(s),\alpha) - b(\bar{x}(s),\alpha) \right) m(\bar{x}(s),d\alpha) \right| \\
&\qquad \le K_t E_{\hat{x}}^{1/2} \int_0^t \sup_\alpha \left| b(\bar{x}_p^\delta(s),\alpha) - b(\bar{x}(s),\alpha) \right| ds,
\end{aligned} \tag{4.6}
$$

where both $\bar{x}(t)$ and $\bar{x}_p^\delta(t)$ are from the uncontrolled process (1.4). For each continuous $\hat{x}$, this expression goes to zero by the continuity condition (A3.1.1) on $b(\cdot)$, as $\bar{x}_p^\delta(t) \to \bar{x}(t)$ for each path, uniformly on any bounded time interval. Thus, uniformly in $m(\cdot)$ and for each continuous $\hat{x}$, as $\delta \to 0$

$$
||P^{\delta,m}(\hat{x},t,\cdot) - P^m(\hat{x},t,\cdot)|| \to 0. \tag{4.7}
$$

The convergence in (4.7) is also uniform on any compact $\hat{x}$-set of initial conditions.

Using (4.7), the method of Theorem 3.2 can be used to show that

$$
||\mu^{\delta,m}(\cdot) - \mu^m(\cdot)|| \to 0 \tag{4.8}
$$

uniformly in $m(\cdot)$. Using the expressions (4.3), (4.7), (4.8), and the representation (4.4) of the finite-time cost (but with control form $m(\bar{x}(t),d\alpha)$ used) yields that the cost component corresponding to $k(\cdot)$ converges as asserted. We omit the details concerning the reflection term. ∎

Suppose that, for a sequence of periodic relaxed feedback controls $\{m^\delta(\cdot)\}$, we apply $m^\delta(\cdot)$ to both the system (4.2b) where the dynamics and cost rate depend on the memory segment $\bar{x}_p^\delta(t)$ and to the system (1.9) or (4.2c) where $\bar{x}(t)$ is used for the memory segment in the dynamics and cost rate. Then the next theorem shows that the difference between the costs goes to zero as $\delta \to 0$, uniformly in the selected sequence of controls.

Theorem 4.4. *Let $m^\delta(\cdot)$ be a periodic relaxed feedback control. Then, uniformly in the sequence $\{m^\delta(\cdot), \delta \to 0\}$, for each $\hat{x}$ and as $\delta \to 0$*

$$
||P^{\delta,m^\delta}(\hat{x},t,\cdot) - P^{m^\delta}(\hat{x},t,\cdot)|| \to 0 \tag{4.9}
$$

uniformly on any bounded time interval. The convergence is uniform in any compact $\hat{x}$-set. Also, as $\delta \to 0$,

$$
\begin{aligned}
&||\mu^{\delta,m^\delta}(\cdot) - \mu^{m^\delta}(\cdot)|| \to 0, \\
&\gamma^\delta(m^\delta) - \gamma(m^\delta) \to 0.
\end{aligned} \tag{4.10}
$$

Proof. The proof of the assertions concerning convergence is like that of similar assertions in Theorems 4.2 and 4.3. We need only replace the middle line of (4.6) by

$$K_t E_{\hat{x}}^{1/2} \int_0^t ds \int_U \left| \left(b(\bar{x}_p^\delta(s), \alpha) - b(\bar{x}(s), \alpha) \right) m^\delta(\bar{x}_p^\delta(s), \tau^\delta(s), d\alpha) \right|,$$

where $x(\cdot)$ and $\bar{x}_p^\delta(\cdot)$ are obtained from the uncontrolled process (1.4). ■

The next theorem implies that the periodic model (4.2b) is a good approximation to (1.1) in that good periodic relaxed feedback controls for (4.2b) are also good for (1.1). Define $\bar{\gamma} = \inf_m \gamma(m)$, where the inf is over all relaxed feedback controls, and $\bar{\gamma}^\delta = \inf_m \gamma^\delta(m)$, where the inf is over all periodic relaxed feedback controls. We will require the following two assumptions.

A4.1. *The infimum $\bar{\gamma}^\delta$ of $\gamma^\delta(m)$ over periodic relaxed feedback controls (for system* (4.2b)*) is at least as small as the infimum over nonperiodic relaxed controls (for system* (4.2a)*).*

A4.2. *The infimum of $\gamma(m)$ over periodic relaxed feedback controls (for system* (4.2c)*) is no smaller than the infimum $\bar{\gamma}$ over nonperiodic relaxed controls (for system* (1.9)*).*

The next assumption is stronger than (A4.1). It will be needed in Section 5 to show that the optimal costs for the memory segment and control processes used there converge to the optimal cost for the periodic relaxed controls and periodic memory segment of this section. It will also be needed to prove the convergence of the numerical algorithms for the ergodic cost criterion in Chapter 8. We expect that both (A4.1) and (A4.3) hold in general, although a proof is lacking.[2] [3]

A4.3. *Consider the class of admissible relaxed controls $r(\cdot)$ such that there is a controlled process $x(\cdot)$ with periodic memory segment $\bar{x}_p^\delta(t)$, and that is stationary in the sense that the distribution of $(x(t+\cdot), r(t+\cdot) - r(t), z(t+$*

[2] Consider (A4.1). Suppose that at time t the dynamics and cost rate depend only on $\bar{x}_p^\delta(t)$ and $\tau^\delta(t)$, as in the periodic approximation defined by (4.2.6). Then is it possible that a control that depends on the full $\bar{x}(t)$ would yield a smaller cost than the best one that depended only on $\bar{x}_p^\delta(t)$ and $\tau^\delta(t)$? The answer to this question is not known at present. But it is highly plausible that it is not the case.

[3] For each relaxed feedback control with values $m(\bar{x}_p^\delta(t), \tau^\delta(t), d\alpha)$ at time t, there is a unique stationary process and invariant measure. This is the process that starts at a (uniformly distributed) random time in $[0, \delta]$, with the initial condition associated with starting at t having the distribution $\int \mu^{\delta,m}(d\hat{v}) P^{\delta,m}(\hat{v}, t, \cdot)$. The starting points of the renewal (still interval δ) cycles depend on the path. The associated cost is still $\gamma^\delta(m)$. The reason for the restriction to stationary processes in (A4.3) is that the limit processes will have this form. Using results for the problem with no delays as a guide, where one cannot do better than with relaxed feedback controls [56, Chapter 4, Theorem 6.1], it seems very likely that the assumption (A4.3) is not restrictive. However, the methods that were used in the proof of [56, Chapter 4, Theorem 6.1] for the problem with no delays do not seem to carry over to the problem with delays.

$\cdot) - z(t))$ does not depend on t. Then the infimum of the costs over periodic relaxed feedback controls is at least as small as the infimum over this class of relaxed controls.

For the relaxed control $r(\cdot)$ in (A4.3), for future use let $\gamma^\delta(r)$ denote the average cost per unit time, which must exist due to the stationarity.

Theorem 4.5. *Assume the conditions of this section and* (A4.1), (A4.2). *Then $\bar{\gamma}^\delta \to \bar{\gamma}$ as $\delta \to 0$. For each $\epsilon > 0$, there is an $\epsilon_1 > 0$ such that an ϵ_1-optimal periodic relaxed feedback control for* (4.2b) *is ϵ-optimal for both* (1.1) *and* (4.2c), *where the memory segment is $\bar{x}(t)$.*

Proof. Let $m^\epsilon(\cdot)$ be an ϵ-optimal relaxed feedback control for (1.9). Then by Theorem 4.3, (A4.1), and the minimality of $\bar{\gamma}^\delta$ and $\bar{\gamma}$ for (4.2a,b) and (1.9), resp.,

$$\bar{\gamma}^\delta \leq \gamma^\delta(m^\epsilon) \to \gamma(m^\epsilon) \leq \bar{\gamma} + \epsilon.$$

Because ϵ is arbitrary, $\limsup_\delta \bar{\gamma}^\delta \leq \bar{\gamma}$. Next, let $m^{\delta,\epsilon}(\cdot)$ be an ϵ-optimal periodic relaxed feedback control for the δ-periodic model (4.2b). Thus $\gamma^\delta(m^{\delta,\epsilon}) \leq \bar{\gamma}^\delta + \epsilon$. By Theorem 4.4, $\gamma^\delta(m^{\delta,\epsilon}) - \gamma(m^{\delta,\epsilon}) \to 0$ as $\delta \to 0$. Since $\gamma(m^{\delta,\epsilon}) \geq \bar{\gamma}$ by (A4.2) and $\epsilon > 0$ is arbitrary, $\liminf_\delta \bar{\gamma}^\delta \geq \bar{\gamma}$, which completes the proof that $\bar{\gamma}^\delta \to \bar{\gamma}$.

The last assertion of the theorem follows from $\lim_{\delta\to 0} |\gamma^\delta(m^{\delta,\epsilon}) - \gamma(m^{\delta,\epsilon})| = 0$ and $\limsup_{\delta\to 0} |\gamma(m^{\delta,\epsilon}) - \bar{\gamma}| \leq \epsilon$, both of which follow from the previous computations. ∎

The next two theorems will be useful for proving convergence of the numerical algorithms.

Discrete approximation to U. Let U^n be a set with finitely many points such that $U^n \to U$. Define $b_n(\hat{x}, \alpha)$ by dividing U into a finite number of disjoint subsets U_i^n whose diameters are less than $1/n$, selecting a point $\alpha_i^n \in U_i^n$ and letting $b_n(\hat{x}, \alpha) = b(\hat{x}, \alpha_i^n)$ for $\alpha \in U_i^n$. Define $k_n(\cdot)$ analogously. The use of $b_n(\cdot)$ and $k_n(\cdot)$ is equivalent to exchanging U for U^n. Then the arguments in Theorems 3.1, 3.2, 4.3, and 4.4 imply that the costs change little if the control-value set U is approximated by U^n. This is stated in the following theorem.

Theorem 4.6. *Let U^n replace U in either the model* (1.9) *or* (4.2a,b). *Then, for large n, the invariant measures and costs change little, uniformly in δ and $m(\cdot)$, for either relaxed feedback or periodic relaxed feedback controls.*

5.4.2 Smoothed Nearly Optimal Controls

Let U contain only finitely many points. For $\alpha \in U$, write $m(\bar{x}_p^\delta(t), \tau^\delta(t), \{\alpha\})$ for the weight that the measure puts on the point α. For $l\delta \leq t < l\delta + \delta$,

$\bar{x}_p^\delta(t)$ is a function of $\tilde{X}^\delta(t)$. Thus, without confusion, we can use the notation $m(\tilde{X}^\delta(t), \tau^\delta(t), d\alpha)$ for $m(\bar{x}_p^\delta(t), \tau^\delta(t), d\alpha)$. Let $\tilde{X}^\delta$ denote the canonical value of $\tilde{X}^\delta(t)$, and τ^δ the canonical value of $\tau^\delta(t)$. For the purposes of proving the convergence of the numerical procedures, it will be helpful to know that for each δ and $\epsilon > 0$, there is an ϵ-optimal periodic relaxed control for (4.2) that is continuous in $\tilde{X}^\delta$ and τ^δ, and this is shown in the next theorem. See [56, Equation 4.3.17] for a related result, where the $\sqrt{2\pi\epsilon}$ should be to the power r, which is the dimension of the state variable x there.

Theorem 4.7. *Let U contain only finitely many points and let $\epsilon > 0$. For any δ and a periodic relaxed feedback control $m(\cdot)$ for (4.2b), there is a periodic relaxed feedback control $m^\epsilon(\cdot)$ such that, for each $\alpha \in U$, $m^\epsilon(\tilde{X}^\delta, \tau^\delta, \{\alpha\})$ is continuous in $\tilde{X}^\delta$ and τ^δ. Also $|\gamma^\delta(m) - \gamma^\delta(m^\epsilon)| \leq \epsilon$.*

Proof. Fix $\delta > 0$ and $m(\cdot)$. Each component of $\tilde{X}^\delta$ takes values in G, and τ^δ takes values in $[0, \delta)$. Let us extend the definition of $m(\tilde{X}^\delta, \tau^\delta, \{\alpha\})$ so that for each $\alpha \in U$ it is defined for all values of the other variables in the Euclidean space of the appropriate dimension. Suppose that there are periodic relaxed feedback controls $m^n(\tilde{X}^\delta, \tau^\delta, d\alpha)$, with associated transition probabilities $P^{\delta, m^n}(\cdot)$, such that for almost all $\tau \in (0, \delta)$ and each $\alpha \in U$, $m^n(\tilde{X}^\delta, \tau^\delta, \{\alpha\}) \to m(\tilde{X}^\delta, \tau^\delta, \{\alpha\})$ for almost all (Lebesgue measure) $\tilde{X}^\delta$ whose components have values in G. By a computation analogous to (4.6), for some constant K_t we have

$$||P^{\delta,m}(\hat{x}, t, \cdot) - P^{\delta,m^n}(\hat{x}, t, \cdot)|| \leq$$

$$K_t E_{\hat{x}}^{1/2} \int_0^t ds \int_U \left| b(\bar{x}_p^\delta(s), \alpha) \left[m(\tilde{X}^\delta(s), \tau^\delta(s), d\alpha) - m^n(\tilde{X}^\delta(s), \tau^\delta(s), d\alpha) \right] \right|. \tag{4.11}$$

By the convergence of $m^n(\tilde{X}^\delta, \tau^\delta, \alpha)$ for almost all (Lebesgue measure) $\tilde{X}^\delta$ whose components have values in G, for almost all $\tau^\delta \in [0, \delta)$ and each $\alpha \in U$, together with the density properties of $\tilde{X}^\delta$ in Theorem 1.4, the expression (4.11) goes to zero as $n \to \infty$. Then the arguments of Theorems 4.3 and 4.4 imply that $\gamma^\delta(m^n) \to \gamma^\delta(m)$ as $n \to \infty$. We have only to show that there are such $m^n(\tilde{X}^\delta, \tau, \alpha)$ that are continuous in $(\tilde{X}^\delta, \tau^\delta)$.

Constructing the smoothed control. The approximations $m^n(\tilde{X}^\delta, \tau^\delta, \alpha)$ are constructed via a smoothing procedure. Let $\rho \to 0$ replace $n \to \infty$ and let d=number of real values needed to specify $\tilde{X}^\delta$. Define the smoothed control $m^\rho(\tilde{X}^\delta, \tau^\delta, \{\alpha\}), \alpha \in U$, by

$$m^\rho(\tilde{X}^\delta, \tau^\delta, \{\alpha\})$$

$$= \frac{1}{(2\pi\rho)^{d/2}} \frac{1}{(2\pi\rho)^{1/2}} \int_{I\!R^d} \int_{-\infty}^{\infty} e^{-|X - \tilde{X}^\delta|^2/2\rho} e^{-|\tau - \tau^\delta|^2/2\rho} m(X, \tau, \{\alpha\}) dX d\tau. \tag{4.12}$$

$m^\rho(\tilde{X}^\delta, \tau^\delta, \{\alpha\})$ is continuous in $\tilde{X}^\delta$ and τ^δ for each $\alpha \in U$. Now restrict the domain to $(\tilde{X}^\delta, \tau^d)$-values such that the components of $\tilde{X}^\delta$ have values in G, and the values of τ^δ are in $[0, \delta)$. For each $\alpha \in U$, it converges to $m(\tilde{X}^\delta, \tau^\delta, \{\alpha\})$ for almost all values in the desired range. ■

5.4.3 Delays in the Variance Term

Up to now, we have assumed that the variance term $\sigma(\cdot)$ did not depend on the delayed path. The reason had to do with the method of constructing the solutions to the controlled problem via the Girsanov transformation method, which allows only the drift term to be changed, so that we would not be able to change $\sigma(\bar{x}(t))$ for $\sigma(\bar{x}_p^\delta(t))$ when deriving the periodic approximation, and similarly for other approximations.

Delay-dependent noise terms can be handled under stronger assumptions. This will be illustrated for the example of periodic delays. Consider the processes

$$dx^\delta(t) = \sigma(\bar{x}_p^\delta(t))dw(t) + dz^\delta(t), \tag{4.13a}$$

$$dx(t) = \sigma(\bar{x}(t))dw(t) + dz(t), \tag{4.13b}$$

where $\bar{x}_p^\delta(t)$ is the periodic approximation to the memory segment corresponding to the solution $x^\delta(\cdot)$, and $z^\delta(\cdot)$ is the reflection term, with the analogous definitions for (4.13b). Suppose that, for any initial condition, (4.13a) and (4.13b) have unique strong-sense solutions that are also weak-sense unique. That is, given a Wiener process $w(\cdot)$, for any $\delta > 0$ and initial condition $\hat{x}$ we can construct solutions that are functionals of $w(\cdot)$ and the initial condition. The density properties in Section 1 can still be shown to hold. Given any relaxed feedback control, whether periodic or nonperiodic, we can construct solutions via the Girsanov transformation method of Subsection 1.1, where $\sigma(x(t))$ is replaced by either $\sigma(\bar{x}(t))$ or $\sigma(\bar{x}_p^\delta(t))$, according to the case. The Doeblin condition continues to hold as do the results on convergence of the costs and of the transition probabilities to the invariant measure, for each fixed $m(\cdot)$ and δ. But the proofs of results such as in Theorems 4.3 and 4.6, which compare solutions with different memory segments, need to be modified.

Consider the proof of Theorem 4.3. Equation (4.5) holds as all uncontrolled processes can be defined in terms of the same Wiener process $w(\cdot)$, no matter what δ is. Using the inequality (3.1) yields the bounds

$$||P^{\delta,m}(\hat{x}, t, \cdot) - P^m(\hat{x}, t, \cdot)|| \leq$$

$$K_t E_{\hat{x}}^{1/2} \int_0^t ds \int_U \left|\left(\sigma^{-1}(\bar{x}_p^\delta(s))b(\bar{x}_p^\delta(s), \alpha) - \sigma^{-1}(\bar{x}(s))b(\bar{x}(s), \alpha)\right) m(\bar{x}(s), d\alpha)\right|. \tag{4.14}$$

Using the uniqueness hypotheses, under appropriate continuity conditions on $b(\cdot)$ and $\sigma(\cdot)$, this expression will go to zero as $\delta \to 0$. Using such results, we can obtain the foregoing convergence and approximation results. For simplicity of development, we will continue with only weak-sense existence and uniqueness of the uncontrolled model and use the form $\sigma(x(t))dw(t)$.

5.5 The Periodic-Erlang Approximation

To use the periodic approximation of the previous section, one needs to keep track of the elapsed time since the beginning of the current cycle, as was the case in Subsection 4.2.2. The periodic-Erlang approximation defined by (4.2.8) in Subsection 4.2.4 was a useful way of approximating this elapsed time, and the method can be carried over to the ergodic cost problem. It will be a convenient basis for the numerical approximations.

Definitions. The general assumptions of this section are those of Section 1 and that $b(\cdot)$ satisfies (A3.1.1). We will use the notation of Subsection 4.2.4. Recall the Erlang state $L^{\delta_0,\delta}(t)$ (which is independent of the $x(\cdot)$ and $w(\cdot)$ in (1.4)), the random times $\bar{\sigma}_l^{\delta_0,\delta}$ and $\sigma_l^{\delta_0}$, and the memory segments $\bar{x}_e^{\delta,\delta_0}(t)$ of (4.2.8) and their limits (as $\bar{L}^{\delta_0,\delta} \to \infty$) $\bar{x}_p^{\delta}(t)$ of (4.2.6). For $\bar{\sigma}_l^{\delta,\delta_0} \leq t < \bar{\sigma}_{l+1}^{\delta,\delta_0}$, recall the definitions $\tilde{X}_e^{\delta,\delta_0}(t) = \left(x(t), x(\bar{\sigma}_l^{\delta,\delta_0}), \ldots, x(\bar{\sigma}_{l-Q_\delta^+}^{\delta,\delta_0})\right)$ with canonical value $\tilde{X}^{\delta,\delta_0}$, and the vector $\tilde{X}_l^{\delta,\delta_0} = \left(x(\bar{\sigma}_l^{\delta,\delta_0}), \ldots, x(\bar{\sigma}_{l-Q_\delta^+}^{\delta,\delta_0})\right)$. We will use relaxed feedback controls of the form $m(\bar{x}(t), d\alpha)$, and also controls of the form $m(\bar{x}_e^{\delta_0,\delta}(t), L^{\delta_0,\delta}(t), d\alpha)$ that depend on the Erlang state and the periodic-Erlang path memory approximation. We suppose that the σ-algebra $\mathcal{F}_t$ has been augmented so that it measures $\{L^{\delta_0,\delta}(s), s \leq t\}$.

Construction of the controlled process. In analogy to (4.1), for a relaxed feedback control $m(\bar{x}(t), d\alpha)$, the process with memory segment $\bar{x}_e^{\delta_0\delta}(t)$ is defined from (1.4) via the measure $P_{\hat{x}}^{\delta_0,\delta.m}$ obtained from the Girsanov transformation with Radon–Nikodym derivative $R^{\delta_0,\delta}(t,m) = \exp[\zeta^{\delta_0,\delta}(t,m)]$, where

$$
\begin{aligned}
\zeta^{\delta_0,\delta}(t,m) = & \int_0^t \left[\sigma^{-1}(x(s)) \int_U b(\bar{x}_e^{\delta_0,\delta}(s),\alpha) m(\bar{x}(s), d\alpha)\right]' dw(s) \\
& - \frac{1}{2} \int_0^t \left|\sigma^{-1}(x(s)) \int_U b(\bar{x}_e^{\delta_0,\delta}(s),\alpha) m(\bar{x}(s), d\alpha)\right|^2 ds,
\end{aligned}
\tag{5.1}
$$

where $x(\cdot)$ is the process (1.4) and $\bar{x}_e^{\delta_0,\delta}(\cdot)$ is obtained from it via the definition (4.2.8). Write the transformed process as

$$
dx(t) = dt \int_U b(\bar{x}_e^{\delta,\delta_0}(t),\alpha) m(\bar{x}(t), d\alpha) + \sigma(x(t)) dw_m(t) + dz(t). \tag{5.2a}
$$

If the control $m(\cdot)$ depends on the memory segment $\bar{x}_e^{\delta_0,\delta}(t)$ and the Erlang state, then the system is

$$
dx(t) = dt \int_U b(\bar{x}_e^{\delta,\delta_0}(t),\alpha) m(\bar{x}_e^{\delta_0\delta}(t), L^{\delta_0\delta}(t), d\alpha) + \sigma(x(t)) dw_m(t) + dz(t), \tag{5.2b}
$$

and $m(\bar{x}_e^{\delta_0,\delta}(s), L^{\delta_0,\delta}(s), d\alpha)$ is used in (5.1) in lieu of $m(\bar{x}(s), d\alpha)$. Finally, if the memory segment in the dynamics and cost rate is $\bar{x}(t)$ but the control depends on the approximation, then the system is

$$dx(t) = dt \int_U b(\bar{x}(t), \alpha) m(\bar{x}_e^{\delta_0 \delta}(t), L^{\delta_0 \delta}(t), d\alpha) + \sigma(x(t)) dw_m(t) + dz(t), \quad (5.2c)$$

and $b(\bar{x}(s), \alpha) m(\bar{x}_e^{\delta_0 \delta}(t), L^{\delta_0 \delta}(t), d\alpha)$ is used in (5.1). In all cases, we use the same expression $P^{\delta_0, \delta, m}(\hat{x}, t, \cdot)$ for the transition function.

The next theorem shows that the properties of $L^{\delta_0, \delta}(t)$ do not change under the measure transformation.

Theorem 5.1. *Under $P_{\hat{x}}^{\delta_0, \delta, m}$ and for all cases of (5.2), for each $t \geq 0$ the process $L^{\delta_0, \delta}(t + \cdot) - L^{\delta_0, \delta}(t)$ is independent of $\mathcal{F}_t$, and its distribution is unchanged.*

Proof. For $t \geq 0$, $t_i > 0$, and $T \geq t + t_i, i = 1, \ldots, k$, and a measurable set A consider

$$\begin{aligned} &E_{\hat{x}}^{\delta_0, \delta, m} I_{\{(L^{\delta_0, \delta}(t+t_i) - L^{\delta_0, \delta}(t), i=1,\ldots,k) \in A\}} \\ &\quad = E_{\hat{x}} R^{\delta_0, \delta}(T, m) I_{\{(L^{\delta_0, \delta}(t+t_i) - L^{\delta_0, \delta}(t), i=1,\ldots,k) \in A\}}. \end{aligned} \quad (5.3)$$

Because $w(\cdot)$ and $x(\cdot)$ (from (1.4)) are independent of $L^{\delta_0, \delta}(\cdot)$ under the original measure, the two factors of the right-hand term of (5.3) are mutually independent. Since the expectation of $R^{\delta_0, \delta}(T, m)$ is unity, the expectation on the right side is just the probability of the event $\{(L^{\delta_0, \delta}(t + t_i) - L^{\delta_0, \delta}(t), i = 1, \ldots, k) \in A\}$ under the original measure. This computation implies that the distribution of $L^{\delta_0, \delta}(\cdot)$ is unchanged. To get the independence of $L^{\delta_0, \delta}(t + \cdot) - L^{\delta_0, \delta}(t)$ and $\mathcal{F}_t$ repeat the computation with a conditioning on $\mathcal{F}_t$. The theorem follows from this and the arbitrariness of the T, t_i and k. ■

The next theorem is an analog of Theorems 1.3 and 1.4 for the uncontrolled process, where t_i is replaced by $\bar{\sigma}_i^{\delta_0, \delta}$. It follows from the comments after Theorem 1.4 concerning random times, as, for each $\delta > 0$, $\{\bar{\sigma}_{l+1}^{\delta, \delta_0} - \bar{\sigma}_l^{\delta, \delta_0}, l < \infty\}$ are mutually independent with uniformly bounded variances, $\lim_{\epsilon \to 0} \sup_{\delta_0} P\{\bar{\sigma}_1^{\delta, \delta_0} \leq \epsilon\} = 0$, and the set is independent of both $w(\cdot)$ and the process $x(\cdot)$ defined by (1.4).

Theorem 5.2. *Let $\epsilon_1 > 0$ and let $x(\cdot)$ be the process defined by (1.4). Then there is $\epsilon_2 > 0$ such that for any $B \in \mathcal{B}(G)$ with $l(B) \geq \epsilon_1$ we have $P_0(x, \bar{\sigma}_1^{\delta, \delta_0}, B) = P_x\{x(\bar{\sigma}_1^{\delta, \delta_0}) \in B\} \geq \epsilon_2$ for all $x \in G$ and all δ_0 such that δ/δ_0 is an integer, and conversely. Thus the $P_0(x, \bar{\sigma}_1^{\delta, \delta_0}, \cdot)$ are mutually absolutely continuous with respect to Lebesgue measure, uniformly in δ_0 and x.*

For $l = Q_\delta^+ + 1$, the analogous result holds for the measures $P_x\{\tilde{X}_l^{\delta, \delta_0} \in A\}$ and the Lebesgue measure $l(A)$, where A is a Borel set in $G^{Q_\delta^+ + 1}$, the range space of $\tilde{X}_l^{\delta, \delta_0}$.

The next result extends Theorem 5.2 to the controlled process. Define $a \wedge b = \min\{a, b\}$, and for arbitrary t define the "stopped" processes $\bar{\sigma}_{t(l)}^{\delta, \delta_0} =$

$\bar{\sigma}_l^{\delta,\delta_0} \wedge t$ and $\tilde{X}_{l(t)}^{\delta_0 \delta}$:

$$\tilde{X}_{t(l)}^{\delta_0 \delta} = \left(x(\bar{\sigma}_l^{\delta,\delta_0} \wedge t), \ldots, x(\bar{\sigma}_{l-Q_\delta^+}^{\delta,\delta_0} \wedge t)\right).$$

These stopped processes will be useful as we cannot use the Girsanov transformation on the unbounded interval $[0, \bar{\sigma}^{\delta_0}\delta_l]$. If the relaxed feedback control depends on both the Erlang state and the periodic-Erlang approximation to the memory segment, then we call it a relaxed periodic-Erlang feedback control.

Theorem 5.3. *Let* $P_{\hat{x}}^{\delta,\delta_0,m}\{\tilde{X}_l^{\delta,\delta_0} \in \cdot\}$ *replace* $P_x\{\tilde{X}_l^{\delta,\delta_0} \in \cdot\}$ *in Theorem 5.2 for* $l = Q_\delta^+ + 1$. *Then the conclusions of that theorem hold, uniformly in* $m(\cdot), \hat{x}$ *and* δ_0, *where* $m(\cdot)$ *is either a relaxed feedback control or a relaxed periodic-Erlang feedback control; i.e., for any of the models* (5.2a,b,c).

Proof. Let $l = Q_\delta^+ + 1$. Because for any Borel set A,

$$P_{\hat{x}}^{\delta,\delta_0,m}\{\tilde{X}_l^{\delta,\delta_0} \in A\} = \lim_{t\to\infty} P_{\hat{x}}^{\delta,\delta_0,m}\{\tilde{X}_{t(l)}^{\delta,\delta_0} \in A\},$$

where the limit is uniform in $\hat{x}$, δ_0 and $m(\cdot)$, it is sufficient to prove the theorem for $\tilde{X}_{t(l)}^{\delta,\delta_0}$ replacing $\tilde{X}_l^{\delta,\delta_0}$ with t large. The proof with this replacement is analogous to that of Theorem 1.7. We have

$$P_{\hat{x}}^{\delta,\delta_0,m}\{\tilde{X}_{t(l)}^{\delta_0,\delta} \in A) = E_{\hat{x}} R^{\delta,\delta_0}(\bar{\sigma}_{t(l)}^{\delta,\delta_0}, m) I_{\{\tilde{X}_{t(l)}^{\delta,\delta_0} \in A\}}.$$

Define the set $B_k^{\delta,\delta_0,m}(t) = \{\zeta(\bar{\sigma}_{t(l)}^{\delta,\delta_0}, m) \geq -k\}$. Analogously to (1.16), we have the following lower bound to the above expression:

$$e^{-k}\left[E_{\hat{x}(0)} I_{\{\tilde{X}_{t(l)}^{\delta,\delta_0} \in A\}} - E_{\hat{x}} I_{\{\Omega - B_k^{\delta,\delta_0,m}(t)\}}\right].$$

Let $l(A) \geq \epsilon_1 > 0$. Then, by Theorem 5.2, there is $\epsilon_2 > 0$, depending only on ϵ_1, such that for large enough t and all δ_0 and $\hat{x}$, $E_{\hat{x}} I_{\{\tilde{X}_{t(l)}^{\delta,\delta_0} \in A\}} \geq \epsilon_2$. Thus

$$P_{\hat{x}}^{\delta,\delta_0,m}\{\tilde{X}_{t(l)}^{\delta,\delta_0} \in A\} \geq e^{-k}\left[\epsilon_2 - E_{\hat{x}} I_{\{\Omega - B_k^{\delta,\delta_0,m}(t)\}}\right].$$

Finally, choose k so that $E_{\hat{x}} I_{\{\Omega - B_k^{\delta,\delta_0,m}(t)\}} \leq \epsilon_2/2$ for all $m(\cdot), \hat{x}$, and all large t. ■

The next theorem is a consequence of Theorem 5.3.

Theorem 5.4. *With Lebesgue measure as the reference, the process* $\tilde{X}_l^{\delta,\delta_0}$ *obtained from* (5.2b) *is recurrent, aperiodic, and irreducible and satisfies a Doeblin condition, uniformly in the initial condition, in the periodic-Erlang* *relaxed feedback control, and in δ_0, for small δ_0. There is a unique invariant measure $\mu^{\delta_0,\delta,m}(\cdot)$ for each periodic-Erlang relaxed feedback control $m(\cdot)$.*

In Theorem 5.4, the invariant measure is defined on the range space of the process of samples $\tilde{X}_l^{\delta,\delta_0}$ and this is the interpretation that we will use unless mentioned otherwise. It could also be considered to be a measure on the set of interpolations of $\tilde{X}_l^{\delta,\delta_0}$ on $[-\bar{\theta},0)$, as defined by the terms in the brackets in (4.2.8).

By Theorems 5.3 and 5.4, there are $C<\infty$ and $\rho\in(0,1)$, not depending on $m(\cdot),\hat{x}$, or on δ_0, for small δ_0, such that

$$||P_{\hat{x}}^{\delta_0,\delta,m}\{\tilde{X}_l^{\delta_0,\delta}\in\cdot\}-\mu^{\delta_0,\delta,m}(\cdot)||\le C\rho^l. \tag{5.4}$$

For periodic-Erlang relaxed feedback controls, define the finite-time costs

$$\gamma^{\delta_0,\delta}(t,m)= \\ \frac{1}{t}E_{\hat{x}}^{\delta_0,\delta.m}\left[\int_0^t ds\int_U k(\bar{x}_e^{\delta_0,\delta}(s),\alpha)m(\bar{x}_e^{\delta_0,\delta}(s),L^{\delta_0,\delta}(s),d\alpha)+q'y(t)\right]. \tag{5.5}$$

Theorem 5.5. *For any periodic-Erlang relaxed feedback control $m(\cdot)$ and system* (5.2b), *the limit*

$$\gamma^{\delta_0,\delta}(m)=\lim_{t\to\infty}\gamma^{\delta_0,\delta}(t,m) \tag{5.6}$$

exists, uniformly in $\hat{x},m(\cdot)$, and small δ_0.

Proof. Consider the cost component involving $k(\cdot)$. Define $n(t)=\min\{i:\bar{\sigma}_i^{\delta_0,\delta}\ge t\}$. It is sufficient to work with the sum, as $t\to\infty$,

$$E_{\hat{x}}^{\delta_0,\delta,m}\frac{n(t)}{t}\frac{1}{n(t)}\sum_{i=0}^{n(t)-1}\int_{\bar{\sigma}_i^{\delta_0,\delta}}^{\bar{\sigma}_{i+1}^{\delta_0,\delta}}ds\int_U k(\bar{x}_e^{\delta_0,\delta}(s),\alpha)m(\bar{x}_e^{\delta_0,\delta}(s),L^{\delta_0,\delta}(s),d\alpha).$$

Because the expression is uniformly bounded and $n(t)/t\to 1/\delta$ in probability as $t\to\infty$, uniformly in δ_0, we need only evaluate the limit as $i\to\infty$ of

$$\frac{1}{\delta}E_{\hat{x}}^{\delta_0,\delta.m}\int_{\bar{\sigma}_i^{\delta_0,\delta}}^{\bar{\sigma}_{i+1}^{\delta_0,\delta}}ds\int_U k(\bar{x}_e^{\delta_0,\delta}(s),\alpha)m(\bar{x}_e^{\delta_0,\delta}(s),L^{\delta_0,\delta}(s),d\alpha).$$

By (5.4) for small δ_0 this goes to the limit

$$\frac{1}{\delta}\int\mu^{\delta_0,\delta,m}(d\hat{v})E_{\hat{v}}^{\delta_0,\delta,m}\int_0^{\bar{\sigma}_1^{\delta_0,\delta}}ds\int_U k(\bar{x}_e^{\delta_0,\delta}(s),\alpha)m(\bar{x}_e^{\delta_0,\delta}(s),L^{\delta_0,\delta}(s),d\alpha)$$

uniformly in the initial condition, $m(\cdot)$, and δ_0, as $i\to\infty$. This is the component of $\gamma^{\delta_0,\delta}(m)$ that involves $k(\cdot)$.

For the component of the cost that involves the reflection term, we will prove the uniform integrability of $\{y(\bar{\sigma}_{i+1}^{\delta_0,\delta}) - y(\bar{\sigma}_i^{\delta_0,\delta})\}$ and leave the rest of the details to the reader. For fixed t, write

$$y(\bar{\sigma}_1^{\delta_0,\delta}) = y(\bar{\sigma}_{t(1)}^{\delta_0,\delta}) + (y(\bar{\sigma}_1^{\delta_0,\delta}) - y(\bar{\sigma}_{t(1)}^{\delta_0,\delta})),$$

where $t(1) = \min\{t, \bar{\sigma}_1^{\delta_0,\delta}\}$. The uniform (in $m(\cdot), \hat{x}, \delta_0$) integrability of the first term follows from Lemma 3.2.1, which also implies that the expectation of the second term can be bounded as

$$\begin{aligned} &E_{\hat{x}}^{\delta_0,\delta,m} \left| y(\bar{\sigma}_1^{\delta_0,\delta}) - y(\bar{\sigma}_{t(1)}^{\delta_0,\delta}) \right| \sum_{p=0}^{\infty} I_{\{t+p \le \bar{\sigma}_1^{\delta_0,\delta} < t+p+1\}} \\ &\le K \sum_{p=0}^{\infty} P^{1/2}\{t + p \le \bar{\sigma}_1^{\delta_0,\delta} < t + p + 1\}, \end{aligned}$$

where K does not depend on $\hat{x}, \delta_0$, or $m(\cdot)$, and the sum goes to zero as $t \to \infty$. These computations together with the fact that $\{\bar{\sigma}_{l+1}^{\delta_0\delta} - \bar{\sigma}_l^{\delta_0\delta}, l\}$ are i.i.d., imply the uniform integrability of the $\{y(\bar{\sigma}_{i+1}^{\delta_0,\delta} - y(\bar{\sigma}_i^{\delta_0,\delta})\}$. ■

The following result is an analog of Theorems 4.6 and 4.7 and the proof is omitted.

Theorem 5.6. *Consider system* (5.2b). *Let U^n contain only a finite number of points with $U^n \to U$ as $n \to \infty$. Then if U^n is used in lieu of U, for large n the invariant measures and costs change little, uniformly in δ, δ_0, and in the periodic-Erlang relaxed feedback control $m(\cdot)$.*

Let $\epsilon > 0$ and suppose that U contains only finitely many points. For any δ, δ_0, and a periodic-Erlang relaxed feedback control $m^{\delta_0,\delta}(\cdot)$, there is a periodic-Erlang relaxed feedback control $m^{\delta_0,\delta,\epsilon}(\cdot)$ such that the expression $m^{\delta_0,\delta,\epsilon}(\tilde{X}^{\delta_0,\delta}, L^{\delta_0,\delta}, \alpha)$ is continuous in $\tilde{X}^{\delta_0,\delta}$, for each $\alpha \in U$ and each value of the Erlang state $L^{\delta_0,\delta}$. Also $|\gamma^{\delta,\delta_0}(m^{\delta_0,\delta}) - \gamma^{\delta,\delta_0}(m^{\delta_0,\delta,\epsilon})| \le \epsilon$.

Define $\bar{\gamma}^{\delta_0,\delta}$ to be the infimum of $\gamma^{\delta_0\delta}(m)$ over the relaxed periodic-Erlang feedback controls. Fix $\delta > 0$. The next result shows that as $\delta_0 \to 0$, the costs $\bar{\gamma}^{\delta_0,\delta}$ are no larger than $\bar{\gamma}^{\delta}$, the infimum of the costs for the periodic model. Let $m(\cdot)$ be a periodic control for the model in Section 4 with values $m(\bar{x}_p^{\delta}(t), \tau^{\delta}(t), d\alpha)$, and suppose that U has only finitely many points. Suppose that $m(\cdot)$ is ϵ-optimal for the periodic model and is continuous in the variables $\bar{x}_p^{\delta}(t), \tau^{\delta}(t)$, as allowed by Theorem 4.7. In the next theorem, we will need to apply this control to the model of this section, where the memory segment is periodic-Erlang and the Erlang state measures the passage of time. This can be done by using $m(\bar{x}_e^{\delta_0,\delta}(t), \tau^{\delta_0,\delta}(t), d\alpha)$ at time t, where we set $\tau^{\delta_0,\delta}(t) = k\delta_0$ if $L^{\delta_0,\delta}(t) = k$.

Theorem 5.7.

$$\limsup_{\delta_0 \to 0} \bar{\gamma}^{\delta_0,\delta} \leq \bar{\gamma}^{\delta}. \tag{5.7}$$

Proof. Let $m(\cdot)$ be the control for the periodic model that was described above the theorem statement, and adapted for use on the periodic-Erlang model as noted there. The first step is to show the convergence $\mu^{\delta_0,\delta,m}(\cdot) \to \mu^{\delta,m}(\cdot)$ as $\delta_0 \to 0$. Restrict the invariant measure $\mu^{\delta,m}(\cdot)$ to the sampled process $\tilde{X}_n^\delta = \big(x(n\delta), \ldots, x(n\delta - Q_\delta^+ \delta)\big)$. Then we can write

$$\begin{aligned} ||\mu^{\delta,m}(\cdot) - \mu^{\delta_0,\delta,m}(\cdot)|| &\leq ||\mu^{\delta_0,\delta,m}(\cdot) - P_{\hat{x}}^{\delta_0,\delta,m}\{\tilde{X}_l^{\delta_0,\delta} \in \cdot\}|| \\ &+||\mu^{\delta,m}(\cdot) - P_{\hat{x}}^{\delta,m}\{\tilde{X}_l^{\delta} \in \cdot\}|| \\ &+||P_{\hat{x}}^{\delta_0,\delta,m}\{\tilde{X}_l^{\delta_0,\delta} \in \cdot\} - P_{\hat{x}}^{\delta,m}\{\tilde{X}_l^{\delta} \in \cdot\}||. \end{aligned} \tag{5.8}$$

For small δ_0 and l large enough, the first two terms on the right side can be made arbitrarily small. For each l, as $t \to \infty$, $P\{\bar{\sigma}_{t(l)}^{\delta_0,\delta} \neq \bar{\sigma}_l^{\delta_0,\delta}\} \to 0$, and as $\delta_0 \to 0$, $\bar{\sigma}_l^{\delta_0,\delta} \to l\delta$ w.p.1. Thus, analogously to what was done in Theorem 5.3, to evaluate the last term on the right side of (5.8), it is sufficient to evaluate

$$\left| E_{\hat{x}} \exp(\zeta^{\delta_0,\delta}(\bar{\sigma}_{t(l)}^{\delta_0,\delta}, m)) - E_{\hat{x}} \exp(\zeta^{\delta}(l\delta, m)) \right|. \tag{5.9}$$

for large l and large $t > l\delta$.

Because $E_{\hat{x}} \left| \exp(\zeta^{\delta_0,\delta}(\bar{\sigma}_{t(l)}^{\delta_0,\delta}, m)) - \exp(\zeta^{\delta_0,\delta}(l\delta, m)) \right| \to 0$ as $\delta_0 \to 0$, it is sufficient to replace $\bar{\sigma}_l^{\delta_0,\delta}$ with $l\delta$ in (5.9). Using (3.1), for each l there is a $K_l < \infty$ and the following upper bound to this revised (5.9):

$$\begin{aligned} K_l \int_0^{l\delta} ds \int_U \Big| & b(\bar{x}_e^{\delta_0,\delta}(s), \alpha) m(\bar{x}_e^{\delta_0,\delta}(s), \tau^{\delta_0,\delta}(s), d\alpha) \\ & - b(\bar{x}_p^{\delta}(s), \alpha) m(\bar{x}_p^{\delta}(s), \tau^{\delta}(s), d\alpha) \Big|, \end{aligned}$$

where $\tau^{\delta_0,\delta}(t)$ is defined above the theorem, and the memory segments $\bar{x}_p^\delta(\cdot)$ and $\bar{x}_e^{\delta_0,\delta}(\cdot)$ are based on the uncontrolled model (1.4). As $\delta_0 \to 0$, $\bar{x}_e^{\delta_0,\delta}(t) \to \bar{x}_p^\delta(t)$ and $\tau^{\delta_0,\delta}(t) \to \tau^\delta(t)$ for almost all t. Then by the continuity of $m(\cdot)$ and $b(\cdot)$ for each of the discrete values of α, the last expression goes to zero as $\delta_0 \to 0$. These computations imply the convergence of the invariant measures, as asserted. We omit the proof of convergence of the associated costs as it is similar to what was done in Theorem 5.5.

Because $m(\cdot)$ is ϵ-optimal for the δ-periodic model and is not necessarily optimal for the periodic-Erlang model, $\bar{\gamma}^\delta \geq \gamma^\delta(m) - \epsilon$ and $\bar{\gamma}^{\delta_0,\delta} \leq \gamma^{\delta_0,\delta}(m) \to \gamma^\delta(m)$. The proof is completed as ϵ is arbitrary. ■

Theorem 5.8. *Assume* (A4.3) *in addition to the assumptions of this section. Then* $\bar{\gamma}^{\delta_0,\delta} \to \bar{\gamma}^\delta$ *as* $\delta_0 \to 0$.

Proof. In view of Theorem 5.7, we need to show that

$$\liminf_{\delta_0 \to 0} \bar{\gamma}^{\delta_0,\delta} \geq \bar{\gamma}^{\delta}. \tag{5.10}$$

For arbitrary $\epsilon > 0$, let $m^{\delta_0\delta}(\cdot)$ be ϵ-optimal periodic-Erlang relaxed feedback controls for model (5.2b), and consider the stationary systems that are associated with these controls. Let $r^{\delta_0\delta}(\cdot)$ denote the relaxed control representation of the $m(\bar{x}_e^{\delta_0,\delta}(\cdot), L^{\delta_0,\delta}(\cdot), d\alpha)$, for the stationary system. The costs for the stationary systems are still $\gamma^{\delta_0\delta}(m)$. Index the path, reflection, and Wiener process by δ_0, δ. Then the set $(x^{\delta_0}\delta(\cdot), r^{\delta_0}\delta(\cdot), z^{\delta_0}\delta(\cdot), w^{\delta_0}\delta(\cdot))$ is stationary in the sense used in (A4.3). It is also tight. Extract a weakly convergent subsequence as $\delta_0 \to 0$, indexed also by δ_0 for notational convenience, with limit denoted by $(x(\cdot), r(\cdot), z(\cdot), w(\cdot))$. Then the set is stationary in the sense used in (A4.3) and satisfies

$$dx(t) = dt \int_U b(\bar{x}_p^{\delta}(t), \alpha) r'(d\alpha, t) + \sigma(x(t))dw(t) + dz(t).$$

The process $\bar{x}_p^{\delta}(\cdot)$ is obtained from the solution $x(\cdot)$ via the definition (4.2.6). The cost rate converges to $k(\bar{x}_p^{\delta}(t), \alpha)$. The limit cost is $\gamma^{\delta}(r)$, and, by the convergence, ϵ-optimality of the $m^{\delta_0\delta}(\cdot)$, and (A4.3),

$$\epsilon + \bar{\gamma}^{\delta_0\delta} \geq \gamma^{\delta_0\delta}(m^{\delta_0}\delta) \to \gamma^{\delta}(r) \geq \bar{\gamma}^{\delta}.$$

This yields the theorem. ■

Theorem 5.9. *Assume* (A4.1)–(A4.3) *in addition to the other assumptions of this section. Then, as* $\delta_0 \to 0$ *and then* $\delta \to 0$,

$$\bar{\gamma}^{\delta_0\delta} \to \bar{\gamma}.$$

Proof. By Theorem 5.8, $\bar{\gamma}^{\delta_0\delta} \to \bar{\gamma}^{\delta}$ as $\delta_0 \to 0$. By Theorem 4.5, $\bar{\gamma}^{\delta} \to \bar{\gamma}$ as $\delta \to 0$. ■

Theorem 5.9 says that the optimal cost for (5.2b) approximates that for (1.9) and justifies using a numerical procedure based on (5.2b) to approximate the costs for (1.9). It is plausible that a nearly optimal control for (5.2b) will be nearly optimal for (1.9), but it has not yet been proved.

6

Markov Chain Approximations: Introduction

6.0 Outline of the Chapter

The Markov chain approximation for the no-delay case will be briefly outlined in this chapter. The chapter is essentially a review of the key parts of [58] that will be needed in the sequel. The reader is referred to that reference for a full development for the no-delay case of the methods for getting the approximations, the numerical algorithms, and the convergence proofs. The method will be adapted to the problem with delays in the following chapters, which will be organized so that the methods and results of [58] can be taken advantage of wherever possible. For the most part, in this chapter we will use the no-delay specialization of the reflected diffusion model (3.2.1), which is the standard diffusion process with boundary reflection. All of the usual process models and cost functions can be handled. For example, if the model of concern is stopped when first hitting the boundary, then ignore the reflections and add a stopping cost. The model is then the no-delay form of (3.1.1).

The numerical methods are based on approximations of the controlled process $x(\cdot)$ by a simpler controlled process, for which the evaluation of either the cost function for a fixed control or of the optimal cost can be done with an acceptable amount of computational work. If the approximating controlled process is close to the original process $x(\cdot)$ in an appropriate statistical sense and the cost function is suitably approximated, then we would expect that the value of the cost function for the approximating process for a fixed control (or its optimal value over all controls), and possibly the optimal control itself, will be close to those for the original model. The most useful approximating process is a Markov chain.

The closeness of the chain to the original diffusion is quantified by the local consistency conditions, and this is described in Section 2, which also gives a dynamical representation of the approximating chain that emphasizes its closeness to the diffusion. The numerical algorithms are based on the finite-state Markov chain approximation. But the convergence proofs are based on continuous-time interpolations of the approximating chains, and two such

H.J. Kushner, *Numerical Methods for Controlled Stochastic Delay Systems*,
doi: 10.1007/978-0-8176-4621-9_6,

(asymptotically equivalent) interpolations are discussed in Section 3. There are two types of approximating chains, called the "explicit" and "implicit" approximations, respectively. They differ in the way that the time variable is treated, each can be obtained from the other, and both will play important roles in treating the delay case. The explicit approximation case is discussed in Section 4 and the implicit in Section 5, which also proves the asymptotic equivalence between them and of the various timescales that are used. Section 6 is concerned with the singular and impulsive control forms, and some comments on the ergodic cost function are in Section 7. The notation will be slightly different from that in the references [58, 31, 50], as we wish to adapt or simplify it for the particular purposes of this book.

6.1 The System Model

Most of the discussion in this chapter will focus on the reflected diffusion model (3.2.1) but without delays, namely,

$$dx(t) = b(x(t), u(t))dt + \sigma(x(t))dw + dz(t), \tag{1.1}$$

where $x(t)$ is confined to a compact polyhedral constraint set G (with boundary ∂G) by the reflection term $z(\cdot)$ and (A3.2.1)–(A3.2.2) hold. The control $u(\cdot)$ takes values in a compact set U, $b(\cdot)$ and $\sigma(\cdot)$ are bounded and continuous, and there is a unique weak-sense solution for each relaxed control. Focusing on (1.1) will allow us to concentrate on the essential ideas without the complications inherent in an exhaustive review, but other models will be commented on from time to time. Let $r(\cdot)$ be an admissible relaxed control with derivative $r'(d\alpha, t)$. Then the relaxed control representation is

$$dx(t) = \int_U b(x(t), \alpha)r'(d\alpha, t)dt + \sigma(x(t))dw + dz(t). \tag{1.2}$$

If the boundary is absorbing, then drop $z(\cdot)$.

A discounted cost function. For a vector q and continuous real-valued $k(\cdot)$, the restriction of the cost function (3.4.3) to the no-delay case is

$$W(x, u) = E_x^u \int_0^\infty e^{-\beta t} \left[k(x(t), u(t))dt + q'dy(t)\right], \tag{1.3}$$

with relaxed control form

$$W(x, r) = E_x^r \int_0^\infty e^{-\beta t} \left[\int_U k(x(t), \alpha)r'(d\alpha, t)dt + q'dy(t)\right]. \tag{1.4}$$

Define $V(x) = \inf_r W(x, r)$, where the infimum is over all admissible controls.

Stopping on hitting a boundary. For most of the book, we will concentrate on the discounted cost function. But all of the standard cost functions that are used in stochastic could be used instead. Consider the form of the discounted cost function for a process that is stopped when first hitting the boundary ∂G. Let $g_0(\cdot)$ be a continuous real-valued stopping cost and define $\tau_G = \inf\{t : x(t) \neq G^0\}$, where G^0 is the interior of G. Then (3.4.1) becomes

$$W(x,r) = E_x^r \int_0^{\tau_G} e^{-\beta t} \int_U k(x(t),\alpha) r'(d\alpha,t)dt + E_x^r e^{-\beta\tau_G} g_0(x(\tau)). \quad (1.5)$$

6.2 Approximating Chains and Local Consistency

Approximating chains. The basis of the approximation is a discrete-time finite-state controlled Markov chain whose "local properties" are "consistent" with those of (1.1), as described below. This chain will be interpolated into a continuous-time process that will turn out to be a good approximation to (1.1). Let h denote the approximation parameter. For simplicity, we suppose that it is real-valued and positive, although a vector-valued parameter could be used as well, as noted in [58]. The reference [58, Chapter 5] describes many convenient methods for constructing chains that satisfy the required properties.

For each $h > 0$, let $\{\xi_n^h, n < \infty\}$ be a controlled discrete-parameter Markov chain on a discrete state space $S_h \subset I\!R^r$ with transition probabilities $p^h(x,\tilde{x}|\alpha)$, where S_h becomes dense in $I\!R^r$ as $h \to 0$. The variable α is the canonical value of the control and takes values in U. More generally, all that is needed is that α take values in a set U^h where the closed convex hull of $(b(x,U^h), k(x,U^h))$ converges to the closed convex hull of $(b(x,U), k(x,U))$ as $h \to 0$. Define $G_h = S_h \cap G$.

Local consistency in G. Let u_n^h denote the actual control action for the chain at discrete time n. Define $\Delta\xi_n^h = \xi_{n+1}^h - \xi_n^h$. Let $E_{x,n}^{h,\alpha}$ denote the conditional expectation given all data to step n and that $\xi_n^h = x$ and $u_n^h = \alpha$. Suppose that the following "local consistency" conditions hold in G_h:[1]

$$E_{x,n}^{h,\alpha} \Delta\xi_n^h \equiv b_h(x,\alpha)\Delta\theta(x,\alpha) = b(x,\alpha)\Delta t^h(x,\alpha) + o(\Delta t^h(x,\alpha)),$$

$$\begin{aligned} E_{x,n}^{h,\alpha}[\Delta\xi_n^h - E_{x,n}^{h,\alpha}\Delta\xi_n^h][\Delta\xi_n^h - E_{x,n}^{h,\alpha}\Delta\xi_n^h]' &\equiv a_h(x,\alpha)\Delta t^h(x,\alpha) \\ &= a(x)\Delta t^h(x,\alpha) + o(\Delta t^h(x,\alpha)), \end{aligned} \quad (2.1)$$

$$a(x) = \sigma(x)\sigma'(x),$$

$$\sup_{n,\omega} |\xi_{n+1}^h - \xi_n^h| \xrightarrow{h} 0,$$

[1] (2.1) defines the functions $b_h(\cdot)$ and $a_h(\cdot)$

for some function $\Delta t^h(x,\alpha) > 0$, which we call an "interpolation interval." We assume that $\lim_{h\to 0}\sup_{x\in G_h,\alpha\in U^h}\Delta\theta(x,\alpha) = 0$, but $\inf_{x\in G_h,\alpha\in U^h}\Delta\theta(x,\alpha) > 0$ for each $h > 0$.

We see that the chain has the local conditional drift and covariance properties of (1.1). The local consistency (2.1) is essentially all that is required of the approximating chain, together with analogous conditions for the reflecting boundary or when dealing with impulsive or singular controls. The interpolation intervals are obtained automatically when the transition functions $p^h(x,\tilde{x}|\alpha)$ are constructed [58, Chapter 5]. The local consistency need not hold everywhere. See the example in [58, Section 5.5].

A control policy $u^h = \{u_n^h,\ n < \infty\}$ for the chain is said to be *admissible* if

$$P\{\xi_{n+1}^h = \tilde{x}|\xi_i^h, u_i^h, i \le n\} = p^h(\xi_n^h, \tilde{x}|u_n^h).$$

It is always assumed that the controls are admissible. Let $E_x^{u^h}$ denote the expectation, given that $\xi_0^h = x$ and that either an admissible control sequence $u^h = \{u_n^h, n < \infty\}$ or a feedback control denoted by $u^h(\cdot)$ is used.

Local consistency on the reflecting boundary. Let $\partial G_h^+ \subset S_h$ denote the set of points not in G_h to which the chain might move from points in G_h; i.e., the set of points $\tilde{x} \notin G_h$ for which $p^h(x,\tilde{x}|\alpha) > 0$ for some $x \in G_h$ and $\alpha \in U^h$. This set is referred to as the "reflecting boundary" for the chain, and the points in it are called reflecting points. Keep in mind that if the process is absorbed on the boundary or when it leaves G, then there is no reflection process.

In analogy to the boundary reflection properties of the process (1.1), we need to specify the transition probabilities of the approximating chain when it escapes the set G; i.e., when at points in ∂G_h^+. Recall the definition of $d(x)$ given below (A3.2.1), as the set of reflection directions at the point x. From points in ∂G_h^+, the transitions of the chain are such that they move to G_h, with the conditional mean direction being a reflection direction at x. This is codified as the local consistency condition on the reflecting boundary as follows. We require $\lim_{h\to 0}\text{distance}(\partial G_h^+, G_h) = 0$, and there are $\theta_1 > 0$ and $\theta_2(h) \to 0$ as $h \to 0$ such that for all $x \in \partial G_h^+$, the conditional mean reflection directions satisfy

$$\begin{gathered} E_{x,n}^{h,\alpha}\left[\xi_{n+1}^h - x\right] \in \{a\gamma : \gamma \in d(x) \text{ and } \theta_2(h) \ge a \ge \theta_1 h\}, \\ \Delta t^h(x,\alpha) = 0 \quad \text{for } x \in \partial G_h^+. \end{gathered} \tag{2.2}$$

The last line of (2.2) says that the reflection from states on ∂G_h^+ is instantaneous.

Figure 2.1 illustrates the sets G_h and ∂G^+ for a two-dimensional problem. The set G is the triangle, G_h are the grid points in G, and it is assumed that from points in G_h, the state can transit one unit in any direction, including the diagonals. The set ∂G_h^+ consists of the points labeled r, and the reflection directions on the three boundaries are given by the arrows.

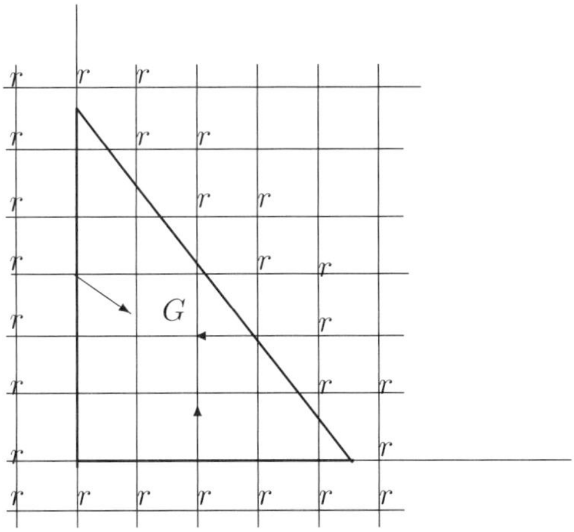

Figure 2.1. Illustration of the sets G_h and ∂G_h^+.

Figure 2.2 illustrates how the reflection direction can be approximated for a simple case. The set G is the rectangle bounded by the boundary lines that are labeled ∂G. The reflection direction vector from the left-hand boundary is d_1, and to attain it from the point x, the chain goes to points $\{a, b\}$ with probabilities such that the average direction is d_1.

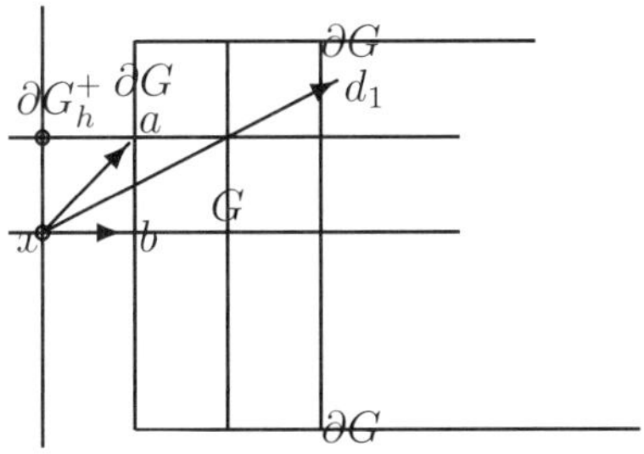

Figure 2.2. Reflection direction and state transitions.

A dynamical representation of the chain. Define $\Delta t_n^h = \Delta t^h(\xi_n^h, u_n^h)$, and $\Delta z_n^h = [\xi_{n+1}^h - \xi_n^h] I_{\{\xi_n^h \notin G\}}$ and let $\mathcal{F}_n^h$ denote the minimal σ-algebra that measures the system data to step n. Let the expectation conditioned on $\mathcal{F}_n^h$ be denoted by E_n^h. Define the vector Δy_n^h with components $\Delta y_{n,i}^h$ by

$\Delta z_n^h = \sum_i d_i \Delta y_{n,i}^h$.[2] Define the martingale difference[3]

$$\beta_n^h = \left[\Delta \xi_n^h - E_n^h \Delta \xi_n^h\right] I_{\{\xi_n^h \in G_h\}}. \tag{2.3}$$

By centering around the conditional expectation, we can write

$$\xi_{n+1}^h = \xi_n^h + \Delta t_n^h b(\xi_n^h, u_n^h) + \beta_n^h + \Delta z_n^h + o(\Delta t_n^h), \tag{2.4}$$

where β_n^h has conditional (on $\mathcal{F}_n^h$) covariance $a_h(\xi_n^h, u_n^h)\Delta t_n^h$. The $o(\Delta t_n^h)$ term is due to the use of $b(\cdot)$ in lieu of $b_h(\cdot)$ (see (2.1)).

A discounted cost function for the approximating chain. By the definition, $d_i \Delta y_{n,i}^h = \Delta z_n^h I_{\{\text{refl is through } i\text{th face}\}}$. If the reflection goes through a corner or edge of G, then the face might not be defined uniquely, and we select any one of the possibilities. Define the interpolated time $t_n^h = \sum_{i=0}^{n-1} \Delta t_i^h$. An analog of (1.3) for the chain is

$$W^h(x, u^h) = E_x^{u^h} \sum_{n=0}^{\infty} e^{-\beta t_n^h} \left[k(\xi_n^h, u_n^h)\Delta t_n^h + q' \Delta y_n^h\right]. \tag{2.5}$$

It follows from Lemma 3.1 in the next section, together with the fact that $k(\cdot)$ is bounded, that the discounted cost functions are well defined. There is a function $Y^h(\cdot)$ such that, for $\xi_n^h \in \partial G_h^+$, $q'[\Delta y_n^h] = Y^h(\xi_n^h, \xi_{n+1}^h)$. If x is on the reflecting boundary, then the transitions are not controlled and the α in $p^h(x, \tilde{x}|\alpha)$ is irrelevant and will usually be dropped. Then, we can write the Bellman equation for the cost function (2.5) as

$$V^h(x) = \begin{cases} \min_{\alpha \in U} \left[\sum_{\tilde{x}} e^{-\beta \Delta t^h(x,\alpha)} p^h(x, \tilde{x}|\alpha) V^h(\tilde{x}) + k(x,\alpha)\Delta t^h(x,\alpha)\right], & x \in G_h, \\ \sum_{\tilde{x}} p^h(x, \tilde{x}) \left[V^h(\tilde{x}) + Y^h(x, \tilde{x})\right], & x \notin G_h. \end{cases} \tag{2.6}$$

The discount factor $e^{-\beta \Delta t^h(x,\alpha)}$ can be expensive to compute. One can simplify by using, for example, $[1 - \beta \Delta t^h(x,\alpha)]$. The exact form of the discount factor is not critical. Various alternatives and numerical simplifications of the discount factor are discussed in [58, Chapter 5]. All are asymptotically equivalent, in that the limits of the solutions as $h \to 0$ are the same.

[2] Strictly speaking, as noted in Chapter 4, for the decomposition to hold for $\xi_n^h \neq \partial G$, the d_i would have to be replaced by d_i plus an error that goes to zero as $h \to 0$, as (A3.2.2) and (2.2) only guarantee that the directions converge as the state converges to the boundary (from the outside). But for notational simplicity we always ignore this difference, which has no effect, asymptotically.

[3] Here and in the sequel, when we say that some process derived from the chain is a martingale or martingale difference, the relevant filtration is that generated by the path and control data.

Constant interpolation interval. An interpolation interval $\Delta t^h(\cdot)$ that does not depend on the state or control is sometimes useful to simplify the coding and will he helpful in the next chapter. This is easily obtained, and the modified transition probabilities and interpolation interval are readily obtained from the $p^h(\cdot)$ and $\Delta t^h(\cdot)$ in the following way. Define the new interpolation interval $\overline{\Delta}^h = \inf_{\alpha \in U^h, \xi \in G_h} \Delta t^h(\xi, \alpha)$. The possibility that $\overline{\Delta}^h < \Delta t^h(x, \alpha)$ at some (x, α) is compensated for by allowing the state x to communicate with itself at that point. Let $\bar{p}^h(x, \tilde{x}|\alpha)$ denote the new transition probabilities. Conditioned on the event that a state does not communicate with itself on the current transition, the transition probabilities are the $p^h(\cdot)$. Thus, the general formula for getting $\bar{p}^h(x, \tilde{x}|\alpha)$ from the $p^h(\cdot)$ is ([58, Section 7.7])

$$\begin{aligned} \bar{p}^h(x, \tilde{x}|\alpha) &= p^h(x, \tilde{x}|\alpha)(1 - \bar{p}^h(x, x|\alpha)), \qquad \text{for } x \neq \tilde{x}, \\ \bar{p}^h(x, x|\alpha) &= 1 - \frac{\overline{\Delta}^h}{\Delta t^h(x, \alpha)}. \end{aligned} \tag{2.7}$$

The transition probabilities from the reflecting states are not affected.

Transforming the transition probabilities so that the interpolation interval is constant has some advantages, notably for proving convergence for the ergodic cost problem. Allowing a state to communicate with itself can slow down the convergence of the numerical computations. But when coding for the Bellman equation and using a policy improvement algorithm, it is possible to normalize such that the actual computations are all in terms of the original data and there is no slow down. See [58, Section 7.7].

6.3 Continuous-Time Interpolations

6.3.1 The Continuous-Time Interpolation $\xi^h(\cdot)$

The numerical algorithms are those for the Markov chain approximation. But the proofs of convergence are based on continuous-time interpolations of the chain that approximate the controlled process $x(\cdot)$. These interpolations are not needed for the numerical algorithms. There are two interpolations that will be needed. The first interpolation, called $\xi^h(\cdot)$, uses the interpolation intervals $\Delta t_n^h = \Delta\theta(\xi_n^h, u_n^h)$. Recall the definition of the interpolated time $t_n^h = \sum_0^{n-1} \Delta t_i^h$. Then define the continuous-parameter interpolations $\xi^h(\cdot)$ by

$$\xi^h(t) = \xi_n^h, \ u^h(t) = u_n^h, \ t \in [t_n^h, t_{n+1}^h). \tag{3.1}$$

The definition in Equation (3.1) is clear if $\xi_n^h \in G_h$, as then $\Delta t_n^h > 0$. If $\xi_n^h \in \partial G_h^+$, the set of reflecting states, then $\Delta t_n^h = 0$ and the interval $[t_n^h, t_{n+1}^h)$ is empty. Thus, in the construction of the interpolation, the reflecting states are ignored and, if ξ_n^h is a reflecting state, then $\xi^h(t_n^h) = \xi_{n+1}^h$, the state that the reflecting state ξ_n^h goes to. The reflections are "instantaneous" in the interpolation. Figure 3.1 illustrates the construction.

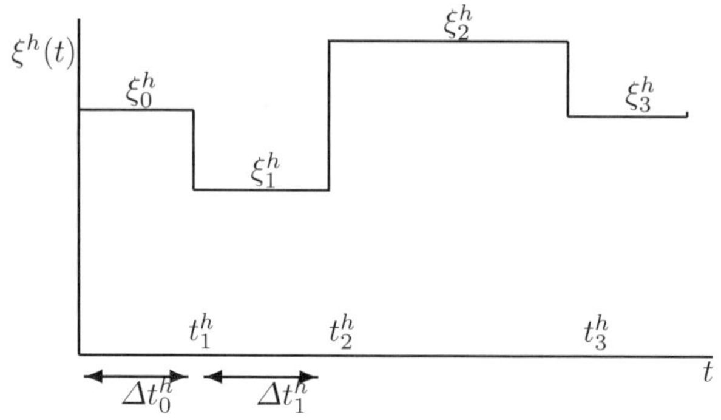

Figure 3.1. The interpolation $\xi^h(\cdot)$.

The interpolated process $\xi^h(\cdot)$ is piecewise-constant. Given the value of the current state and control action, the current interval is known.

Discussion. The similarity of the cost function (2.5) to (1.3) and the similarity of the local properties of the interpolation $\xi^h(\cdot)$ to those of the original controlled diffusion $x(\cdot)$ suggest that the $V^h(x)$ defined by (2.6) would be a good approximation to $V(x)$ for small values of h. This turns out to be true. Any sequence $\xi^h(\cdot)$ has a subsequence that converges to a controlled diffusion of the type (1.1). Suppose that $\bar{u}^h(\cdot)$ is the optimal control for the chain $\{\xi_n^h, n < \infty\}$ with cost function (for example) (2.5), and suppose that some subsequence (denoted by $\xi^h(\cdot)$) converges to a limit diffusion $x(\cdot)$ with admissible control $\tilde{u}(\cdot)$. Then the cost functionals $V^h(x)$ for the sequence of chains will converge to the cost functional $W(x, \tilde{u})$ for the limit process. Because $V(x)$ is the optimal value function, we have that $W(x, \tilde{u}) \geq V(x)$ and, hence, $\liminf_h V^h(x) \geq V(x)$. The reverse inequality can be proved by another approximation procedure [58, Chapters 10,11], which uses the fact that $V^h(x)$ is the minimal cost function for the controlled chain. See also the details of the proofs for the models with delays in Chapter 8. The Markov chain approximation method is thus quite straightforward: (a) get a locally consistent chain; (b) get a suitable approximation to the original cost function for the chain; (c) solve the Bellman equation for the optimal value and control for the approximating chain. Generally, there would not be a subsequence of the $\bar{u}^h(\cdot)$ that converges, but there is always a subsequence of the set of relaxed control representations that converges, and this is sufficient for the proof.

A dynamical representation. For each $t \geq 0$, define the time index, where $\sum_0^{-1} = 0$,

$$d^h(t) = \max\left\{n : \sum_{i=0}^{n-1} \Delta t_i^h = t_n^h \leq t\right\}.$$

Note that $d^h(t)$ will *never* be the index of a reflecting state, as the time intervals for those are zero. Then

$$\xi^h(t) = \xi^h(0) + \sum_{i=0}^{d^h(t)-1} b_h(\xi_i^h, u_i^h)\Delta t_i^h + B^h(t) + z^h(t), \tag{3.2}$$

where

$$B^h(t) = \sum_{i=0}^{d^h(t)-1} \beta_i^h, \quad z^h(t) = \sum_{i=0}^{d^h(t)-1} \Delta z_i^h, \quad y^h(t) = \sum_{i=0}^{d^h(t)-1} \Delta y_i^h.$$

Let $r(\cdot)$ denote the relaxed control representation of $u^h(\cdot)$. Then in interpolated and relaxed control form, and modulo an asymptotically negligible error,

$$\xi^h(t) = x(0) + \int_0^t \int_{U^h} b_h(\xi^h(s), \alpha) r^h(d\alpha\, ds) + B^h(t) + z^h(t). \tag{3.3}$$

Although the fact will not be used in the sequel, it is interesting to note that $B^h(\cdot)$ is an approximation to a stochastic integral in the following sense. There are martingale differences Δw_n^h whose continuous-time interpolation (intervals Δt_n^h) converges weakly to a standard Wiener process and $\beta_n^h \approx \sigma(\xi_n^h)\Delta w_n^h$ in the sense that [53, Section 6.6]

$$\sum_{n=1}^{d^h(t)-1} \beta_n^h = \sum_{n=1}^{d^h(t)-1} \sigma(\xi_n^h)\Delta w_n^h + \text{asymptotically negligible error.}$$

6.3.2 A Markov Continuous-Time Interpolation

By the construction of $\xi^h(\cdot)$, its interpolation intervals are $\Delta\theta(\xi_n^h, u_n^h)$, which are known when ξ_n^h and u_n^h are known. With the cost function (2.5) and Bellman equation (2.6), the control u_n^h depends only on ξ_n^h, in which case the discrete parameter process $\{\xi_n^h\}$ is Markov, but $\xi^h(\cdot)$ is not. A very useful (for the convergence proofs) alternative continuous-parameter interpolation of $\{\xi_n^h, n < \infty\}$ is a Markov (if the control is feedback) process itself. This is constructed as follows. Let $\{\nu_n\}$ be random variables that are independent of the $\{\xi_n^h, u_n^h\}$ and are mutually independent and identically distributed, with ν_n being exponentially distributed with mean unity. Define $\Delta\tau_n^h = \Delta t_n^h \nu_n$ and $\tau_n^h = \sum_{i=0}^{n-1} \Delta\tau_i^h$. Define $\psi^h(\cdot)$ by

$$\psi^h(t) = \sum_{i:\tau_{i+1}^h \le t} \Delta\xi_i^h + \xi_0^h. \tag{3.4}$$

Thus $\psi^h(t) = \xi_n^h$ on $[\tau_n^h, \tau_{n+1}^h)$. $\psi^h(\cdot)$ is a continuous-time Markov chain, whose holding times $\Delta\tau_n^h$, given ξ_n^h, u_n^h, are exponentially distributed with mean Δt_n^h. The construction is illustrated in Figure 3.2. Let $u_\tau^h(\cdot)$ denote the interpolation of the u_n^h with intervals $\Delta\tau_n^h$, and relaxed control representation $r_\tau^h(\cdot)$.

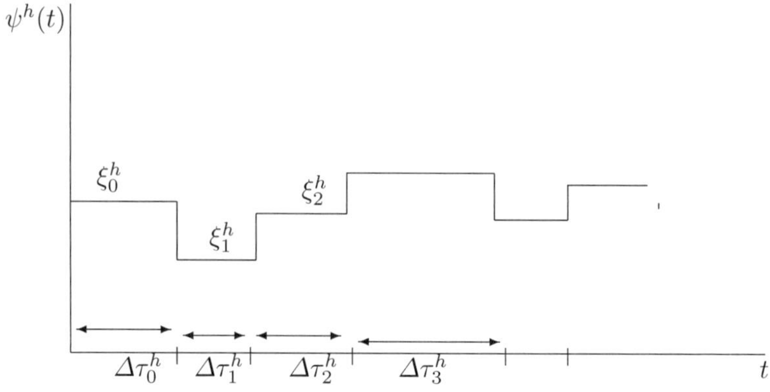

Figure 3.2. Illustration of the interpolation $\psi^h(\cdot)$.

Local properties of $\psi^h(\cdot)$ in G_h. Let $E_{x,t}^{h,\alpha}$ (with associated conditional probability $P_{x,t}^{h,\alpha}$) denote the expectation given the data

$$\left\{\psi^h(s), u_\tau^h(s), s \le t; \tau_n^h : \tau_n^h \le t; \psi^h(t) = x, u_\tau^h(t) = \alpha\right\},$$

where $x \in G_h$. Because the intervals between jumps are $\Delta t_n^h \nu_n$, where ν_n is exponentially distributed and independent of $\mathcal{F}_n^h$, the jump rate of $\psi^h(\cdot)$, when it is in state x and the control value is α, is $1/\Delta t^h(x,\alpha)$. Given a jump, the distribution of the next state is given by the $p^h(x,\tilde{x}|\alpha)$, and the conditional mean change, for $x \in G_h$ and control value α used, is $b_h(x,\alpha)\Delta t^h(x,\alpha)$. Thus

$$P_{x,t}^{h,\alpha}\{\text{ jump on } [t,t+\delta)\} = \frac{\delta}{\Delta\theta(x,\alpha)} + o(\delta).$$

For $\delta > 0$, define $\Delta\psi^h(t) = \psi^h(t+\delta) - \psi^h(t)$. The local properties of $\psi^h(\cdot)$ are

$$P_{x,t}^{h,\alpha}\{\psi^h(t+\delta) = \tilde{x}, \text{ jump on } [t,t+\delta)\} = \frac{\delta}{\Delta\theta(x,\alpha)} p^h(x,\tilde{x}|\alpha) + o(\delta), \quad (3.5)$$

$$\begin{aligned} E_{x,t}^{h,\alpha}\Delta\psi^h(t) &= P_{x,t}^{h,\alpha}\{\text{ jump on } [t,t+\delta)\} \sum_{\tilde{x}} p^h(x,\tilde{x}|\alpha)(\tilde{x}-x) \\ &= P_{x,t}^{h,\alpha}\{\text{ jump on } [t,t+\delta)\} b_h(x,\alpha)\Delta\theta(x,\alpha) \\ &= \frac{\delta}{\Delta t^h(x,\alpha)} b(x,\alpha)\Delta\theta(x,\alpha) + \delta o(\Delta\theta(x,\alpha)) + o(\delta) \\ &= \delta\, b(x,\alpha) + \delta o(\Delta\theta(x,\alpha)) + o(\delta), \end{aligned} \quad (3.6)$$

$$E_{x,t}^{h,\alpha}[\Delta\psi^h(t)][\Delta\psi^h(t)]' = a(x)\delta + \delta o(\Delta\theta(x,\alpha)) + o(\delta). \quad (3.7)$$

The behavior at the reflection states is the same as that for $\xi^h(\cdot)$. Thus, from the point of view of control, the discrete parameter chain and the two interpolations are asymptotically equivalent.

Let $z_\tau^h(\cdot)$ denote the interpolation of $\sum_{i=0}^{n-1} \Delta z_i^h$ with intervals $\Delta \tau_n^h$. We can decompose $\psi^h(\cdot)$ in terms of the continuous-time compensator, reflection term, and martingale as

$$\psi^h(t) = x(0) + \int_0^t b_h(\psi^h(s), u_\tau^h(s))ds + B_\tau^h(t) + z_\tau^h(t), \tag{3.8}$$

where the quadratic variation process of the martingale[4] $B_\tau^h(t)$ is

$$\int_0^t a_h(\psi^h(s), u_\tau^h(s))ds.$$

In terms of the relaxed controls,

$$\int_0^t b_h(\psi^h(s), u_\tau^h(s))ds = \int_0^t \int_{U^h} b_h(\psi^h(s), \alpha) r_\tau^{h,\prime}(s, d\alpha)ds.$$

It can be shown that ([58, Section 10.4.1]) there is a martingale $w^h(\cdot)$ (with respect to the filtration generated by the path and control processes, possibly augmented by an "independent" Wiener process) such that

$$B_\tau^h(t) = \int_0^t \sigma_h(\psi^h(s))dw^h(s) = \int_0^t \sigma(\psi^h(s))dw^h(s) + \epsilon^h(t),$$

where $\sigma_h(\cdot)[\sigma_h(\cdot)]' = a_h(\cdot)$ (recall the definition of $a_h(\cdot)$ in (2.1)), $w^h(\cdot)$ has quadratic variation process It and converges weakly to a standard (real or vector-valued, according to the case) Wiener process. The martingale $\epsilon^h(\cdot)$ is due to the difference between $\sigma(x)$ and $\sigma_h(x, \alpha)$ and

$$\lim_{h\to 0} \sup_{u^h} E \sup_{s\le t} |\epsilon^h(s)|^2 = 0 \tag{3.9}$$

for each t. Thus

$$\begin{aligned}\psi^h(t) = x(0) &+ \int_0^t \int_U b_h(\psi^h(s), \alpha) r_\tau^{h,\prime}(d\alpha, s)ds \\ &+ \int_0^t \sigma(\psi^h(s))dw^h(s) + z_\tau^h(t) + \epsilon^h(t).\end{aligned} \tag{3.10}$$

Estimates of the reflection terms. The following result [58, Theorem 1.3, Chapter 11] is an analog of Lemma 3.2.1 for application to the approximating chains.

---

[4] In [58], $M^h(\cdot)$ was used for $B_\tau^h(\cdot)$.

Lemma 3.1. *Assume* (A3.2.1)–(A3.2.2). *Suppose that the chain is modified so that we still have* $\lim_{h\to 0}\sup_n \Delta\xi_n^h = 0$, *but in* (2.1), $b_h(\xi_n^h, u_n^h)$ *and* $a_h(\xi_n^h, u_n^h)$ *are replaced by general measurable processes* $\tilde{b}_h(n)$ *and* $\tilde{a}_h(n)$, *resp., that are* $\mathcal{F}_n^h$ *measurable and bounded in norm by some constant* K. *Suppose that local consistency continues to hold on the reflecting boundary. Then the corresponding reflection terms satisfy*

$$\lim_{T\to 0}\limsup_{h\to 0}\sup_{\tilde{b}_h,\tilde{a}_h,x(0)} E\left|z^h\right|^2(T) = 0.$$

For any $T < \infty$,

$$\limsup_{h\to 0}\sup_{\tilde{b}_h,\tilde{a}_h,x(0)} E\left|z^h\right|^2(T) < \infty.$$

The conclusions hold if $z_\tau^h(\cdot)$ *replaces* $z^h(\cdot)$.

Note on convergence. For any subsequence $h \to 0$, there is a further subsequence (also indexed by h for simplicity) such that $(\psi^h(\cdot), r_\tau^h(\cdot), w^h(\cdot), z_\tau^h(\cdot))$ converges weakly to random processes $(x(\cdot), r(\cdot), w(\cdot), z(\cdot))$, where $r(\cdot)$ is a relaxed control, $(x(\cdot), r(\cdot), w(\cdot), z(\cdot))$ is nonanticipative with respect to the standard vector-valued Wiener process $w(\cdot)$, and the set satisfies

$$x(t) = x(0) + \int_0^t\int_U b(x(s),\alpha)r'(d\alpha, s)ds + \int_0^t \sigma(x(s))dw(s) + z(t),$$

where $z(\cdot)$ is the reflection term. Along the selected subsequence, $W^h(x, r^h) \to W(x, r)$. The proofs of these facts are in [58, Chapters 10, 11]. It is sufficient to require local consistency, continuity of the various dynamical and cost rate functions (this can be weakened, see the reference), and weak-sense uniqueness of the solution for any relaxed control. For the problem with boundary reflections, (A3.2.1) and (A3.2.2) are needed, whereas for the absorbing boundary case, we need (A3.4.1) and (A3.4.2).

The dynamic programming equation for cost (2.5) and process $\psi^h(\cdot)$. One natural analog of the cost function (2.5) for the process $\psi^h(\cdot)$, with continuous discounting and using the fact that $\psi^h(\cdot)$ is constant on the intervals $[\tau_n^h, \tau_{n+1}^h)$, is

$$\begin{aligned} W^h(x, u^h) &= E_x^{u^h}\sum_{n=0}^{\infty} e^{-\beta\tau_n^h}\left[\left(\int_0^{\tau_{n+1}^h-\tau_n^h} e^{-\beta t}dt\right) k(\xi_n^h, u_n^h) + q'\Delta y_n^h\right] \\ &= E_x^{u^h}\int_0^\infty e^{-\beta t}\left[k(\psi^h(t), u_\tau^h(t))dt + q'dy_\tau^h(t)\right]. \end{aligned} \tag{3.11}$$

The dynamic programming equation is the same as (2.6) except for an asymptotically negligible difference in the discount factor. Because

$$E_{x,n}^{h,\alpha} \int_0^{\Delta\tau_n^h} e^{-\beta s} ds = \frac{\Delta\theta(x,\alpha)}{1+\beta\Delta\theta(x,\alpha)}$$

and

$$E_{x,n}^{h,\alpha} e^{-\beta\Delta\tau_n^h} = \frac{1}{1+\beta\Delta\theta(x,\alpha)},$$

the *effective average discount factor* from time τ_n^h to time τ_{n+1}^h, given that $\xi_n^h = x$ and $u_n^h = \alpha$, is

$$\frac{1}{1+\beta\Delta\theta(x,\alpha)} = \exp[-\beta\Delta\theta(x,\alpha)](1+O(\Delta\theta(x,\alpha))).$$

Then the dynamic programming equation for the controlled chain $\{\xi_n^h, n < \infty\}$ and cost (3.11) is

$$V^h(x) = \min_{\alpha\in U}\left[\frac{1}{1+\beta\Delta\theta(x,\alpha)}\sum_{\tilde{x}} p^h(x,\tilde{x}|\alpha)V^h(\tilde{x}) + k(x,\alpha)\frac{\Delta\theta(x,\alpha)}{1+\beta\Delta\theta(x,\alpha)}\right], \tag{3.12}$$

for $x \in G_h$. For $x \in \partial G_h^+$, it is the same as in (2.6). As $h \to 0$, the differences between the solutions to (2.6) and (3.12) goes to zero.

6.4 The "Explicit" Approximation Procedure

Numerous approaches to the construction of the transition probabilities $p^h(x,\tilde{x}|\alpha)$ of the approximating chains are discussed in [58, Chapters 5 and 12] and the reader is referred to that reference for full details. However, to motivate some of the terminology that is to be used, a brief description of a simple procedure for the construction in G will be given, for a one-dimensional model (hence G is an interval). The procedure is intended to be illustrative of one possibility. The method is based on finite-difference approximations. But these are used in a purely formal way to get the transition probabilities. The methods of proof are purely probabilistic and no analysis based on finite differences is used. What matters is only that the end result satisfies (2.1).

Example of the construction of the transition probabilities. Consider the one-dimensional problem $dx = b(x,u)dt + \sigma(x)dw$. For an arbitrary continuous function $k(\cdot)$, let us formally consider the partial differential equation

$$\mathcal{L}^\alpha W(x,\alpha) + k(x,\alpha) = 0,\ x \in (0,B), \tag{4.1}$$

where $\mathcal{L}^\alpha$ is the differential operator of $x(\cdot)$ when the control is fixed at α; namely,

$$\mathcal{L}^\alpha f(x) = b(x,\alpha)\frac{\partial f(x)}{\partial x} + \frac{1}{2}\sigma^2(x)\frac{\partial^2 f(x)}{\partial x^2}.$$

Suppose that

$$\inf_{\alpha \in U^h, x \in G} [\sigma^2(x) - h|b(x,\alpha)|] > 0 \tag{4.2}$$

for the values of h of concern. The function $k(\cdot)$ is just a "place holder." Its exact values are irrelevant.

Let us use the finite-difference approximations

$$f_x(x) \to \frac{f(x+h) - f(x-h)}{2h} \tag{4.3}$$

for the first derivative, and

$$f_{xx}(x) \to \frac{f(x+h) + f(x-h) - 2f(x)}{h^2} \tag{4.4}$$

for the second derivative. Letting $W(x,\alpha)$ replace $f(x)$, substitute (4.3) and (4.4) into (4.1) and collect terms to yield the approximation to (4.1) with solution $W^h(x,\alpha)$:

$$\begin{aligned} &\frac{W^h(x+h,\alpha) - W^h(x-h,\alpha)}{2h} b(x,\alpha) \\ &\quad + \frac{W^h(x+h,a) + W^h(x-h,\alpha) - 2W^h(x,\alpha)}{h^2} \frac{\sigma^2(x)}{2} + k(x,a) = 0, \end{aligned}$$

or (which defines the functions $p^h(\cdot)$ and $\Delta t^h(\cdot)$)

$$\begin{aligned} W^h(x,\alpha) &= \frac{\sigma^2(x) + hb(x)}{2\sigma^2(x)} W^h(x+h,\alpha) + \frac{\sigma^2(x) - hb(x,a)}{2\sigma^2(x)} W^h(x-h,\alpha) \\ &\quad + k(x,\alpha) \frac{h^2}{\sigma^2(x)} \\ &= p^h(x, x+h|\alpha) W^h(x+h,\alpha) + p^h(x, x-h|\alpha) W^h(x-h,\alpha) \\ &\quad + k(x,a) \Delta t^h(x,\alpha). \end{aligned} \tag{4.5}$$

For $\tilde{x} \neq x \pm h$, set $p^h(x, \tilde{x}|\alpha) = 0$. The $p^h(\cdot)$ are transition probabilities for a Markov chain and are locally consistent with the process $x(\cdot)$, in the sense that (2.1) holds. We can see that the interpolation interval $\Delta t^h(\cdot)$ appeared as a consequence of the derivation of the transition probabilities.

If the condition (4.2) does not hold, then use the noncentral ("upwind") differences

$$\begin{aligned} f_x(x) &\to \frac{f(x+h) - f(x)}{h}, \quad \text{if } b(x,\alpha) \geq 0, \\ f_x(x) &\to \frac{f(x) - f(x-h)}{h}, \quad \text{if } b(x,\alpha) < 0. \end{aligned} \tag{4.6}$$

That is, if the velocity at a point is nonnegative, then use the forward difference, and if the velocity at a point is negative, then use the backward difference.

Schemes such as (4.6) are known as the "upwind" approximation method in numerical analysis. Define the *positive and negative parts* of a real number by: $a^+ = \max[a, 0]$, $a^- = \max[-a, 0]$. Using (4.6) in lieu of (4.3) and collecting terms yields

$$\begin{aligned} p^h(x, x+h|\alpha) &= \frac{\sigma^2(x)/2 + hb^+(x,\alpha)}{\sigma^2(x) + h|b(x,\alpha)|}, \\ p^h(x, x-h|\alpha) &= \frac{\sigma^2(x)/2 + hb^-(x,\alpha)}{\sigma^2(x) + h|b(x,\alpha)|}, \\ \Delta t^h(x,\alpha) &= \frac{h^2}{\sigma^2(x) + h|b(x,\alpha)|}. \end{aligned} \tag{4.7}$$

For $\tilde{x} \neq x \pm h$, set $p^h(x, \tilde{x}|\alpha) = 0$. Then the constructed $p^h(\cdot)$ are locally consistent transition probabilities in G for a controlled Markov chain. We see that in either case, the formal finite-difference approximation can be written as

$$W^h(x,\alpha) = \sum_{\tilde{x}} p^h(x,\tilde{x}|\alpha) W^h(\tilde{x},\alpha) + k(x,\alpha)\Delta\theta(x,\alpha) \tag{4.8}$$

for $x \in G_h$.

We emphasize that no claim is made that the convergence of the finite-difference approximations can be proved via the classical methods of numerical analysis. The finite-difference approximation is used only to get the transition probabilities of a Markov chain that is locally consistent in G.

A multidimensional example. For illustrative purposes, here is another special case from [58, Chapter 5]. Let the matrix $a(\cdot)$ be diagonal with entries $a_{ii}(x)$. First suppose that

$$\inf_{x,\alpha} [a_{ii}(x) - h|b_i(x,\alpha)|] \geq 0, \qquad \text{for all } i. \tag{4.9}$$

Let e_i denote the unit vector in the ith coordinate direction and define $A(x) = \sum_j a_{jj}(x)$. Then a development analogous to what led from (4.1) to (4.5) yields the locally consistent transition probabilities. for $x \in G_h$,

$$p^h(x, x \pm e_i h|\alpha) = \frac{a_{ii}(x) \pm hb_i(x,\alpha)}{2A(x)}, \quad \Delta t^h(x,\alpha) = \frac{h^2}{A(x)}. \tag{4.10}$$

If for a point $x \in G_h$, (4.9) fails at coordinates $i \in I$, then "upwind" forms analogous to (4.6) can be used for the first derivatives at coordinates $i \in I$ to yield the locally consistent values

$$\begin{aligned} \Delta t^h(x,\alpha) &= \frac{h^2}{A(x) + \sum_{j \in I} h|b_j(x,\alpha)|}, \\ p^h(x, x \pm e_i h|\alpha) &= \begin{cases} \dfrac{a_{ii}(x)/2 \pm hb_i(x,\alpha)/2}{A(x) + \sum_{j\in I} h|b_j(x,\alpha)|}, & i \notin I, \\ \dfrac{a_{ii}(x)/2 + hb_i^{\pm}(x,\alpha)}{A(x) + \sum_{j\in I} h|b_j(x,\alpha)|}, & i \in I, \end{cases} \end{aligned} \tag{4.11}$$

where the nonlisted $p^h(x, \tilde{x}|\alpha)$ are zero.

6.5 The "Implicit" Approximating Processes

We will call processes that satisfy the local consistency condition (2.1) *explicit approximating processes.* The name derives partly from the so-called explicit finite-difference approximations that were used in the example of the previous subsection. But the name also derives from the fact that the advance of time is explicit: at step n, interpolated time advances by $\Delta t^h(\xi_n^h, u_n^h)$ or $\Delta\tau^h(\xi_n^h, u_n^h)$, according to whether the interpolation $\xi^h(\cdot)$ or $\psi^h(\cdot)$ is used.

There is another approach, leading to what we will call an *implicit approximating procedure.* The fundamental difference between the explicit and implicit approaches to the Markov chain approximation lies in the fact that in the latter the time variable is treated as just another state variable. It is discretized in the same manner as are the other state variables. For the no-delay case, the approximating Markov chain has a state space that is a discretization of the (x, t)-space, and the component of the state of the chain that comes from the original time variable does not necessarily increase its value at each step. The idea is analogous when there are delays, and leads to some interesting and possibly more efficient numerical schemes, as will be seen in the next three chapters. The idea will be motivated by an example based on finite-difference approximations. Keep in mind that the finite-difference method is used only as one convenient method of constructing the transition probabilities. As for the explicit method, the proofs of convergence are purely probabilistic, and no "finite-difference" analysis is used. A form of the so-called implicit finite-difference approximation will be used to construct the transition probabilities in the example below, which provides another motivation for calling the general procedure "implicit."

An example. As noted above, the essential difference between what are called the explicit and implicit approximation approaches to the Markov chain approximation lies in the fact that in the former the time variable is treated differently than the state variables: It records interpolated time and its value increases by Δt_n^h or $\Delta\tau_n^h$ at step n if $\xi_n^h \in G_h$. In the implicit approximation approach, the time variable is just another state variable, and its value does not necessarily increase at each step where $\xi_n^h \in G_h$. The following example, analogous to the one based on finite differences in the previous section, will illustrate the differences.

Continue to use the special one-dimensional model $x(\cdot)$ in the first example of the previous section. Fix the control at the value α and for fixed $T < \infty$ and any $0 \le t < T$, consider the cost function

$$W(x, t, \alpha) = E_{x,t}^{\alpha} \int_t^T k(x(s), \alpha) ds + E_{x,t}^{\alpha} g(x(T)), \tag{5.1}$$

where $g(\cdot)$ is an arbitrary continuous function (which will serve only as a place holder). Formally, $W(\cdot)$ satisfies the partial differential equation:

$$W_t(x,t,\alpha) + \mathcal{L}^\alpha W(x,t,\alpha) + k(x,\alpha) = 0,\ t < T, \tag{5.2}$$

with the boundary condition $W(x,T,\alpha) = g(x)$. As usual in the dynamic computation of cost functions, (5.2) has a terminal condition and is solved backwards in time.[5]

Let $\delta > 0$ denote the discretization interval for time, and suppose that T is an integral multiple of δ. Use the approximations

$$\begin{aligned} f_t(x,t) &\to \frac{f(x,t+\delta) - f(x,t)}{\delta}, \\ f_x(x,t) &\to \frac{f(x+h,t) - f(x-h,t)}{2h}, \\ f_{xx}(x,t) &\to \frac{f(x+h,t) + f(x-h,t) - 2f(x,t)}{h^2}. \end{aligned} \tag{5.3}$$

Note that the last two equations of (5.3) use the argument t (rather than $t+\delta$) on the right side.

Suppose that $\sigma^2(x) > h|b(x,\alpha)|$ for all (x,α). Using (5.3) to approximate the partial differential equation (5.2), we obtain the finite-difference approximation

$$\begin{aligned} &\left[1 + \sigma^2(x)\frac{\delta}{h^2}\right] W^{h,\delta}(x,n\delta,\alpha) \\ &\quad = \left[\frac{\sigma^2(x)}{2}\frac{\delta}{h^2} + \frac{b(x,\alpha)\delta}{2h}\right] W^{h,\delta}(x+h,n\delta,\alpha) \\ &\quad + \left[\frac{\sigma^2(x)}{2}\frac{\delta}{h^2} - \frac{b(x,\alpha)\delta}{2h}\right] W^{h,\delta}(x-h,n\delta,\alpha) \\ &\quad + W^{h,\delta}(x,n\delta+\delta,\alpha) + k(x,\alpha)\delta. \end{aligned}$$

With the obvious definitions of $p^{h,\delta}(\cdot)$ and $\Delta t^{h,\delta}(\cdot)$, let us divide the terms in the above expression by the coefficient of $W^{h,\delta}(x,n\delta,\alpha)$ and rewrite it as

$$\begin{aligned} W^{h,\delta}(x,n\delta,\alpha) =& \sum_{\tilde{x}} p^{h,\delta}(x,n\delta;\tilde{x},n\delta|\alpha) W^{h,\delta}(\tilde{x},n\delta,\alpha) \\ &+ p^{h,\delta}(x,n\delta;x,n\delta+\delta|\alpha) W^{h,\delta}(x,n\delta+\delta,\alpha) \\ &+ k(x,\alpha)\Delta t^{h,\delta}(x,\alpha), \end{aligned} \tag{5.4}$$

with boundary condition $W^{h,\delta}(x,T,\alpha) = g(x)$. The values of the undefined $p^{h,\delta}(x,n\delta;\tilde{x},n\delta|\alpha)$ are set to zero. The $p^{h,\delta}(\cdot)$ are nonnegative and satisfy

[5] (5.2) is an equation on a finite time interval $[0,T]$, but the derived transition probabilities will be useable on $[0,\infty)$.

$$\sum_{\tilde{x}} p^{h,\delta}(x, n\delta; \tilde{x}, n\delta|\alpha) + p^{h,\delta}(x, n\delta; x, n\delta + \delta|\alpha) = 1.$$

It can be seen from the last two expressions that we can consider the $p^{h,\delta}(\cdot|\alpha)$ to be one-step transition probabilities of a Markov chain $\{\zeta_n^{h,\delta}, n < \infty\}$ on the "(x,t)-state space"

$$\{0, \pm h, \pm 2h, \ldots\} \times \{0, \delta, 2\delta, \ldots\},$$

under control value α. The $p^{h,\delta}(x, n\delta; \tilde{x}, n\delta|\alpha)$ is the probability that the path state goes from x to $\tilde{x}$ and the time state does not advance, and $p^{h,\delta}(x, n\delta; x, n\delta + \delta|\alpha)$ is the probability that the path state does not change but the time state advances by δ, all under control value α. The value of T does not appear in (5.4) and its value is irrelevant.

It is evident that time is being treated as just another state variable. It can be shown that for $x \neq \tilde{x}$ we have

$$p^{h,\delta}(x, n\delta; \tilde{x}, n\delta|\alpha) = p^h(x, \tilde{x}|\alpha) \times \text{normalization}(x), \tag{5.5}$$

where the $p^h(x, \tilde{x}|\alpha)$ are the transition probabilities defined by (4.5).

Write $\zeta_n^{h,\delta} = (\phi_n^{h,\delta}, \xi_n^{h,\delta})$, where $\phi_n^{h,\delta}$ represents the time variable, and $\xi_n^{h,\delta}$ represents the "spatial" state. Let $u_n^{h,\delta}$ denote the control that is used at step n. Let $E_{x,n}^{h,\delta,\alpha}$ denote the expectation conditioned on the data to step n, with $\xi_n^{h,\delta} = x$ and $u_n^{h,\delta} = \alpha$. Define $\Delta\xi_n^{h,\delta} = \xi_{n+1}^{h,\delta} - \xi_n^{h,\delta}$ and $\Delta\phi_n^{h,\delta} = \phi_{n+1}^{h,\delta} - \phi_n^{h,\delta}$. Then for the above example we have

$$\begin{aligned}
E_{x,n}^{h,\delta,\alpha} \Delta\xi_n^{h,\delta} &= b(x,\alpha)\Delta t^{h,\delta}(x,\alpha), \\
\text{cov}_{x,n}^{h,\delta,\alpha} \Delta\xi_n^{h,\delta} &= \sigma^2(x)\Delta t^{h,\delta}(x,\alpha) + \Delta t^{h,\delta}(x,\alpha)O(h), \\
E_{x,n}^{h,\delta,\alpha} \Delta\phi_n^{h,\delta} &= \Delta t^{h,\delta}(x,\alpha).
\end{aligned}$$

Thus, the "spatial" component of the controlled chain is locally consistent in the sense of (2.1), but with interpolation intervals $\Delta t_n^{h,\delta} = \Delta t^{h,\delta}(\xi_n^{h,\delta}, u_n^{h,\delta})$. The conditional mean increment of the "time" component of the state is $\Delta t^{h,\delta}(x,\alpha)$, where $\xi_n^{h,\delta} = x$ and $u_n^{h,\delta} = \alpha$. We have constructed an approximating Markov chain via an "implicit" method. In the traditional use of the approximation (5.3) in numerical analysis, it was called an implicit approximation method because (5.4) cannot be solved by a simple backward iteration. At each n, (5.4) determines the values of the $\{W^{h,\delta}(x, n\delta, u)\}$ implicitly.

6.5.1 The General Implicit Approximation Method

The above special case illustrated one method of getting the implicit approximation. However, there is a general method for getting the transition probabilities for the implicit approximation method that starts with any set $p^h(\cdot), \Delta t^h(\cdot)$ satisfying (2.1) and does not require the use of any particular method of construction. It is based simply on an extension of the representation (5.5). Suppose that at the current step the time variable does not advance. Then, conditioned on this event and on the value of the current spatial state, the distribution of the next spatial state is just $p^h(x,\tilde{x}|\alpha)$. So one needs only determine the probability that the time variable advances, conditioned on the current state. This is obtained by a "local consistency" argument, and no matter how the $p^h(\cdot)$ were derived, the (no-delay) transition probabilities $p^{h,\delta}(\cdot)$ and interpolation interval $\Delta t^{h,\delta}(\cdot)$ for the implicit approximation procedure can be determined from the $p^h(\cdot)$ and $\Delta t^h(\cdot)$ by the formulas [58, Section 12.4], for $x \in G_h$,

$$\begin{aligned} p^h(x,\tilde{x}|\alpha) &= \frac{p^{h,\delta}(x,n\delta;\tilde{x},n\delta|\alpha)}{1-p^{h,\delta}(x,n\delta;x,n\delta+\delta|\alpha)}, \\ p^{h,\delta}(x,n\delta;x,n\delta+\delta|\alpha) &= \frac{\Delta t^h(x,\alpha)}{\Delta t^h(x,\alpha)+\delta}, \\ \Delta t^{h,\delta}(x,\alpha) &= \frac{\delta \Delta t^h(x,\alpha)}{\Delta t^h(x,\alpha)+\delta}. \end{aligned} \tag{5.6}$$

The reader is referred to the reference for the full details of the derivation.

For $x \in G_h$, the general local consistency equations for the implicit approximation are (these formulas define $b_{h,\delta}(\cdot)$ and $a_{h,\delta}(\cdot)$)

$$\begin{aligned} E^{h,\delta,\alpha}_{x,n} \Delta\xi^{h,\delta}_n &= b_{h,\delta}(x,\alpha)\Delta t^{h,\delta}(x,\alpha) \\ &= b(x,\alpha)\Delta t^{h,\delta}(x,\alpha) + o(\Delta t^{h,\delta}(x,\alpha)), \end{aligned}$$

$$\begin{aligned} \operatorname{cov}^{h,\delta,\alpha}_{x,n} \Delta\xi^{h,\delta}_n(x,\alpha) &= a_{h,\delta}(x,\alpha)\Delta t^{h,\delta}(x,\alpha) \\ &= a(x)\Delta t^{h,\delta}(x,\alpha) + o(\Delta t^{h,\delta}(x,\alpha)), \end{aligned} \tag{5.7a}$$

$$E^{h,\delta,\alpha}_{x,n} \Delta\phi^{h,\delta}_n = \Delta t^{h,\delta}(x,\alpha),$$

$$\sup_{x \in G_h,\alpha} \Delta t^h(x,\alpha)/\delta \to 0 \text{ as } h,\delta \to 0. \tag{5.7b}$$

Because the reflection is instantaneous if the spatial state leaves G, the transition probabilities at the reflecting states are unchanged and the local consistency condition for the reflecting states is the analog of (2.2). In the no-delay case, the implicit approximation procedure was used in [58] largely to deal with control problems that were defined over a fixed finite time interval. It will be used in a quite different way in the delay case in the following chapters, where it will be helpful in dealing with the memory requirements.

Figure 5.1 illustrates the explicit and the implicit state transitions for a one-dimensional example. The horizontal axis denotes interpolated time, and the vertical axis denotes the spatial variable. For the explicit approximation procedure, the time interval at point x and control value α is $\Delta t^h(x,\alpha)$, and the point x transits to either point a or b. For the implicit approximation procedure, the point (space, time)= (x, ϕ) goes to either (a, ϕ) or (b, ϕ) if time does not advance and to $(x, \phi + \delta)$ if time advances.

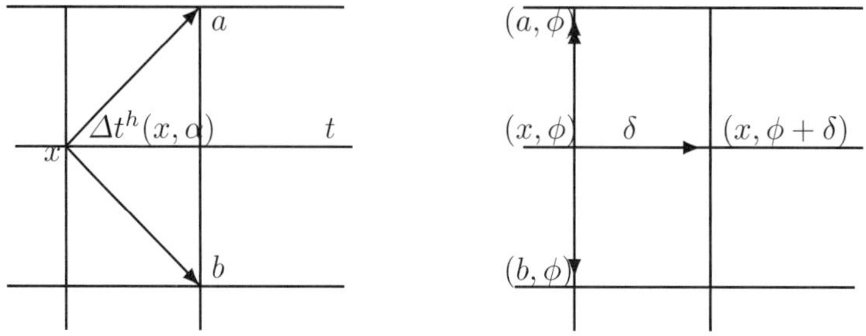

(a) Explicit approximation. (b) Implicit approximation.

Figure 5.1.

6.5.2 Continuous-Time Interpolations

Let $\Delta z_n^{h,\delta}$ denote the reflection term for the $\xi_n^{h,\delta}$ process, and define the components $\Delta y_{i,n}^{h,\delta}$ by $\Delta z_n^{h,\delta} = \sum_i d_i \Delta y_{i,n}^{h,\delta}$. Define $\Delta t_n^{h,\delta} = \Delta t^{h,\delta}(\xi_n^{h,\delta}, u_n^{h,\delta})$ and the martingale differences

$$
\begin{aligned}
\beta_n^{h,\delta} &= \left[\left(\xi_{n+1}^{h,\delta} - \xi_n^{h,\delta} \right) - b_{h,\delta}(\xi^{h,\delta}, u_n^{h,\delta}) \Delta t_n^{h,\delta} \right] I_{\{\xi_n^{h,\delta} \in G_h\}}, \\
\beta_{0,n}^{h,\delta} &= \left(\phi_{n+1}^{h,\delta} - \phi_n^{h,\delta} \right) - \Delta t_n^{h,\delta}.
\end{aligned}
$$

The conditional covariance of $\beta_{0,n}^{h,\delta}$ is $o(\Delta t_n^{h,\delta})$ and that of $\beta_n^{h,\delta}$ is $a(\xi_n^{h,\delta})\Delta t_n^{h,\delta} + o(\Delta t_n^{h,\delta})$. We can write

$$
\begin{aligned}
\xi_{n+1}^{h,\delta} &= \xi_n^{h,\delta} + b(\xi_n^{h,\delta}, u_n^{h,\delta}) \Delta t_n^{h,\delta} + o(\Delta t_n^{h,\delta}) + \beta_n^{h,\delta} + \Delta z_n^{h,\delta}, \\
\phi_{n+1}^{h,\delta} &= \phi_n^{h,\delta} + \Delta t_n^{h,\delta} + \beta_{0,n}^{h,\delta}.
\end{aligned}
\tag{5.8}
$$

Define the "interpolated" times $t_n^{h,\delta} = \sum_0^{n-1} \Delta t_i^{h,\delta}$. Define the continuous-parameter interpolations $\xi^{h,\delta}(\cdot)$, etc., as follows. For $t \in [t_n^{h,\delta}, t_{n+1}^{h,\delta})$, set

$$
\begin{gathered}
\xi^{h,\delta}(t) = \xi_n^{h,\delta}, \quad \phi^{h,\delta}(t) = \phi_n^{h,\delta} \quad u^{h,\delta}(t) = u_n^{h,\delta}, \\
z^{h,\delta}(t) = \sum_{i=0}^{n-1} \Delta z_i^{h,\delta}, \quad y^{h,\delta}(t) = \sum_{i=0}^{n-1} \Delta y_i^{h,\delta}, \quad B^{h,\delta}(t) = \sum_{i=0}^{n-1} \beta_i^{h,\delta}.
\end{gathered}
\tag{5.9}
$$

Set $\zeta^{h,\delta}(\cdot) = (\xi^{h,\delta}(\cdot), \phi^{h,\delta}(\cdot))$. With $r^{h,\delta}(\cdot)$ denoting the relaxed control representation of $u^{h,\delta}(\cdot)$, we have the analog of (3.3), modulo an asymptotically negligible error,

$$\xi^{h,\delta}(t) = x(0) + \int_0^t \int_{U^h} b_{h,\delta}(\xi^{h,\delta}(s), \alpha) r^{h,\delta}(d\alpha\, ds) + B^{h,\delta}(t) + z^{h,\delta}(t). \quad (5.10)$$

A Markov process interpolation. Analogously to what was done in Section 3 for the explicit approximation case, let $\{\nu_n\}$ be random variables that are independent of the $\{\xi_n^{h,\delta}, u_n^{h,\delta}, n < \infty\}$ and are mutually independent and identically distributed, with ν_n being exponentially distributed with mean unity. Define $\Delta\tau_n^{h,\delta} = \nu_n \Delta t_n^{h,\delta}$, and $\tau_n^{h,\delta} = \sum_{i=0}^{n-1} \Delta\tau_i^{h,\delta}$. Define the continuous-parameter interpolations: For $t \in [\tau_n^{h,\delta}, \tau_{n+1}^{h,\delta})$,

$$\begin{aligned} &\psi^{h,\delta}(t) = \xi_n^{h,\delta}, \ \phi_\tau^{h,\delta}(t) = \phi_n^{h,\delta} \ \ u_\tau^{h,\delta}(t) = u_n^{h,\delta}, \\ &z_\tau^{h,\delta}(t) = \sum_{i=0}^{n-1} \Delta z_i^{h,\delta}, \ y_\tau^{h,\delta}(t) = \sum_{i=0}^{n-1} \Delta y_i^{h,\delta}, \ B_\tau^{h,\delta}(t) = \sum_{i=0}^{n-1} \beta_i^{h,\delta}. \end{aligned} \quad (5.11)$$

Set $\zeta_\tau^{h,\delta}(\cdot) = (\psi^{h,\delta}(\cdot), \phi_\tau^{h,\delta}(\cdot))$. With $r_\tau^{h,\delta}(\cdot)$ denoting the relaxed control representation of $u_\tau^{h,\delta}(\cdot)$, we have the analog of (3.10):

$$\psi^{h,\delta}(t) = x(0) + \int_0^t \int_{U^h} b_{h,\delta}(\psi^{h,\delta}(s), \alpha) r_\tau^{h,\delta}(d\alpha\, ds) + B_\tau^{h,\delta}(t) + z_\tau^{h,\delta}(t). \quad (5.12)$$

If the control is feedback, then $\psi^{h,\delta}(\cdot)$ is a continuous-time Markov chain.

As noted in connection with (3.10), there is a martingale ([58, Section 10.4.1]) $w^{h,\delta}(\cdot)$ (with respect to the filtration generated by the state and control processes, possibly augmented by an "independent" Wiener process) such that

$$B_\tau^{h,\delta}(t) = \int_0^t \sigma(\psi^{h,\delta}(s)) dw^{h,\delta}(s) + \epsilon^{h,\delta}(t),$$

where $w^{h,\delta}(\cdot)$ has quadratic variation process It and converges weakly to a standard Wiener process. The martingale $\epsilon^{h,\delta}(\cdot)$ satisfies (3.9), with h, δ replacing h.

An alternative approximating chain. There is an alternative way of interpolating the $\xi_n^{h,\delta}$ that will be used in Chapters 7 and 8, which looks at the process only at those times that the time variable $\phi_n^{h,\delta}$ advances. Define $v_0^{h,\delta} = 0$ and for $n > 0$ define

$$v_n^{h,\delta} = \min\{i > v_{n-1}^{h,\delta} : \phi_i^{h,\delta} - \phi_{i-1}^{h,\delta} = \delta\}. \quad (5.13)$$

Then define $\tilde{\xi}_n^{h,\delta} = \xi_{v_n^{h,\delta}}^{h,\delta}$. Define the continuous-parameter interpolation $\tilde{\xi}^{h,\delta}(t) = \tilde{\xi}_n^{h,\delta}$ for $t \in [n\delta, n\delta + \delta)$. Define $\tilde{u}^{h,\delta}(\cdot)$ analogously. It will be seen in Theorem 5.1 that $\tilde{\xi}^{h,\delta}(\cdot)$ is asymptotically equal to $\xi^{h,\delta}(\cdot)$ in the sense that the difference converges to zero.

6.5.3 Representations of the Cost Function

The timescale based on $\phi^{h,\delta}(\cdot)$ will be useful in Chapters 7 and 8. To prepare for that, we now give some representations of the cost functions in terms of it.

Control on $[0, T]$. Recall the cost criterion (5.1) and the equation (5.4) whose solution formally approximates (5.1). The solution to (5.4) can be represented in terms of the path up to the first time that the time component $\phi_n^{h,\delta}$ reaches or exceeds the value T. To do this we start by defining the stopping time

$$N^{h,\delta}(T) = \min\{n : \phi_n^{h,\delta} \geq T\}. \tag{5.14}$$

Let $u^{h,\delta} = \{u_n^{h,\delta}, n < \infty\}$ be an admissible control sequence and let $E_{x,n}^{h,\delta,u^{h,\delta}}$ denote the expectation given the data to step n, that $u^{h,\delta}$ is used, and that $\xi_n^{h,\delta} = x$. Then the solution to (5.4) for the general vector-valued state case, with use of the control sequence $u^{h,\delta}$ in place of the constant value α, and with the boundary condition $W^{h,\delta}(x, T, u) = g(x)$ and $\phi_n^{h,\delta} = t$, can be written as

$$\begin{aligned} W^{h,\delta}(x, t, u^{h,\delta}) &= E_{x,n}^{h,\delta,u^{h,\delta}} \left[\sum_{i=n}^{N^{h,\delta}(T)-1} k(\xi_i^{h,\delta}, u_i^{h,\delta}) \Delta t_i^{h,\delta} + g(\xi_{N^{h,\delta}(T)}^{h,\delta}) \right] \\ &= E_{x,n}^{h,\delta,u^{h,\delta}} \left[\sum_{i=n}^{\infty} k(\xi_i^{h,\delta}, u_i^{h,\delta}) \Delta t_i^{h,\delta} I_{\{\phi_i^{h,\delta} < T\}} + g(\xi_{N^{h,\delta}(T)}^{h,\delta}) \right]. \end{aligned} \tag{5.15}$$

Because $\Delta t^{h,\delta}(x, \alpha) = E_{x,i}^{h,\delta,\alpha}[\phi_{i+1}^{h,\delta} - \phi_i^{h,\delta}]$, (5.15) equals

$$E_{x,n}^{h,\delta,u^{h,\delta}} \left[\sum_{i=n}^{N^{h,\delta}(T)-1} k(\xi_i^{h,\delta}, u_i^{h,\delta})[\phi_{i+1}^{h,\delta} - \phi_i^{h,\delta}] + g(\xi_{N^{h,\delta}(T)}^{h,\delta}) \right]. \tag{5.16}$$

We can write this last expression as

$$E_{x,t}^{h,\delta,u^{h,\delta}} \left[\int_t^T k(\tilde{\xi}^{h,\delta}(s), \tilde{u}^{h,\delta}(s)) ds + g(\tilde{\xi}^{h,\delta}(T)) \right].$$

The discounted cost function: Reflecting diffusion. Consider the discounted cost

$$W^{h,\delta}(x, u^{h,\delta}) = E_x^{h,\delta,u^{h,\delta}} \sum_{n=0}^{\infty} e^{-\beta t_n^{h,\delta}} \left[k(\xi_n^{h,\delta}, u_n^{h,\delta}) \Delta t_n^{h,\delta} + q' \Delta y_n^{h,\delta} \right]. \tag{5.17}$$

By the results of the next subsection, (5.17) is asymptotically equal to

$$E_x^{h,\delta,u^{h,\delta}} \sum_{n=0}^{\infty} e^{-\beta\phi_n^{h,\delta}} \left[k(\xi_n^{h,\delta}, u_n^{h,\delta}) \left(\phi_{n+1}^{h,\delta} - \phi_n^{h,\delta} \right) + q' \Delta y_n^{h,\delta} \right]. \tag{5.18}$$

Modulo an asymptotically negligible interpolation error and in relaxed control notation, (5.17) is also equal to

$$E_x^{h,\delta,u^{h,\delta}} \int_0^\infty \int_{U^h} e^{-\beta t} \left[k(\xi^{h,\delta}(s), \alpha) r^{h,\delta}(d\alpha\, ds) + q' dy^{h,\delta}(s) \right]. \tag{5.19}$$

By the results of the next subsection, this is asymptotically equal to

$$E_x^{h,\delta,u^{h,\delta}} \int_0^\infty \int_{U^h} e^{-\beta t} \left[k(\psi^{h,\delta}(s), \alpha) r_\tau^{h,\delta}(d\alpha\, ds) + q' dy_\tau^{h,\delta}(s) \right]. \tag{5.20}$$

The errors in all cases go to zero as $h, \delta \to 0$, uniformly in the initial data and control. Recalling from (5.6) that $p^{h,\delta}(x, n\delta, x, n\delta + \delta | \alpha)\delta = \Delta t^{h,\delta}(x, \alpha)$, we can write the Bellman equation for (5.18) as follows (the current value of the time variable $\phi^{h,\delta}$ is irrelevant): For $x \in G_h$,

$$\begin{aligned} V^{h,\delta}(x) = \min_\alpha \Bigg[& \sum_{\tilde{x}} p^{h,\delta}(x, n\delta, \tilde{x}, n\delta | \alpha) V^{h,\delta}(\tilde{x}) \\ & + e^{-\beta\delta} p^{h,\delta}(x, n\delta, x, n\delta + \delta | \alpha) V^{h,\delta}(x) + k(x, \alpha) \Delta t^{h,\delta}(x, \alpha) \Bigg]. \end{aligned} \tag{5.21}$$

Recall the definition of the function $Y^h(\cdot)$ above (2.6) and that the transitions from the reflecting states are not controlled. Then, for $x \in \partial G_h^+$, the set of reflecting states,

$$V^{h,\delta}(x) = \sum_{\tilde{x}} p^h(x, \tilde{x}) \left[V^{h,\delta}(\tilde{x}) + Y^h(x, \tilde{x}) \right]. \tag{5.22}$$

In (5.22), we use the expression $p^h(x, \tilde{x})$ for the transition probabilities as the transitions for the reflecting states do not depend on δ or α. Theorem 5.1 implies that the optimal values determined by (2.6) and ((5.21), (5.22)) are asymptotically equal.

6.5.4 Asymptotic Equivalence of the Timescales

For each $t \geq 0$, define the time indices:

$$\begin{aligned} d^h(t) &= \max \left\{ n : \sum_{i=0}^{n-1} \Delta t_i^h = t_n^h \leq t \right\}, \\ d^{h,\delta}(t) &= \max \left\{ n : \sum_{i=0}^{n-1} \Delta t_i^{h,\delta} = t_n^{h,\delta} \leq t \right\}, \\ d_\tau^h(t) &= \max \left\{ n : \tau_n^h \leq t \right\}, \quad d_\tau^{h,\delta}(t) = \max \left\{ n : \tau_n^{h,\delta} \leq t \right\}. \end{aligned} \tag{5.23a}$$

As noted above (3.2), $d^h(t)$ will never be the index of a reflecting state, as the interpolation intervals for those are zero. Define the stopping times

$$\tilde{d}^h(t) = \min\left\{n : \sum_{i=0}^{n-1} \Delta t_i^h = t_n^h \geq t\right\},$$
$$\tilde{d}^{h,\delta}(t) = \min\left\{n : \sum_{i=0}^{n-1} \Delta t_i^{h,\delta} = t_n^{h,\delta} \geq t\right\}, \tag{5.23b}$$
$$\tilde{d}_\tau^h(t) = \min\left\{n : \tau_n^h \geq t\right\}, \quad \tilde{d}_\tau^{h,\delta}(t) = \min\left\{n : \tau_n^{h,\delta} \geq t\right\}.$$

Theorem 5.1. *For each* $t > 0$,

$$\lim_{h\to 0} \sup_{u^h,x} E_x^{h,u^h} \sup_{s\leq t} \left[\sum_{i=0}^{\tilde{d}^h(s)} (\Delta\tau_i^h - \Delta t_i^h)\right]^2 = 0. \tag{5.24}$$

Also,

$$\lim_{h\to 0,\, \delta\to 0} \sup_{u^{h,\delta},x} E_x^{h,\delta,u^{h,\delta}} \sup_{s\leq t} \left[\sum_{i=0}^{\tilde{d}^{h,\delta}(s)} (\Delta\tau_i^{h,\delta} - \Delta t_i^{h,\delta})\right]^2 = 0. \tag{5.25}$$

Equation (5.24) *holds with* $\tilde{d}_\tau^h(\cdot)$ *replacing* $\tilde{d}^h(\cdot)$, *and* (5.25) *holds with* $\tilde{d}_\tau^{h,\delta}(\cdot)$ *replacing* $\tilde{d}^{h,\delta}(\cdot)$. *Let* $I_n^{h,\delta}$ *be the indicator function of the event that time advances for the implicit approximation procedure at step* n. *Then* (5.25) *also holds with* $\tilde{d}_\tau^{h,\delta}(\cdot)$ *used and* $\Delta t_n^h(1 - I_n^{h,\delta})$ *replacing* $\Delta t_n^{h,\delta}$ *and* $\Delta\tau_n^h(1 - I_n^{h,\delta})$ *replacing* $\Delta\tau_n^{h,\delta}$.

Let $\phi^{h,\delta}(\cdot)$ *denote the interpolation of the* $\phi_n^{h,\delta}$ *with the intervals* $\Delta t_n^{h,\delta}$. *Then* $\phi^{h,\delta}(\cdot)$ *converges weakly and in mean square (uniformly on any finite time interval) to the process with value* t *at time* t. *The result of the last sentence holds if the intervals are* $\Delta\tau_n^{h,\delta}$.

Proof. Owing to the mutual independence of the exponentially distributed random variables $\{\nu_n\}$ and their independence of everything else, the discrete parameter process $L_n = \sum_{i=0}^n (\Delta\tau_i^h - \Delta t_i^h)$ is a martingale. By Doob's inequality for martingales, the expectation of the $\sup_{s\leq t}$ of the squared term in (5.24), conditioned on $\{\Delta t_i^h\}$, satisfies

$$E_x^{h,u^h} \sup_{s\leq t} \left[\sum_{i=0}^{\tilde{d}^h(s)} \left[\Delta\tau_i^h - \Delta t_i^h\right]^2 \middle| \Delta t_i^h, i < \infty\right]$$
$$\leq 4E_x^{h,u^h} \left[\sum_{i=0}^{\tilde{d}^h(t)} \left[\Delta\tau_i^h - \Delta t_i^h\right]^2 \middle| \Delta t_i^h, i < \infty\right]$$
$$= 4 \sum_{i=0}^{\tilde{d}^h(t)} [\Delta t_i^h]^2 \leq 4(t + \sup_n \Delta t_n^h) \sup_n \Delta t_n^h \xrightarrow{h} 0,$$

which yields (5.24). Equation (5.25) and the rest of the first paragraph of the theorem are proved in the same way.

To prove the assertions concerning the asymptotic behavior of $\phi^{h,\delta}(\cdot)$, write

$$\phi^{h,\delta}(t) = \sum_{i=0}^{d^{h,\delta}(t)-1} \Delta t_i^{h,\delta} + \sum_{i=0}^{d^{h,\delta}(t)-1} \beta_{0,i}^{h,\delta}.$$

The first sum equals t, modulo $\sup_n \Delta t_n^{h,\delta}$. The variance of the martingale term is bounded by δt, modulo $\delta + \sup_n \Delta t_n^{h,\delta}$, and the term converges weakly to the zero process. This yields the next to last assertion of the theorem. This and (5.25) yield the last assertion of the theorem. ■

6.5.5 Convergence

The proofs of convergence for the explicit approximation procedure for the no-delay case of this chapter are in [58, Chapter 10 and 11]. The main difference in the proofs of the explicit and implicit approximation procedures lies in the fact that the timescales are different. These are shown to be asymptotically equal by Theorem 5.1, and the results are summarized in the next theorem. See the proofs of convergence for the delay case in Chapter 8 for more detail.

Theorem 5.2. *Let* $\{\xi_n^h\}$ *and* $\{\xi_n^{h,\delta}\}$ *be locally consistent. Assume the model* ((1.1), (1.3)), *or* ((1.2), (1.4)) *in relaxed control notation. Assume* (A3.2.1), (A3.2.2) *and* (A3.4.3). *Let* $b(\cdot), \sigma(\cdot), k(\cdot)$ *be continuous,* U *compact, and* $U^h \to U$ *as* $h \to 0$. *Let* (1.2) *have a unique weak-sense solution for each admissible relaxed control. Then* $V^h(\cdot)$ *given by* (2.6) *and* $V^{h,\delta}(\cdot)$ *given by* (5.21) *converge to the optimal cost for* ((1.2), (1.4)), *as* $h \to 0$ *and* $(h,\delta) \to 0$, *respectively. The analogous results hold for the cost function* (1.5) *if* (A3.4.1) *and* (A3.4.2) *replace the conditions* (A3.2.1) *and* (A3.2.2).

6.6 Singular and Impulsive Controls

6.6.1 Singular Controls

Consider the specialization of the model (3.6.1):

$$dx(t) = b(x(t))dt + q_1(x(t-))d\lambda(t) + \sigma(x(t))dw(t) + dz(t), \tag{6.1}$$

where

$$d\lambda(t) = \lambda(t) - \lambda(t-dt) = \sum_i v_i d\lambda_i(t),$$

where the v_i are direction vectors and the $\lambda_i(\cdot)$ are real-valued and right continuous. Let the cost function be

$$W(x,\lambda) = E_x^\lambda \int_0^\infty e^{-\beta t}\left[k(x(t))dt + q'dy(t) + q'_\lambda d\lambda(t)\right]. \tag{6.2}$$

The components $q_{\lambda,i}$ of the vector q_λ are all positive. The development of the Markov chain approximation for the singular control model is in [58, Section 8.3] and the proof of convergence is in [58, Section 11.2].[6] One can use ordinary and singular controls together, but we confine the discussion to the strictly singular control problem for expositional simplicity. Let $p^h(x,\tilde{x})$ and $\Delta t^h(x)$ denote a transition probability and interpolation interval that are locally consistent for the uncontrolled problem, which is (6.1) with $\lambda(\cdot) = 0$. Let $\Delta\lambda_n^h$, with components $\Delta\lambda_{i,n}^h$, denote the increment in the control at step n. The cost function for the chain is

$$W^h(x,\lambda^h) = E_x^{h,\lambda^h} \sum_{n=0}^\infty e^{-\beta\theta_n}\left[k(\xi_n^h)\Delta\theta_n + q'_\lambda \Delta\lambda_n^h + q'\Delta y_n^h\right]. \tag{6.3}$$

The case where $q_1(\cdot)$ is constant. Until further notice, suppose that $q_1(\cdot)$ does not depend on x, which is the case considered in [58]. Then the controls in (6.1) and (6.2) appear in a purely additive fashion. Hence, if we wish to realize some $\Delta\lambda_n^h > 0$ at step n, whatever its value, we can consider it to be the sum of a sequence of very small impulses occurring instantaneously. This idea will be used for the construction of the approximating Markov chain for the controlled problem. Divide the behavior of the chain at a typical iterate n into the following three mutually exclusive types:

1. Suppose that $\xi_n^h = x \in \partial G_h^+$. Then we have a "reflection step," and $\Delta\theta(x) = 0$. The next value ξ_{n+1}^h is determined in accordance with the local consistency condition (2.2).

   Otherwise, we are at $x \in G_h \subset G$ and there are the following two choices, only one of which can be exercised at a time:

2. Do not exercise control and get ξ_{n+1}^h from $x = \xi_n^h$ by using the uncontrolled transition probability $p^h(x,\tilde{x})$ with the associated interpolation interval being $\Delta\theta(x)$.
3. Exercise control and choose the increment in the singular control, with the interpolation interval being $\Delta t^h(x) = 0$.

   For $x \in G_h$, the Bellman equation is (replacing the first equation in (2.6))

$$V^h(x) = \min\left\{e^{-\beta\Delta\theta(x)}\sum_{\tilde{x}} p^h(x,\tilde{x})V^h(\tilde{x}) + k(x)\Delta\theta(x),\right.$$
$$\left.\min_{\Delta\lambda}\left[\sum_{\tilde{x}} p^h(x,\tilde{x}|\Delta\lambda)V^h(\tilde{x}) + \sum_i q_{i,\lambda}\Delta\lambda_i\right]\right\}. \tag{6.4}$$

[6] To help reference to the book [58], note that $F(\cdot)$ was used there for our $\lambda(\cdot)$.

In (6.4), $\Delta\lambda = (\Delta\lambda_1, \ldots)$ denotes the increment in the singular control at the current step, and $p^h(x, \tilde{x}|\Delta\lambda)$ is the transition probability under control action $\Delta\lambda$. Owing to the current assumption that the $q_1(\cdot)$ in (6.1) does not depend on x, we need only increment one component of $\lambda(\cdot)$ at a time and can suppose that the increments $\Delta\lambda_i^h$ in the components go to zero as $h \to 0$. Then we can replace the Bellman equation by

$$\begin{aligned} V^h(x) = \min \Bigg\{ & e^{-\beta\Delta\theta(x)} \sum_{\tilde{x}} p^h(x, \tilde{x}) V^h(\tilde{x}) + k(x)\Delta\theta(x), \\ & \min_{i, \Delta\lambda_i} \left[\sum_{\tilde{x}} p^h(x, \tilde{x}|\Delta\lambda_i) V^h(\tilde{x}) + q_{i,\lambda}\Delta\lambda_i \right] \Bigg\}. \end{aligned} \tag{6.5}$$

The $p^h(x, \tilde{x}|\Delta\lambda_i)$ in (6.5) is the probability that the next state is $\tilde{x}$, given the increment $\Delta\lambda_i$ in the ith component of the singular control, with the increments in the other components being zero. So, first one checks to see which component of the control it is best to increment, with an interpolation interval zero, and then compares this with the value of not having any singular control increment, in which case the process evolves with $p^h(x, \tilde{x})$ and $\Delta t^h(x)$. The values of $\Delta\lambda_i$ will depend on h. For $x \in \partial G_h^+$, the Bellman equation is the second expression in (2.6).

The possible values for the $\Delta\lambda_i$ were not specified. The actual values are not important for convergence provided only that they go to zero as $h \to 0$ and take the current state to another grid point. Convenience of coding is a primary concern, and we will illustrate one useful procedure with a two-dimensional problem.

Choosing the control when $q_1(\cdot)$ is constant: An example. Because $q_1(\cdot)$ is now constant, replace it by unity in (6.1). A general procedure will be illustrated via a two-dimensional example. Suppose that there are two directions for the control effects, namely, v_1 and v_2. A control value in only one of the directions will be chosen at each control step. This is not required for the convergence but is convenient for the coding. Thus we must choose the direction and the magnitude in that direction. It is best if the state moves only "locally." But a direction v_i might not take any point $x \in G_h$ through any nearby point on the grid. Because of this the choice of the control value and transition probability is a little indirect. Refer to Figure 6.1. Let us scale the vector $v_i, i = 1, 2$, so that the vector hv_i takes the point x to the point x_i on the left hand or lower lines. The v_i and $q_{i,\lambda}$ can always be scaled so that there is no loss of generality in making this supposition. In the example, neither point x_1 or x_2 is in G_h.

Suppose that the direction v_2 is selected and denote the corresponding transition probabilities by $p^h(x, \tilde{x}|hv_2)$. Then to attain the mean value x_2, $p^h(x, \tilde{x}|hv_2)$ would randomize among the points y_1 and y_2, and the value of $\Delta\lambda_2$ that takes x to x_2 would be used for the increment in the cost.

On the randomization to attain the desired conditional mean. Let J_n^h denote the indicator function of the event that step n is a control step, and augment $\mathcal{F}_n^h$ so that it also measures $\{J_i^h, i \le n\}$. With n being a control step and $\Delta\lambda_n^h$ being the chosen value of the conditional (given the system data to step n) expectation of the control increment at the nth step, we have $E_n^h \Delta\xi_n^h J_n^h = \Delta\lambda_n^h J_n^h$. In the limit, as $h \to 0$, the effects of the randomizations that are used to attain the desired conditional means disappear, as we will now see. The randomization errors $(\Delta\xi_n^h - \Delta\lambda_n^h)J_n^h$ are a martingale difference sequence and

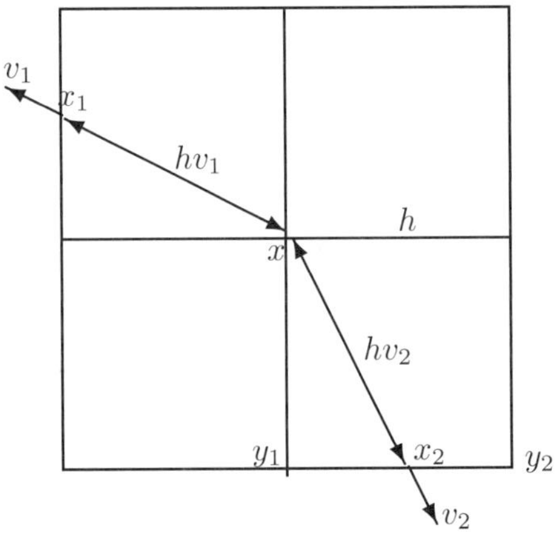

Figure 6.1. Choosing the singular control values.

$$E \sup_{n \le N} \left| \sum_{j=0}^{n-1} [\Delta\xi_j^h J_j^h - \Delta\lambda_j^h J_j^h] \right|^2 = O(h) E \sum_{j=0}^{N-1} |\Delta\lambda_n^h|.$$

This implies that the effects of the randomization errors go to zero if the sequence of the costs that are due to the control are bounded, which must be the case in any optimization problem.

The procedure with $q_1(\cdot)$ being x-dependent. Recall the discussion in Section 3.6 concerning the problems that can arise when $q_1(\cdot)$ depends on x. In particular, there was an example that illustrated the difference in the effects of a single jump of magnitude J in some component of $\lambda(\cdot)$ on one hand, and what would happen if that jump were the limit of many small jumps occurring in a sequence, one after the other. Because of this potential complication, when $q_1(\cdot)$ depends on x, the Bellman equation (6.4) must be used in lieu of (6.5). The main complication from a coding point of view is that the transitions will not always be local. To date, for the bulk of problems of interest, the singular control does not depend on the state.

6.6.2 Impulsive Control

The numerical algorithms for the no-delay form of the impulsively controlled model (3.6.17)

$$dx(t) = b(x(t))dt + DdF(t) + \sigma(x(t))dw(t) + dz(t)$$

with cost function (3.1.18)

$$W(\hat{x}, F) = E_{\hat{x}}^F \left[\int_0^\infty e^{-\beta t} k(x(t))dt + \sum_i e^{-\beta \tau_i} g(x(\tau_i -), \nu_i) \right]$$

is handled as is the general singular control problem, and similarly for the boundary absorption version. What distinguishes the impulsively controlled problem is the strict positivity of the function $g(x, \nu)$ for $\nu \neq 0$. This "setup" cost ensures that the problem is well defined and that there are only a finite number of impulses on any finite time interval.

6.7 The Ergodic Cost Function

6.7.1 Introduction

A brief overview of the setup for the ergodic cost criterion will be presented. A fuller treatment is in [58, Chapters 7 and 11]. Let $\{X_n\}$ be a finite-state controlled Markov chain, with transition probabilities $p(x, \tilde{x}|\alpha)$, where the control parameter α takes values in the compact set U. Suppose that, for each feedback control $u(\cdot)$, the state space consists of a single ergodic set and a transient set. The cost criterion is taken to be

$$\gamma(u) = \lim_{n \to \infty} \frac{1}{n} E_x^u \sum_{i=0}^{n-1} C(X_i, u(X_i)). \tag{7.1}$$

The limit exists and does not depend on the initial condition x.

There is a vector-valued function $W(u)$ with values $W(x, u)$ such that

$$W(x, u) = \sum_{\tilde{x}} p(x, \tilde{x}|u(x)) W(\tilde{x}, u) + C(x, u(x)) - \gamma(u). \tag{7.2}$$

The solution to (7.2) is not unique, because if we added a constant k to each component of $W(u)$, the result would also solve (7.2). One candidate for $W(u)$ is

$$W(x, u) = \sum_{n=0}^{\infty} E_x^u [C(\xi_n, u(\xi_n)) - \gamma(u)],$$

where the sum is well defined as the summands go to zero geometrically.

Suppose that $p(x,\tilde{x}|\alpha)$ and $C(x,\alpha)$ are continuous functions of α for each x in the state space. Then the Bellman equation is

$$V(x) = \min_{\alpha \in U} \left[\sum_{\tilde{x}} p(x,\tilde{x}|\alpha)V(\tilde{x}) + C(x,\alpha) - \bar{\gamma} \right]. \tag{7.3}$$

Under our conditions, there is a solution $(V(\cdot),\bar{\gamma})$, where $\bar{\gamma}$ is unique and is the minimal value of the ergodic cost.

Approximation in policy space. A standard way of solving (7.3) is via the approximation in policy space algorithm. One recursively computes a minimizing sequence of feedback controls $u_n(\cdot), n = 0,1,\ldots$. Suppose that $u_n(\cdot)$ has been computed. Then solve (7.2) with $u(\cdot) = u_n(\cdot)$ to obtain the cost $\gamma(u_n)$ and the function $W(u_n)$. Then compute $u_{n+1}(\cdot)$ by

$$u_{n+1}(x) = \arg\min_{\alpha \in U} \left[\sum_{\tilde{x}} p(x,\tilde{x}|\alpha)W(\tilde{x},u_n) + C(x,\alpha) \right]. \tag{7.4}$$

A difficulty in solving (7.3) is that there is no contraction operator, as there would be for the discounted cost problem. There are convenient ways of circumventing this difficulty by "centering" procedures, and methods of solution are discussed in [58], as are methods for accelerating the convergence. Under the given conditions $\gamma(u_n) \downarrow \bar{\gamma}$, the minimal cost.

6.7.2 The Markov Chain Approximation Method

Consider the model (1.1), where $a(x) = \sigma(x)\sigma'(x)$ is strictly positive definite and has a continuous inverse, and let $\{\xi_n^h\}$ denote a locally consistent Markov chain approximation. Recall that the state space G is a convex polyhedron. With the usual methods of construction of the chain, it is an ergodic process for each feedback control, and the transition probabilities are continuous in the control, and we make these assumptions. The system is (1.1) with cost function, under a relaxed control,

$$\gamma(x,r) = \limsup_T \frac{1}{T} E_x^{h,r} \int_0^T \left[\int_U k(x(t),\alpha)r'(d\alpha,t)dt + q'dy(t) \right]. \tag{7.5}$$

The limit might depend on x. For a relaxed feedback control the form is

$$\gamma(m) = \limsup_T \frac{1}{T} E_x^{h,m} \int_0^T \left[\int_U k(x(t),\alpha)m(x(t),d\alpha)dt + q'dy(t) \right]. \tag{7.6}$$

If the relaxed feedback control is equivalent to an ordinary feedback control $u(\cdot)$, then write $\gamma(u)$.

Let $\bar{\gamma}$ denote the infimum of the cost over all admissible controls, and suppose that it does not depend on the initial condition. Conditions for this to be the case, as well as the proof that we cannot do better by using relaxed controls over relaxed feedback controls and that there is an optimal relaxed feedback control, are in [56, Chapter 4].

The cost for the approximating chain. If the current state ξ_n^h is reflecting, then $\Delta z_n^h = O(h)$. It is convenient to represent the associated cost $q' E_{\xi_n^h}^h \Delta y_n^h$ in the form $k_0^h(\xi_n^h)h$, where $k_0^h(x) = 0$ for $x \in G$

For an arbitrary admissible control sequence $u = \{u_n^h\}$, the cost for the chain is

$$\gamma^h(x,u) = \limsup_n \frac{E_x^{h,u} \sum_{i=0}^n \left[k(\xi_i^h, u_i^h)\Delta t_i^h + k_0^h(\xi_i^h)h\right]}{E_x^{h,u} \sum_{i=0}^n \Delta t_i^h}. \tag{7.7}$$

Suppose that the feedback control $u(\cdot)$ is used. Then there is a unique invariant measure that we denote by $\pi^h(u)$, with values $\pi^h(x,u)$. Then $\gamma^h(x,u)$ does not depend on the initial condition x, and by the ergodic theorem for Markov chains [14] there is a limit in (7.7) and it equals

$$\gamma^h(u) = \frac{\sum_x \left[k(x,u(x))\Delta\theta(x,u(x)) + k_0^h(x)h\right]\pi^h(x,u)}{\sum_x \Delta\theta(x,u(x))\pi^h(x,u)}. \tag{7.8}$$

Weighing the invariant measure with respect to the time $\Delta t^h(x,\alpha)$ that the chain spends in a state x yields the measure $\mu^h(u)$ with values

$$\mu^h(x,u) = \frac{\Delta\theta(x,u(x))\pi^h(x,u)}{\sum_{\tilde{x}} \Delta\theta(\tilde{x},u(\tilde{x}))\pi^h(\tilde{x},u)}. \tag{7.9}$$

Using (7.9), the component of (7.8) that involves $k(\cdot)$ can be written as

$$\sum_x k(x,u(x))\mu^h(x,u). \tag{7.10}$$

It turns out that if the feedback control $u(\cdot)$ is continuous and the diffusion is well-defined under it, then $\mu^h(\cdot,u)$ converges weakly to the invariant measure of the diffusion. Note that $\mu^h(x,u) = 0$ if x is a reflecting state. The measure on the set of reflecting states ∂G_h^+ that is defined by $\mu_R^h(x,u) = h\pi^h(x,u)/\sum_{\tilde{x}} \Delta\theta(\tilde{x},u(\tilde{x}))\pi^h(\tilde{x},u)$ converges to the local time measure on the boundary.

For any feedback control $u(\cdot)$, there is a vector-valued function $W^h(u)$ with values $W^h(x,u)$ such that

$$\begin{aligned} &W^h(x,u) = \\ &\sum_{\tilde{x}} p^h(x,\tilde{x}|u(x))W^h(\tilde{x},u) + \left[k(x,u(x)) - \gamma^h(u)\right]\Delta\theta(x,u(x)) + k_0^h(x)h. \end{aligned} \tag{7.11}$$

As for the situation at the beginning of the section, $W(u)$ is unique only up to additive constant. One solution is

$$W^h(x,u) = \sum_{n=0}^{\infty} E_x^{h,u} \left(\left[k(\xi_n^h, u_n^h) - \gamma^h(u) \right] \Delta\theta_n + k_0^h(\xi_n^h) h \right).$$

To see that any pair $(W^h(u), \gamma)$ that solves (7.11) yields that $\gamma = \gamma^h(u)$, multiply each side by $\pi^h(x,u)$, sum over x, and use the invariance of $\pi^h(u)$ (which cancels the terms involving $W^h(u)$). The result is that γ equals the expression in (7.8).

With $\bar{\gamma}^h$ denoting the infimum of the costs, the Bellman equation is

$$V^h(x) = \min_{\alpha \in U} \left[\sum_{\tilde{x}} p^h(x, \tilde{x}|\alpha) V^h(\tilde{x}) + \left[k(x,\alpha) - \bar{\gamma}^h \right] \Delta\theta(x,\alpha) + k_0^h(x) h \right]. \tag{7.12}$$

Methods for solving this equation for the optimal control and cost are covered in [58, Chapter 7].

Convergence. For a feedback control $u(\cdot)$, the ergodic theorem for Markov chains implies the following (the first four limits are w.p.1):

$$\begin{aligned} \gamma^h(u) &= \lim_n \frac{1}{t_n^h} \sum_{i=0}^{n-1} \left[k(\xi_i^h, u_i^h) \Delta t_i^h + k_0^h(\xi_i^h) h \right] \\ &= \lim_n \frac{1}{\tau_n^h} \sum_{i=0}^{n-1} \left[k(\xi_i^h, u_i^h) \Delta \tau_i^h + k_0^h(\xi_i^h) h \right] \\ &= \lim_n \frac{1}{\tau_n^h} \int_0^{\tau_n^h} \left[k(\psi^h(s), u(\psi^h(s))) ds + q' dy_\tau^h(s) \right] \\ &= \lim_t \frac{1}{t} \int_0^t \left[k(\psi^h(s), u(\psi^h(s))) ds + q' dy_\tau^h(s) \right] \\ &= \lim_t \frac{1}{t} \int_0^t E_x^{h,u} \left[k(\psi^h(s), u(\psi^h(s))) ds + q' dy_\tau^h(s) \right]. \end{aligned} \tag{7.13}$$

The first two lines of (7.13) equal the cost for the chain that would be obtained if we started at time zero with the stationary distribution of the chain ξ_n^h. The last line is the cost that would be obtained for the process $\psi^h(\cdot)$, if we started with its stationary distribution, given by (7.9). These costs are equal.

Now let $\bar{u}^h(\cdot)$ be the optimal control and consider the cost for the associated stationary process. This can be written as

$$\bar{\gamma}^h = \gamma^h(\bar{u}^h) = E^{h,\bar{u}^h} \int_0^1 \left[k(\psi^h(t), u^h(\psi^h(t))) dt + q' dy_\tau^h(t) \right],$$

where $\psi^h(0)$ has the stationary distribution of the continuous parameter process $\psi^h(\cdot)$. Let $r_\tau^h(\cdot)$ denote the relaxed control representation of the control $u^h(\psi^h(\cdot))$ for the stationary process, and rewrite the above expression as

$$\bar{\gamma}^h = \gamma^h(\bar{u}^h) = E^{h,\bar{u}^h} \int_0^1 \left[\int_U k(\psi^h(t), \alpha) r_\tau^{h,\prime}(d\alpha, t) dt + q' dy_\tau^h(t) \right],$$

We will now give an outline of the proof of convergence. More details on the weak convergence are in Section 8.5. Let $w^h(\cdot)$ be defined as below (3.8). The set of processes $(\psi^h(\cdot), r^h_\tau(\cdot), z^h_\tau(\cdot), w^h(\cdot))$ associated with the stationary system is tight. Let $(x(\cdot), r(\cdot), z(\cdot), w(\cdot))$ denote the limit of a weakly convergent subsequence, indexed by h for notational convenience. Then the limit is stationary in the sense that the probability law of $(x(t+\cdot), r(t+\cdot)-r(t), z(t+\cdot)-z(t), w(t+\cdot)-w(t))$ does not depend on t. The process $w(\cdot)$ is Wiener, and the other processes are nonanticipative with respect to $w(\cdot)$. The limit set satisfies

$$dx(t) = \int_U b(x(t), \alpha) r'(d\alpha, t)dt + \sigma(x(t))dw(t) + dz(t).$$

We have $\bar{\gamma}^h \to \gamma(r) \geq \bar{\gamma}$, where $\gamma(r)$ is the stationary cost associated with the limit system. Thus

$$\liminf_h \bar{\gamma}^h \geq \bar{\gamma}. \tag{7.14}$$

To prove that $\bar{\gamma}^h \to \bar{\gamma}$, we need to show that

$$\limsup_h \bar{\gamma}^h \leq \bar{\gamma}. \tag{7.15}$$

To prove this, we need the fact that for any $\epsilon > 0$, there is an ϵ-optimal feedback control $u^\epsilon(\cdot)$ for the diffusion that is continuous in x.[7] Apply this control to the chain, and consider the resulting stationary system. Let $(\psi^h(\cdot), z^h_\tau(\cdot), w^h(\cdot))$ denote the interpolated processes for this system. The set is tight, and any weak-sense limit $(x(\cdot), z(\cdot), w(\cdot))$ is stationary in the sense used above, and satisfies

$$dx(t) = b(x(t), u^\epsilon(x(t)))dt + \sigma(x(t))dw(t) + dz(t).$$

Then by the minimality of $\bar{\gamma}^h$, the ϵ-optimality of $u^\epsilon(\cdot)$ for the original system, the weak convergence and the stationarity of the limit, we have

$$\begin{aligned}
\bar{\gamma}^h &\leq \gamma^h(u^\epsilon) \\
&= E^{h,u^\epsilon} \int_0^1 \left[k(\psi^h(t), u^\epsilon(\psi^h(t)))dt + q' dy^h_\tau(t)\right] \\
&\to E^{u^\epsilon} \int_0^1 \left[k(x(t), u^\epsilon(x(t)))dt + q' dy(t)\right] \\
&= \gamma(u^\epsilon) \leq \bar{\gamma} + \epsilon.
\end{aligned} \tag{7.16}$$

This and the arbitrariness of ϵ yield (7.15). Hence $\bar{\gamma}^h \to \bar{\gamma}$.

---

[7] A proof of this fact is in [56, Chapter 4].

7

Markov Chain Approximations: Path and Control Delayed

7.0 Outline of the Chapter

This chapter adapts the Markov chain approximation methods that were introduced in Chapter 6 to the problem with delays. The approximating chains are constructed almost exactly as they are for the no-delay case, except that the transition probabilities must take the delays into account. Various numerical approximations are developed. They are reasonable and well motivated. But in view of the fact that rather little is known about either approximation or numerical methods for delay equations, the algorithms are to be viewed as a first step and will hopefully encourage additional work. When constructing an algorithm, there are two large issues of concern, and both must be kept in mind. One is numerical feasibility. The other concerns the proof of convergence, as the approximating parameter goes to zero.

Because the basic state space of the problem with delays is infinite-dimensional, one must work with approximations. One can devise "Markov chain like" approximations that converge to the original model and for which the optimal value functions converge to that for the original model. Alternatively, one can first approximate the original model, say along the lines done in Chapter 4, so that the resulting problem is finite-dimensional. Then approximate the result for numerical purposes. Both approaches are taken in this chapter, although the latter one is more realistic, as the memory requirements are much less. As seen in Chapter 4, the suggested finite-dimensional approximations are often quite good. Further approximations are developed in Chapters 8 and 9 when the path or control values are delayed, and they will often be advantageous.

The validity of an approximation to the original model depends on the relative insensitivity of the values and controls to the quantities that are being approximated, whether it is the path, path and delay, control, and so forth. The greater the sensitivity, the finer the approximation needs to be. This is a particularly difficult problem for the delay model, as the behavior can be quite sensitive to the delay, and little is known about this in general.

H.J. Kushner, *Numerical Methods for Controlled Stochastic Delay Systems*,
doi: 10.1007/978-0-8176-4621-9_7,

The proofs of convergence in [58] are purely probabilistic, being based on weak convergence methods. The idea is to interpolate the chain to a continuous-time process in a suitable manner, show that the Bellman equation for the interpolation is the same as for the chain, and then show that the interpolated processes converge to an optimal diffusion as the approximating parameter goes to zero. The approach is parallel to this for the problem with delays, and we try to arrange the development with an eye to using the powerful methods and results of [58] to the extent possible, so as to simplify the proof of convergence.

Section 1 introduces the unapproximated model and the main assumptions. As for the nondelay case, the main assumption is local consistency. This condition is the same as that for the nondelay problem, with the appropriate delay-dependent drift and diffusion terms used. The state of the problem, as needed for the numerical procedure, consists of a segment of the path (over the delay interval) and of the control path as well (if the control is also delayed). The only change in the local consistency condition is the use of the "memory segment" arguments in the drift and diffusion functions. As in Chapter 6, the local consistency condition says no more than that the conditional mean change (resp, covariance) in the state of the approximating chain is proportional to the drift (resp, covariance) of the original diffusion process, modulo small errors. It need not hold everywhere (see, e.g., [58, Section 5.5]). Transition probabilities for the approximating chain are readily obtained from the formulas that are used for the nondelay case in [58].

For pedagogical purposes, in much of the development, we divide the discussion into a part where only the path is delayed in the dynamics and a part where both the control and path are delayed, for which the algorithms are much more complicated The delay system analogs of all of the cost functions covered in [58] can be dealt with. But for simplicity of exposition, most of the discussion is confined to the discounted case, with boundary reflection. If the process is stopped on reaching a boundary then, with the model of Section 3.1 and the cost function (3.4.1), all of the approximation methods and convergence results will hold, and the necessary theorems are stated. Section 1 concludes with the discussion of the continuous time interpolations. These interpolations, which are used for the convergence proofs only and not for the numerical algorithms, are a little more complicated than those used for the no-delay case in Chapter 6, owing to the need to represent the "memory segment" argument in a way that can be used in the development of efficient approximation methods.

In Section 2, some particular Markov chain approximations are introduced, with the aim of efficiency in the use of memory. The implicit approximation method of Chapter 5 has some advantages in dealing with the memory problem, and this is discussed in Section 3. Section 4 deals with various details concerning the relation between the implicit approximation procedure and the model with randomly varying delays in Subsection 4.2.3. Keep in mind that these randomly varying delays are not a feature of the original model, but appear in the numerical approximation as a consequence of the use of the implicit approximation procedure to simplify the memory problem. One could treat the case where the original model has randomly time varying delays as well, as noted in the comments below (3.1.8), but at the expense of increased memory requirements.

Chapter 8 continues the development of the ideas in this chapter and contains the proofs of the convergence theorems.

7.1 The Model and Local Consistency

The approach to numerical approximation is analogous to what was done for the no-delay case in Chapter 6. The main new issues concern accounting for the fact that $b(\cdot), \sigma(\cdot)$, and $k(\cdot)$ depend on the "memory" segments of the solution path and/or the control, whichever is delayed in the dynamics. We will construct an "approximating" controlled finite-state process $\{\xi_n^h, n \geq 0\}$ and interpolation intervals $\{\Delta t_n^h, n \geq 0\}$ in much the same way as was done in Chapter 6. This approximating process will serve as the basis of the numerical procedure. It will be seen that these processes are constructed as easily as they are for the no-delay problem in [58]. Although $\{\xi_n^h\}$ itself is not a Markov chain due to the memory, one can embed it into a finite-state Markov chain. It is the Bellman equation for the embedded chain that needs to be solved to get the optimal cost. Indeed a main concern are representations for such Markov chains that are efficient from the point of view of computation. In this section, a generic approximation will be constructed. Although it often requires too much memory to be of practical use, it will provide the foundation for the alternative and more practical approximations in Section 3 and in the next chapter.

7.1.1 The Models

The model is the controlled reflected diffusion of Section 3.2. Assumptions (A3.2.1) and (A3.2.2) on the constraint set G are always used. Other conditions will be given when needed. Rewriting the equations for convenience, when both the path and control are delayed and in terms of ordinary controls, the model is (3.2.3):

$$x(t) = x(0) + \int_0^t b(\bar{x}(s), \bar{u}(s))ds + \int_0^t \sigma(\bar{x}(s))dw(s) + z(t), \tag{1.1}$$

where the conditions (A3.1.2) and (A3.1.3) hold. For notational simplicity, we suppose that, if both the path and control are delayed, then the maximum delay is the same for both. The case where they are not the same is a simple and obvious modification. In relaxed control notation, (1.1) is

$$x(t) = x(0) + \int_0^t \bar{b}(\bar{x}(s), \bar{r}(t))ds + \int_0^t \sigma(\bar{x}(s))dw(s) + z(t), \tag{1.2}$$

where, as in (3.1.6),

$$\bar{b}(\bar{x}(t), \bar{r}(t)) = \int_{-\bar{\theta}}^0 \int_U b(\bar{x}(t), \alpha, \theta)r'(d\alpha, t+\theta)\mu_c(d\theta), \tag{1.3}$$

and

$$\int_0^t \bar{b}(\bar{x}(s), \bar{r}(s))ds = \int_{-\bar{\theta}}^0 \left[\int_0^t ds \int_U b(\bar{x}(s), \alpha, \theta)r'(d\alpha, s+\theta)\right] \mu_c(d\theta).$$

The discounted cost function (3.4.4) is

$$W(\hat{x}, \hat{r}, r) = E_{\hat{x},\hat{r}}^r \int_0^\infty ds \int_{-\bar{\theta}}^0 \int_U e^{-\beta t} \left[k(\bar{x}(t), \alpha, \theta)r'(d\alpha, t+\theta)\mu_c(d\theta)dt + q'dy(t)\right], \tag{1.4}$$

where $\hat{x}$ and $\hat{r}$ denote the initial memory segments of the path and control, resp. The existence of an optimal control was shown in Theorem 3.5.1.

If the path only is delayed, then we drop the control memory segment term, and the model specializes to

$$x(t) = x(0) + \int_0^t ds \int_U b(\bar{x}(s), \alpha)r'(d\alpha, s) + \int_0^t \sigma(\bar{x}(s))dw(s) + z(t), \tag{1.5}$$

$$W(\hat{x}, r) = E_{\hat{x}}^r \int_0^\infty \int_{-\bar{\theta}}^0 \int_U e^{-\beta t} \left[k(\bar{x}(t), \alpha)r'(d\alpha, t)dt + q'dy(t)\right]. \tag{1.6}$$

As usual, if the process stops on hitting the boundary, then drop (A3.1.2) and (A3.1.3) and add (A3.4.1) and (A3.4.2).

7.1.2 Delay in Path Only: Local Consistency and Interpolations

The approximating chain ξ_n^h takes values in the set S_h, and the definitions of $S_h, G_h = S_h \cap G$ and ∂G_h^+ from the beginning of Section 6.2 are used. As for the no-delay problem, the key requirement that is placed on the approximating chain is that it satisfy a local consistency condition analogous to (6.2.1). The dynamics of (1.5) at time t involve the memory segment $\bar{x}(t)$ of the path on the delay interval $[t - \bar{\theta}, t]$. An analogous dependence must hold for the dynamics of the ξ_n^h process. The definition of the memory segment of the approximating chain will depend on the particular continuous-time interpolation of the ξ_n^h values that is used, and several useful forms will be developed in the sequel and in the next chapter. For simplicity, we will start by using an analog of the explicit approximation procedure of Sections 6.2–6.4. This will not usually yield the best form of the memory segment, but it provides a convenient introduction to the overall approximation method. Suppose that $\xi_n^h, \Delta t_n^h$ are available (these will be constructed below) and, as in Section 6.2, define the interpolated time $t_n^h = \sum_{i=0}^{n-1} \Delta t_i^h$. The process $\xi^h(\cdot)$ is defined to be the piecewise-constant continuous-time interpolation of $\{\xi_n^h\}$ with intervals $\{\Delta t_n^h\}$, as in (6.3.1). Recall the discussion below (6.3.1) concerning the interpolation at the reflecting states. In particular, if ξ_n^h is a reflecting state, then $\xi^h(t_n^h) = \xi_{n+1}^h$, which is the state that the reflecting state ξ_n^h is instantaneously sent to.

Path memory segments. Define the segment $\bar{\xi}_n^h$ of the path $\xi^h(\cdot)$ by

$$\bar{\xi}_n^h(\theta) = \xi^h(t_n^h + \theta) \quad \text{for } \theta \in [-\bar{\theta}, 0), \quad \text{and } \bar{\xi}_n^h(0) = \xi_n^h. \tag{1.7a}$$

This is the segment of the interpolated path on $[t_n^h - \bar{\theta}, t_n^h)$ with the value ξ_n^h at $\theta = 0$. If $\xi_n^h \in G_h$, then $\bar{\xi}_n^h(\theta) = \xi^h(t_n^h + \theta)$ for all $\theta \in [-\bar{\theta}, 0]$. Define the process $\bar{\xi}^h(t)$ by

$$\bar{\xi}^h(t) = \bar{\xi}_n^h, \qquad \text{for } t_n^h \le t < t_{n+1}^h. \tag{1.7b}$$

Let $\hat{\xi}$ denote the canonical value of $\bar{\xi}_n^h$.

To construct the dynamics of the approximating chain, we will need to define a path memory segment that plays the role of $\bar{x}(t)$. There is a great deal of flexibility in the way that this approximation is constructed from the $\{\xi_n^h\}$. The choice influences the computational complexity, and we return to this issue in subsequent sections. Until further notice, we use $\bar{\xi}_n^h$. This choice is not always suitable for numerical purposes, and will later be modified in various ways to simplify the numerical computations. The exact form of the approximation is not important at this point.

The initial condition $\bar{x}(0) = \{x(t) : -\bar{\theta} \le t \le 0\}$ for (1.5) is an arbitrary continuous function. This will have to be approximated for numerical convenience. Until further notice, we simply assume that we use a sequence $\bar{\xi}_0^h \in D(G_h; [-\bar{\theta}, 0])$, that is piecewise-constant and that converges to $\bar{x}(0)$ uniformly on $[-\bar{\theta}, 0]$ as $h \to 0$.

Local consistency in G_h. For numerical purposes it is often useful to approximate the set U. Thus, as in Section 6.2, let U^h be a sequence of compact sets that converges to U as $h \to 0$ in the sense that the closed convex hull of $(b(x, U^h), k(x, U^h))$ converges to the closed convex hull of $(b(x, U), k(x, U))$ as $h \to 0$. Each U^h might contain only a finite set of points.

Let $\xi_n^h \in G_h$. Analogously to the no-delay case in Section 6.2, the chain and intervals are assumed to satisfy the following local consistency properties. Let u_n^h (with values in U^h) denote the control applied at time n. The distribution of ξ_{n+1}^h, given the initial data and $\{\xi_i^h, u_i^h, i \le n\}$, will depend only on the current path memory segment $\bar{\xi}_n^h$ and current control u_n^h and not otherwise on n, analogously to the case in Chapter 6. Recall the definition $\Delta \xi_n^h = \xi_{n+1}^h - \xi_n^h$ and that of the martingale difference β_n^h in (6.2.3), and let $E_{\hat{\xi},n}^{h,\alpha}$ denote the expectation given all data to time n, with $u_n^h = \alpha$ and $\bar{\xi}_n^h = \hat{\xi}$. Analogously to the definition for the no-delay case in (6.2.1), local consistency is said to hold if there is a function $\Delta t^h(\cdot)$ such that, for $\hat{\xi}(0) = \xi_n^h \in G_h$,[1]

$$\begin{aligned} &E_{\hat{\xi},n}^{h,\alpha} \Delta\xi_n^h = b_h(\hat{\xi},\alpha)\Delta t^h(\hat{\xi},\alpha) = b(\hat{\xi},\alpha)\Delta t^h(\hat{\xi},\alpha) + o(\Delta t^h(\hat{\xi},\alpha)), \\ &E_{\hat{\xi},n}^{h,\alpha} \beta_n^h[\beta_n^h]' = a_h(\hat{\xi})\Delta t^h(\hat{\xi},\alpha) = a(\hat{\xi})\Delta t^h(\hat{\xi},\alpha) + o(\Delta t^h(\hat{\xi},\alpha)), \\ &a(\hat{\xi}) = \sigma(\hat{\xi})\sigma'(\hat{\xi}), \\ &\sup_{n,\omega} |\xi_{n+1}^h - \xi_n^h| \xrightarrow{h} 0, \quad \sup_{\hat{\xi},\alpha} \Delta t^h(\hat{\xi},\alpha) \xrightarrow{h} 0. \end{aligned} \tag{1.8}$$

The reflecting boundary is treated the same as in Section 6.2. If ξ_n^h is a reflecting state, then it is sent to a state in G_h, with no control applied. The mean of $\xi_{n+1}^h - \xi_n^h$, conditioned on the data to time n, is a reflection direction at the point ξ_n^h. In particular, (6.2.2) holds. Define $\Delta t_n^h = \Delta t^h(\bar{\xi}_n^h, u_n^h)$.

We have the analog of (6.2.4)

$$\xi_{n+1}^h = \xi_n^h + \Delta t_n^h b(\bar{\xi}_n^h, u_n^h) + \beta_n^h + \Delta z_n^h + o(\Delta t_n^h). \tag{1.9}$$

Constructing the transition probabilities. For simplicity in the development, we will suppose that S_h is a regular h-grid. Hence the points in G_h are h units apart in each direction. This is done only to simplify the notation. Any of the state spaces G_h that are allowed in [58] can be used here. In particular, the state space approximation parameter can depend on the coordinate direction. The simple example of the construction in Section 6.4 and, indeed, any of the methods in [58] for obtaining the transition probabilities and interpolation intervals for the no-delay case can be readily adapted to the delay case.

For the no-delay problem and $x \in G_h$, all of the methods in [58] for generating the controlled transition probabilities $p^h(x,\tilde{x}|\alpha)$ when the grid spacing was uniform in each coordinate direction gave results that depended only on the grid spacings, the "next state" $\tilde{x}$, and on the drift and covariance functions $b(x,\alpha)$ and $a(x) = \sigma(x)\sigma'(x)$, resp. They did not depend on the state and control values in any other way. In addition the transition probability for the chains in [58] for the no-delay case could be written as a ratio in the following way. There are functions $N^h(\cdot)$ and $D^h(\cdot)$ such that for $x \in G_h$,

$$\begin{aligned} P\{\xi_1^h = \tilde{x} | \xi_0^h = x, u_0^h = \alpha\} = p^h(x,\tilde{x}|\alpha) &= \frac{N^h(b(x,\alpha),a(x),\tilde{x})}{D^h(b(x,\alpha),a(x))}, \\ \Delta t^h(x,\alpha) = T^h(b(x,\alpha),a(x)) &= \frac{h^2}{D^h(b(x,\alpha),a(x))}. \end{aligned} \tag{1.10}$$

[1] (1.8) defines $b_h(\cdot)$ and $a_h(\cdot)$.

The particular forms of $N^h(\cdot)$ and $D^h(\cdot)$ depend on the actual approximation method.[2] The function $D^h(\cdot)$ is simply a normalization, so that the sum of the probabilities over $\tilde{x}$ is unity. The transition probability from a state x to a state $\tilde{x}$ must be a function of $b(\cdot)$, $a(\cdot)$, and $\tilde{x}$, only, because that is the only information that is available. Hence, the representation (1.10) is unrestrictive. The values of $N^h(\cdot)$ and $D^h(\cdot)$ for the two examples of construction in Section 6.4 are obvious from (6.4.5) or (6.4.7).

For the delay case, we can use the identical forms. For any of the approximation methods in [58] or elsewhere for getting the $N^h(\cdot)$, $D^h(\cdot)$ in (1.10) that yield locally consistency in the sense of (6.2.1), for $\hat{\xi}(0) \in G_h$ we can use the forms

$$\begin{aligned} p^h(\hat{\xi}, \tilde{x}|\alpha) = P\{\xi_1^h = \tilde{x} | \bar{\xi}_0^h = \hat{\xi}, u_0^h = \alpha\} &= \frac{N^h(b(\hat{\xi},\alpha), a(\hat{\xi}), \tilde{x})}{D^h(b(\hat{\xi},\alpha), a(\hat{\xi}))}, \\ \Delta t^h(\hat{\xi},\alpha) = \frac{h^2}{D^h(b(\hat{\xi},\alpha), a(\hat{\xi}))}&. \end{aligned} \tag{1.11}$$

In particular,

$$p^h(\bar{\xi}_n^h, \tilde{x}|u_n^h) = P\{\xi_{n+1}^h = \tilde{x} | \bar{\xi}_n^h, u_n^h\} = \frac{N^h(b(\bar{\xi}_n^h, u_n^h), a(\bar{\xi}_n^h), \tilde{x})}{D^h(b(\bar{\xi}_n^h, u_n^h), a(\bar{\xi}_n^h))}. \tag{1.12}$$

With the use of (1.11), local consistency in the sense of (6.2.1) implies the local consistency (1.8). It is only the dependence on $b(\cdot)$, $a(\cdot)$, $\tilde{x}$, and h that matters, no matter what the form of the memory segment $\hat{\xi}$. The above discussion is formalized by the following assumption. The assumption is not needed if local consistency is otherwise assured.

A1.1. *The transition probabilities and interpolation intervals are given in the form* (1.10) *with the delay dependencies incorporated* (*yielding* (1.11)), *where* (1.10) *is locally consistent for the no-delay case.*[3]

A discounted cost function. Let $E_{\hat{\xi}}^{h,u^h}$ denote the expectation given initial condition[4] $\hat{\xi} = \bar{\xi}_0^h$ and control sequence $u^h = \{u_n^h, 0 \le n < \infty\}$. Define Δz_n^h

[2] The form of $T^h(\cdot)$ in (1.10) supposes that S_h is a grid with the same spacing in each coordinate direction, so that h is real-valued. This is chosen for simplicity in the development. For more general forms of S_h, the functions $T^h(\cdot)$, $N^h(\cdot)$ and $D^h(\cdot)$ might also depend on the current state x and the local spacing of the states. But whatever they are, they are functions of the drift and diffusion functions. See [58, Section 5.2]. With the delay-dependencies of these functions incorporated, the resulting transition probabilities and interpolation interval would yield the desired local consistency. All that is needed is local consistency.

[3] The form is usually $N(hb(x,\alpha), a(x), \tilde{x})/D(hb(x,\alpha), a(x))$ for some functions $N(\cdot), D(\cdot)$.

[4] The approximation $\hat{\xi}$ of the initial condition will depend on h in general.

and Δy_n^h as above (6.2.3). An approximation to the discounted cost function (3.4.3) for the chain is

$$W^h(\hat{\xi}, u^h) = E_{\hat{\xi}}^{h,u^h} \sum_{n=0}^{\infty} e^{-\beta t_n^h} \left[k(\bar{\xi}_n^h, u_n^h) \Delta t^h(\bar{\xi}_n^h, u_n^h) + q' \Delta y_n^h \right], \qquad (1.13)$$
$$V^h(\hat{\xi}) = \inf_{u^h} W^h(\hat{\xi}, u^h).$$

By Lemma 6.3.1 (which is [58, Theorem 11.1.3]), the costs are well defined. Let $y^h(\cdot)$ denote the continuous-time interpolation of $\{\Delta y_n^h\}$ with intervals $\{\Delta t_n^h\}$.

A "Markov" continuous-time interpolation. One continuous-time interpolation, namely $\xi^h(\cdot)$, has already been defined. We will now define the analog of the interpolation $\psi^h(\cdot)$ that was defined in Subsection 6.3.2. Let the random variables $\{\nu_n\}$, the interval $\Delta\tau_n^h = \nu_n \Delta t_n^h$, and $\tau_n^h = \sum_{i=0}^{n-1} \Delta\tau_n^h$, be defined as in the first paragraph of Subsection 6.3.2. Then define $\psi^h(t)$ by (6.3.4) or (6.3.8), using the intervals $\Delta\tau_n^h$, all based the processes ξ_n^h and Δt_n^h of this chapter. Because the timescale of the $\psi^h(\cdot)$ uses the intervals $\Delta\tau_n^h$, and that of the memory segment $\bar{\xi}_n^h$ uses the intervals Δt_n^h, the dynamical equation for $\psi^h(\cdot)$ will be a little awkward. But keep in mind that this dynamical equation will be used only in the proofs of convergence and not for the numerical computations. As for the no-delay case, the chains ξ_n^h are used for the numerics, with whatever approximation to the path memory segment is used.

Recall the definitions of the interpolations $u_\tau^h(\cdot)$, $B_\tau^h(\cdot)$, $r_\tau^h(\cdot)$, and $z_\tau^h(\cdot)$, in Subsection 6.3.2, and the definition $d_\tau^h(s) = \max\{n : \tau_n^h \le s\}$ in (6.5.23). Define the function $q_\tau^h(s) = t_{d_\tau^h(s)}^h$. Given interpolated time s in the scale determined by the $\Delta\tau_n^h$, the function $d_\tau^h(s)$ is the index of the process ξ_n^h that gives $\psi^h(s)$ in the sense that we have $\psi^h(s) = \xi_{d_\tau^h(s)}^h = \xi^h(q_\tau^h(s))$. With this notation, the conditional drift rate of the process $\psi^h(\cdot)$ at time s is $b_h(\bar{\xi}^h(q_\tau^h(s)), u_\tau^h(s))$ ($b_h(\cdot)$ was defined in (1.8)). Decomposing the process $\psi^h(\cdot)$ into a compensator, martingale, and reflection term as in (6.3.8), and using relaxed control terminology, leads to the representation

$$\psi^h(t) = \xi_0^h + \int_0^t \int_{U^h} b_h(\bar{\xi}^h(q_\tau^h(s)), \alpha) r_\tau^h(d\alpha\, ds) + B_\tau^h(t) + z_\tau^h(t), \qquad (1.14)$$

where $\xi_0^h = \bar{\xi}_0^h(0)$ and $B_\tau^h(\cdot)$ is a martingale with quadratic variation process

$$\int_0^t a_h(\bar{\xi}^h(q_\tau^h(s)))ds.$$

As noted below (6.3.8), there is a martingale $w^h(\cdot)$ with quadratic variation It and that converges weakly to a Wiener process such that

$$B_\tau^h(t) = \int_0^t \sigma(\bar{\xi}^h(q_\tau^h(s)))dw^h(s) + \epsilon^h(t)$$

where $\lim_{h\to 0} E \sup_{s\le t} |\epsilon^h(s)|^2 \to 0$ for each $t < \infty$

Modulo an asymptotically negligible error due to the "continuous time" approximation of the discount factor, the cost function (1.13) can be written as

$$W^h(\hat{\xi}, u^h) = E_{\hat{\xi}}^{h,u^h} \int_0^\infty \int_{U^h} e^{-\beta t} \left[k(\bar{\xi}^h(q_\tau^h(s)), \alpha) r_\tau^h(d\alpha\, ds) + q' dy_\tau^h(s)\right]. \tag{1.15}$$

The following theorem says that any method for solving the control problem for any locally consistent approximation will yield an approximation to the value for the original model (1.5). The proof is in Section 5 of the next chapter. The absorbing boundary case is dealt with in Theorem 1.3.

Theorem 1.1. *Let $\xi_n^h, \Delta t_n^h$ be locally consistent with the model* (1.5) *whose initial condition is $\bar{x}(0)$, a continuous function, and with cost function* (1.6) *and its approximation*

$$E_{\hat{\xi}}^{h,u^h} \sum_{n=0}^{\infty} e^{-\beta t_n^h} \left[k(\bar{\xi}_n^h, u_n^h)\Delta t_n^h + q' \Delta y_n^h\right] \tag{1.16}$$

being used. Let $\bar{\xi}_0^h \in D(G_h; [-\bar{\theta}, 0])$ be any piecewise-constant sequence that converges to $\bar{x}(0)$ uniformly on $[-\bar{\theta}, 0]$. Assume (A3.1.1), (A3.1.2), (A3.2.1)–(A3.2.3), *and* (A3.4.3). *Then $V^h(\bar{\xi}_0^h) \to V(\bar{x}(0))$ as $h \to 0$.*

7.1.3 Delay in the Path and Control

Now consider the model (relaxed control form) (1.2), with cost (1.4), where both the path and control are delayed. As for the case where only the path is delayed, one constructs an approximating chain $\{\xi_n^h, n \ge 0\}$ and interpolation intervals $\{\Delta t_n^h, n \ge 0\}$. The initial data for (1.2) is $\bar{x}(0) = \{x(s), -\bar{\theta} \le s \le 0\} \in C(G; [-\bar{\theta}, 0])$ and $\bar{u}(0) = \{u(s), -\bar{\theta} \le s \le 0\} \in L_2(U; [-\bar{\theta}, 0])$, where the control segment is needed due to the delay in the control. The control memory segment for the approximating chain is slightly different. For the process (1.2), either the segment $\{u(s), s \in [-\bar{\theta}, 0]\}$ or the segment $\{u(s), s \in [-\bar{\theta}, 0)\}$ will do for the initial control data. But for the chain, the control at time 0, namely, u_0^h, which is used to get ξ_1^h, is to be determined at time 0, and should not be given as part of the initial data. This fact accounts for our use of the control segment on the half open $[-\bar{\theta}, 0)$ as the initial data. Let $\hat{u}$ denote the canonical value of the control memory segment on the half open interval. With α denoting the canonical value of the current value of the control, we can write terms such as $\bar{b}(\hat{\xi}, \hat{u}, \alpha)$ without ambiguity, depending on the memory segments and the current control value.

Definitions of the control memory segments. In the remainder of this section, we continue to use the full path memory segment $\bar{\xi}_n^h$ from the previous subsection. Given the initial control data $\bar{u}(0)$, we need to approximate it for use on the chain, and, analogously, obtain a control memory segment for each step of the chain. In this subsection, we will use a form for the control memory segment that is analogous to $\bar{\xi}_n^h$. It will usually be very costly in terms of the required memory, but serves as a useful introduction. Alternative, and more efficient, approximations will be discussed in the next chapter. The control memory segment will be denoted by $\bar{u}_n^h$, with canonical value $\hat{u}$, and is defined in terms of the continuous-time interpolation of the control process, as follows. Let u_n^h denote the control that is used on step n. An interpolation interval $\Delta t^h(\hat{\xi}, \hat{u}, \alpha)$ will be defined in the local consistency condition (1.23). Redefine $\Delta t_n^h = \Delta t^h(\bar{\xi}_n^h, \bar{u}_n^h, u_n^h)$ and $t_n^h = \sum_{i=0}^{n-1} \Delta t_i^h$, and define the interpolation $u^h(\cdot)$ of $\{u_n^h\}$ with intervals $\{\Delta t_n^h\}$. Then define the full control memory segment $\bar{u}_n^h = \{u^h(t_n^h + \theta), \theta \in [-\bar{\theta}, 0)\}$. It is the segment of $u^h(\cdot)$ on $[t_n^h - \bar{\theta}, t_n^h)$. Then $(\hat{u}, \alpha)$ denotes the canonical value of the control on a closed interval $[t - \bar{\theta}, t]$ for any t. Let $\bar{u}_0^h$ be any piecewise-constant function in $D(U^h; [-\bar{\theta}, 0))$ that converges to the function $\bar{u}(0)$ in the L_2-sense as $h \to 0$ and extend the definition of $u^h(\cdot)$ to $[-\bar{\theta}, \infty)$.

Summarizing, in this section the memory state at time n of the approximating chain and the associated dynamic program is $\bar{\xi}_n^h, \bar{u}_n^h$, the value of $\xi^h(\cdot)$ on the closed interval $[t_n^h - \bar{\theta}, t_n^h]$, together with the segment of $u^h(\cdot)$ on $[t_n^h - \bar{\theta}, t_n^h)$.

The distribution of ξ_{n+1}^h, given the initial data and $\{\xi_i^h, u_i^h, i \le n\}$, will depend only on the current memory segments $\bar{\xi}_n^h, \bar{u}_n^h$, and the current control u_n^h, and not on n otherwise. Let $E_{\hat{\xi},\hat{u},n}^{h,\alpha}$ denote the expectation given all data to step n, and that $\bar{\xi}_n^h = \hat{\xi}, \bar{u}_n^h = \hat{u}$, with control value α used at time n. Keep in mind that if control value α is used at step n for the chain, then it is used on $[t_n^h, t_{n+1}^h)$ for the interpolation $\xi^h(\cdot)$. Letting $r^h(\cdot)$ denote the relaxed control representation of $u^h(\cdot)$ and with the memory segment $\bar{u}_n^h$ being used, we can write the drift term as

$$\bar{b}(\bar{\xi}_n^h, \bar{u}_n^h, u_n^h) = \int_{-\bar{\theta}}^{0} \int_{U^h} b(\bar{\xi}_n^h, \alpha, \theta) r^{h,\prime}(d\alpha, t_n^h + \theta) \mu_c(d\theta). \tag{1.17}$$

Example. Before proceeding with the general definition of local consistency when the control is delayed, which is essentially that used in Chapter 6 and in the previous subsection, let us consider a simple example. In (1.1), let

$$b(\bar{x}(t), \bar{u}(t)) = b_1(\bar{x}(t), u(t - \bar{\theta})) + b_0(\bar{x}(t), u(t)). \tag{1.18}$$

Then the measure $\mu_c(\cdot)$ is concentrated on the points $-\bar{\theta}$ and 0. The analog of the first line of (1.8) will be

$$E_{\bar{\xi}_n^h, \bar{u}_n^h, n}^{h, u_n^h} \Delta \xi_n^h = \left[b_1(\bar{\xi}_n^h, u(t_n^h - \bar{\theta})) + b_0(\bar{\xi}_n^h, u_n^h) \right] \Delta t_n^h + o(\Delta t_n^h). \tag{1.19}$$

Notation. Recall the definition of $\tilde{r}'(d\alpha, t, \theta)$ above (3.1.8) and in (4.4.2) and its role in the development of the approximating models in Chapter 4. Analogous definitions will be useful in the proofs of convergence in dealing with the various approximations to the piecewise constant control memory segments, as it will be the control memory segment at each t that is being approximated. For this purpose, define the relaxed control derivatives $\tilde{r}_\tau^{h,\prime}(d\alpha, t, \theta)$ and $\tilde{r}^{h,\prime}(d\alpha, t, \theta)$, for $t \in [0, \infty)$ and $\theta \in [-\bar{\theta}, 0]$, by

$$\begin{aligned} \tilde{r}_\tau^{h,\prime}(d\alpha, t, \theta) &= r^{h,\prime}(d\alpha, \tau_n^h + \theta), \quad \text{for } t \in [\tau_n^h, \tau_{n+1}^h), \\ \tilde{r}^{h,\prime}(d\alpha, t, \theta) &= r^{h,\prime}(d\alpha, t_n^h + \theta), \quad \text{for } t \in [t_n^h, t_{n+1}^h). \end{aligned} \tag{1.20}$$

Define the relaxed control derivative $\bar{r}_n^{h,\prime}$ with values $\bar{r}_n^{h,\prime}(d\alpha, \theta)$, for $\theta \in [-\bar{\theta}, 0)$, by

$$\bar{r}_n^{h,\prime}(d\alpha, \theta) = r^{h,\prime}(d\alpha, t_n^h + \theta). \tag{1.21}$$

The $\bar{r}_n^{h,\prime}$ is a representation of the control memory segment in terms of the derivative of its relaxed control representation, which we will find to be very useful. Using (1.20) and the fact that $\tilde{r}_\tau^{h,\prime}(d\alpha, s, \theta)$ is constant for $s \in [\tau_n^h, \tau_{n+1}^h)$, we can write

$$\begin{aligned} \bar{b}(\bar{\xi}_n^h, \bar{u}_n^h, u_n^h)\Delta\tau_n^h &= \bar{b}(\bar{\xi}_n^h, \bar{r}_n^{h,\prime}, u_n^h)\Delta\tau_n^h \\ &= \int_{\tau_n^h}^{\tau_{n+1}^h} \int_{-\bar{\theta}}^0 \int_{U^h} b(\bar{\xi}_n^h, \alpha, \theta)\tilde{r}_\tau^{h,\prime}(d\alpha, s, \theta)\mu_c(d\theta)ds. \end{aligned} \tag{1.22}$$

Equation (1.22) defines $\bar{b}(\bar{\xi}_n^h, \bar{r}_n^{h,\prime}, u_n^h)$, and we will use this notation when working in terms of relaxed controls.

The general definition of local consistency when the control is delayed. The local consistency condition for the chain is that there exists a function $\Delta t^h(\cdot)$ such that for $\hat{\xi} = \bar{\xi}_n^h$, with $\hat{\xi}(0) \in G_h$, and $\hat{u} = \bar{u}_n^h, \alpha = u_n^h$,

$$\begin{aligned} &E_{\hat{\xi},\hat{u},n}^{h,\alpha} \Delta\xi_n^h = \bar{b}_h(\hat{\xi}, \hat{u}, \alpha)\Delta t^h(\hat{\xi}, \hat{u}, \alpha) = \bar{b}(\hat{\xi}, \hat{u}, \alpha)\Delta t^h(\hat{\xi}, \hat{u}, \alpha) + o(\Delta t^h(\hat{\xi}, \hat{u}, \alpha)), \\ &E_{\hat{\xi},\hat{u},n}^{h,\alpha} \beta_n^h[\beta_n^h]' = a_h(\hat{\xi}, \hat{u}, \alpha)\Delta t^h(\hat{\xi}, \hat{u}, \alpha) = a(\hat{\xi})\Delta t^h(\hat{\xi}, \hat{u}, \alpha) + o(\Delta t^h(\hat{\xi}, \hat{u}, \alpha)), \\ &a(\hat{\xi}) = \sigma(\hat{\xi})\sigma'(\hat{\xi}), \\ &\sup_{n,\omega} |\xi_{n+1}^h - \xi_n^h| \xrightarrow{h} 0, \quad \sup_{\hat{\xi},\hat{u},\alpha} \Delta t^h(\hat{\xi}, \hat{u}, \alpha) \xrightarrow{h} 0. \end{aligned} \tag{1.23}$$

The relations in (1.23) define the functions $b_h(\cdot)$ and $a_h(\cdot)$. The reflecting boundary is treated exactly as it was when only the path was delayed, using transition probabilities satisfying (6.2.2).[5]

[5] Recall that $\Delta t_n^h = \Delta t^h(\bar{\xi}_n^h, \bar{u}_n^h, u_n^h)$ when the control is delayed.

The transition probabilities. The following analogs of (1.11) and (1.12) ensure the local consistency, if the same functions $N^h(\cdot)$ and $D^h(\cdot)$ are used. As for (1.11) and (1.12), it is only the dependence on $b(\cdot), a(\cdot), \tilde{x}$, and h that matters, no matter what the form of the memory segments $\hat{\xi}, \hat{u}$.

$$\begin{aligned} p^h(\hat{\xi}, \hat{u}, \tilde{x}|\alpha) = P\{\xi_1^h = \tilde{x}|\bar{\xi}_0^h = \hat{\xi}, \bar{u}_0^h = \hat{u}, u_0^h = \alpha\} = \frac{N^h(\bar{b}(\hat{\xi}, \hat{u}, \alpha), a(\hat{\xi}), \tilde{x})}{D^h(\bar{b}(\hat{\xi}, \hat{u}, \alpha), a(\hat{\xi}))}, \\ \Delta t^h(\hat{\xi}, \hat{u}, \alpha) = T^h(\bar{b}(\hat{\xi}, \hat{u}, \alpha), a(\hat{\xi})) = \frac{h^2}{D^h(\bar{b}(\hat{\xi}, \hat{u}, \alpha), a(\hat{\xi}))}, \end{aligned} \tag{1.24}$$

and

$$p^h(\bar{\xi}_n^h, \bar{u}_n^h, \tilde{x}|u_n^h) = P\{\xi_{n+1}^h = \tilde{x}|\bar{\xi}_n^h, \bar{u}_n^h, u_n^h\} = \frac{N^h(\bar{b}(\bar{\xi}_n^h, \bar{u}_n^h, u_n^h), a(\bar{\xi}_n^h), \tilde{x})}{D^h(\bar{b}(\bar{\xi}_n^h, \bar{u}_n^h, u_n^h), a(\bar{\xi}_n^h))}.$$

The notation $p^h(\bar{\xi}_n^h, \bar{r}_n^{h,\prime}, \tilde{x}|u_n^h)$ will also be used for $p^h(\bar{\xi}_n^h, \bar{u}_n^h, \tilde{x}|u_n^h)$. We formalize the above discussion as follows. The assumption is not needed if local consistency is otherwise ensured.

A1.2. *The transition probabilities and interpolation intervals are given in the form* (1.24)*, where* (1.10) *is locally consistent for the nondelay case.*

Continuous-time interpolations. The continuous-time interpolations are defined as for the case where only the path is delayed, dealt with in the previous subsection. We will write out the expressions for the interpolation $\psi^h(\cdot)$ and the associated discounted cost that are analogous to (1.14) and (1.15). Extend the definition of $u_\tau^h(t)$ to the interval $[-\bar{\theta}, \infty)$ by letting it equal $\bar{u}_0^h(\theta)$ for $\theta \in [-\bar{\theta}, 0)$, and let $r_\tau^h(\cdot)$ denote the relaxed control representation of this extended $u_\tau^h(\cdot)$. Recalling the definition (1.20), for $\xi_0^h \in G_h$ the continuous time interpolation (1.14) is replaced by

$$\begin{aligned} \psi^h(t) = \xi_0^h + \int_{-\bar{\theta}}^0 \left[\int_0^t \int_{U^h} b_h(\bar{\xi}^h(q_\tau^h(s)), \alpha, \theta) \bar{r}_\tau^{h,\prime}(d\alpha, s, \theta) ds \right] \mu_c(d\theta) \\ + B_\tau^h(t) + z_\tau^h(t). \end{aligned} \tag{1.25}$$

Let $E_{\hat{\xi},\hat{u}}^{h,u^h}$ denote the expectation under initial data $\bar{\xi}_0^h = \hat{\xi}$ and control sequence $u^h = \{u_n^h, n \geq 0\}$, with initial control segment (on $[-\bar{\theta}, 0)$) being $\hat{u}$. The analog of the cost function (1.13) is

$$\begin{aligned} W^h(\hat{\xi}, \hat{u}, u^h) = E_{\hat{\xi},\hat{u}}^{h,u^h} \sum_{n=0}^{\infty} e^{-\beta t_n^h} \left[\bar{k}(\bar{\xi}_n^h, \bar{u}_n^h, u_n^h) \Delta t_n^h + q' \Delta y_n^h \right], \\ V^h(\hat{\xi}, \hat{u}) = \inf_{u^h} W^h(\hat{\xi}, \hat{u}, u^h), \end{aligned} \tag{1.26}$$

where $\bar{k}(\cdot)$ is defined analogously to $\bar{b}(\cdot)$ in (1.17).

In integral and relaxed control form, and modulo an asymptotically negligible error due to the approximation of the discount factor, (1.26) equals

$$
\begin{aligned}
&W^h(\hat{\xi}, \hat{u}, u^h) \\
&= E_{\hat{\xi},\hat{u}}^{h,u^h} \int_{-\bar{\theta}}^{0} \mu_c(d\theta) \left[\int_0^\infty dt \int_{U^h} e^{-\beta t} k(\bar{\xi}^h(q_\tau^h(t)), \alpha, \theta) \tilde{r}_\tau^{h,\prime}(d\alpha, t, \theta) \right] \\
&+ E_{\hat{\xi},\hat{u}}^{h,u^h} \int_0^\infty e^{-\beta t} q' dy_\tau^h(t).
\end{aligned}
\tag{1.27}
$$

The following convergence theorem, whose proof is in Section 5 of the next chapter, says that any method for solving the control problem for any locally consistent approximation will yield an approximation to the value for the original model (1.1) or (1.2).

Theorem 1.2. *Let* $\xi_n^h, \Delta t_n^h$ *be locally consistent with* (1.1) *or* (1.2) *in the sense of* (1.23), *with initial data* $\bar{x}(0)$, *a continuous function on* $[-\bar{\theta}, 0]$, *and* $\bar{u}(0) \in L_2(U; [-\bar{\theta}, 0))$. *The cost function for* (1.2) *is* (1.4) *and that for the approximating chain is* (1.26). *Let* $\bar{\xi}_0^h \in D(G_h; [-\bar{\theta}, 0])$ *be piecewise-constant, and converge to* $\bar{x}(0)$ *uniformly on* $[-\bar{\theta}, 0]$. *Let* $\bar{u}_0^h \in D(U^h; [-\bar{\theta}, 0))$ *be piecewise-constant and converge to* $\bar{u}(0)$ *in the sense of* L_2. *Assume* (A3.1.2), (A3.1.3), *and* (A3.2.1)–(A3.2.3), (A3.4.3). *Then* $V^h(\bar{\xi}_0^h, \bar{u}_0^h) \to V(\bar{x}(0), \bar{u}(0))$ *as* $h \to 0$.

7.1.4 Absorbing Boundaries and Other Cost Functions

The next theorem covers the case where the boundary is absorbing rather than reflecting. The proof will be discussed in Section 5 of the next chapter.

Theorem 1.3. *Assume the conditions of either Theorems* 1.1 *or* 1.2, *except those on the reflection directions. Use the cost function* (3.4.1) *if the control is not delayed and* (3.4.2) *if the control is delayed. Assume* (A3.4.1) *and* (A3.4.2). *For the chain let* N_G^h *denote the first time that it leaves* G^0, *the interior of* G, *and use either the cost function*

$$
W^h(\hat{\xi}, u^h) = E_{\hat{\xi}}^{h,u^h} \left[\sum_{n=0}^{N_G^h - 1} e^{-\beta t_n^h} k(\bar{\xi}_n^h, u_n^h) \Delta t^h(\bar{\xi}_n^h, u_n^h) + e^{-\beta N_G^h} g_0(\xi_{N_G^h}^h) \right],
\tag{1.28}
$$

or

$$
\begin{aligned}
&W^h(\hat{\xi}, \hat{u}, u^h) \\
&= E_{\hat{\xi},\hat{u}}^{h,u^h} \left[\sum_{n=0}^{N_G^h - 1} e^{-\beta t_n^h} \bar{k}(\bar{\xi}_n^h, \bar{u}_n^h, u_n^h) \Delta t^h(\bar{\xi}_n^h, \bar{u}_n^h, u_n^h) + e^{-\beta N_G^h} g_0(\xi_{N_G^h}^h) \right],
\end{aligned}
\tag{1.29}
$$

according to the case. Then, according to the case, as $h \to 0$, $V^h(\bar{\xi}_0^h) \to V(\bar{x}(0))$ *or* $V^h(\bar{\xi}_0^h, \bar{u}_0^h) \to V(\bar{x}(0), \bar{u}(0))$.

Optimal stopping. Suppose that we have the option of stopping before G^0 is exited. Then replace N_G^h by the minimum of N_G^h and the stopping time. The theorem continues to hold. Similarly, Theorems 1.1 and 1.2 hold if we allow stopping with a continuous stopping cost. See the development of the optimal stopping problem in [58].

7.1.5 Approximations to the Memory Segments

In applications, keeping the full computed memory segments $\bar{\xi}_n^h, \bar{u}_n^h$ (or $\bar{r}_n^{h,\prime}$) might be too costly in terms of memory. Specific approximations based on truncations and discretizations will be discussed in the next chapter, and an approximation if only the path is delayed is discussed in Section 3. Considerable flexibility is possible in the modeling of the memory segments. It is preferable to use relaxed control notation for the control memory segments, and this will be done in terms of its derivative. So, following the notation for the control memory segment in (1.21), when the full memory segments[6] are used, let us rewrite the equation below (1.24):

$$p^h(\bar{\xi}_n^h, \bar{r}_n^{h,\prime}, \tilde{x}|u_n^h) = \frac{N^h(\bar{b}(\bar{\xi}_n^h, \bar{r}_n^{h,\prime}, u_n^h), a(\bar{\xi}_n^h), \tilde{x})}{D^h(\bar{b}(\bar{\xi}_n^h, \bar{r}_n^{h,\prime}, u_n^h), a(\bar{\xi}_n^h))}, \tag{1.30a}$$

where $\bar{b}(\bar{\xi}_n^h, \bar{r}_n^{h,\prime}, u_n^h)$ is defined in (1.22).

Approximations: Definitions. The approximations to the full memory segments $(\bar{\xi}_n^h, \bar{r}_n^{h,\prime})$ will be denoted by $(\bar{\xi}_{a,n}^{h,\kappa}, \bar{r}_{a,n}^{h,\kappa,\prime})$, where, for $\theta \in [-\bar{\theta}, 0)$, $\bar{r}_{a,n}^{h,\kappa,\prime}(\theta)$ is a probability measure on U^h, and $\bar{\xi}_{a,n}^{h,\kappa}(\theta)$ is G_h-valued for $\theta \in [-\bar{\theta}, 0)$, and $\bar{\xi}_{a,n}^{h,\kappa}(0)$ will have values either in G_h or in the set of reflecting states ∂G_h^+. The variable $\kappa \to 0$ is a parameter of the approximation. It will also be used to index the associated chain, control, interpolation interval, and so forth, and in the applications will generally take the values δ or (δ_0, δ), analogously to the parameters of the approximations used in Chapter 4. The subscript "a" denotes the type of memory segment approximation, analogously to the usage with the approximations in Chapter 4 (e.g., random, periodic, periodic-Erlang), and, unless noted otherwise, it will be used to index only the approximating memory segments and the relaxed control representation of the approximating control memory segment.

With these approximations used, the true transition probabilities are

[6] The full memory segments at iterate n are the interpolations (with intervals $\{\Delta t_n^h\}$) of the paths and control, resp., over the intervals $[t_n^h - \bar{\theta}, t_n^h]$ and $[t_n^h - \bar{\theta}, t_n^h)$, resp.

$$p^h(\bar{\xi}_{a,n}^{h,\kappa}, \bar{r}_{a,n}^{h,\kappa,\prime}, \tilde{x}|\alpha = u_n^{h,\kappa}) = \frac{N^h(\bar{b}(\bar{\xi}_{a,n}^{h,\kappa}, \bar{r}_{a,n}^{h,\kappa,\prime}, u_n^{h,\kappa}), a(\bar{\xi}_{a,n}^{h,\kappa}), \tilde{x})}{D^h(\bar{b}(\bar{\xi}_{a,n}^{h,\kappa}, \bar{r}_{a,n}^{h,\kappa,\prime}, u_n^{h,\kappa}), a(\bar{\xi}_{a,n}^{h,\kappa}))}, \tag{1.30b}$$

where $\bar{b}(\bar{\xi}_{a,n}^{h,\kappa}, \bar{r}_{a,n}^{h,\kappa,\prime}, u_n^{h,\kappa})$ is defined by (1.22) with $(\bar{\xi}_{a,n}^{h,\kappa}, \bar{r}_{a,n}^{h,\kappa,\prime}, u_n^{h,\kappa})$ replacing $(\bar{\xi}_n^h, \bar{r}_n^{h,\prime}, u_n^h)$. Define $\Delta t_n^{h,\kappa} = \Delta t^h(\bar{\xi}_{a,n}^{h,\kappa}, \bar{u}_{a,n}^{h,\kappa}, u_n^{h,\kappa})$, $t_n^{h,\kappa} = \sum_{i=0}^{n-1} \Delta t_i^{h,\kappa}$, with analogous definitions for $\Delta\tau_n^{h,\kappa}$ and $\tau_n^{h,\kappa}$.

For whatever the type "a" of the approximation, let the relaxed control that is defined by the controls $\{u_n^{h,\kappa}\}$ with interpolation intervals $\{\Delta\tau_n^{h,\kappa}\}$ be denoted by $r_\tau^{h,\kappa}(\cdot)$, and let that defined by the interpolation with intervals $\{\Delta t_n^{h,\kappa}\}$ be denoted by $r^{h,\kappa}(\cdot)$. Define the following function of α, t and θ, where $\theta \in [-\bar{\theta}, 0)$:

$$\bar{r}_n^{h,\kappa,\prime}(d\alpha, \theta) = r^{h,\kappa,\prime}(d\alpha, t_n^{h,\kappa} + \theta). \tag{1.31a}$$

$\bar{r}_n^{h,\kappa,\prime}(\cdot)$ is the full memory segment defined by the actual realized control on the interval $[t_n^{h,\kappa} - \bar{\theta}, t_n^{h,\kappa})$. Keep in mind that it is not necessarily equal to the approximating memory segment $\bar{r}_{a,n}^{h,\kappa,\prime}(\cdot)$, which is the one that is actually used in the dynamics and cost function at step n of the chain when the approximation type is "a."

The following functions of α, t and θ, where $\theta \in [-\bar{\theta}, 0]$, will be useful in analyzing the approximations and their convergence:

$$\left.\begin{aligned} \tilde{r}_a^{h,\kappa,\prime}(d\alpha, t, \theta) &= \bar{r}_{a,n}^{h,\kappa,\prime}(d\alpha, \theta), \quad \text{for } \theta \in [-\bar{\theta}, 0) \\ \tilde{r}_a^{h,\kappa,\prime}(d\alpha, t, 0) &= I_{\{u_n^{h,\kappa} \in d\alpha\}}, \end{aligned}\right\} \text{ for } t \in [t_n^{h,\kappa}, t_{n+1}^{h,\kappa}),$$

$$\left.\begin{aligned} \tilde{r}_{a,\tau}^{h,\kappa,\prime}(d\alpha, t, \theta) &= \bar{r}_{a,n}^{h,\kappa,\prime}(d\alpha, \theta), \quad \text{for } \theta \in [-\bar{\theta}, 0) \\ \tilde{r}_{a,\tau}^{h,\kappa,\prime}(d\alpha, t, 0) &= I_{\{u_n^{h,\kappa} \in d\alpha\}}, \end{aligned}\right\} \text{ for } t \in [\tau_n^{h,\kappa}, \tau_{n+1}^{h,\kappa}). \tag{1.31b}$$

We will always use the definitions:

$$\begin{aligned} &\bar{\xi}_a^{h,\kappa}(\cdot) \text{ is the interpolation of } \{\bar{\xi}_{a,n}^{h,\kappa}\}, \text{ with intervals } \{\Delta t_n^{h,\kappa}\}, \\ &\bar{\xi}_n^{h,\kappa} \text{ is the full memory path segment } \{\xi^{h,\kappa}(t_n^{h,\kappa} + \theta), \theta \in [-\bar{\theta}, 0]\}. \end{aligned} \tag{1.32}$$

The interpolated process $\psi^{h,\kappa}(\cdot)$ with the memory segment approximation. With the above definitions, we can write the analog of the interpolation (1.14) with the approximating memory segments used as

$$\begin{aligned} \psi^{h,\kappa}(t) = \xi_0^h &+ \int_{-\bar{\theta}}^0 \left[\int_0^t \int_{U^h} b_h(\bar{\xi}_a^{h,\kappa}(q_\tau^{h,\kappa}(s)), \alpha, \theta) \tilde{r}_{a,\tau}^{h,\kappa,\prime}(d\alpha, s, \theta) \right] \mu_c(d\theta) ds \\ &+ B_\tau^{h,\kappa}(t) + z_\tau^{h,\kappa}(t), \end{aligned} \tag{1.33}$$

where the martingale $B_\tau^{h,\kappa}(\cdot)$ has quadratic variation process

$$\int_0^t a_h(\bar{\xi}_a^{h,\kappa}(q_\tau^{h,\kappa}(s)))ds.$$

General assumptions on the approximating memory segments and a convergence theorem. In subsequent sections and in Chapter 8, particular approximations will be proposed. But for maximum usefulness and simplicity of the proofs, it is convenient to state a convergence theorem in terms of some general properties. Suppose that

$$\lim_{h\to 0} \sup_{\text{control}} \sup_n E \sup_{-\bar{\theta}\le\theta\le 0} \left|\bar{\xi}_{a,n}^{h,\kappa}(\theta) - \bar{\xi}_n^{h,\kappa}(\theta)\right| = 0 \tag{1.34}$$

and (note that the upper limit of integration is $0-$)

$$\sup_{\text{control}} \sup_n E \left| \int_{-\bar{\theta}}^{0-} \int_{U^h} f(\alpha,\theta) \left[r^{h,\kappa,\prime}(d\alpha, t_n^h + \theta) - \bar{r}_{a,n}^{h,\kappa,\prime}(d\alpha,\theta) \right] \mu_c(d\theta) \right| \to 0 \tag{1.35}$$

for each bounded and continuous real-valued function $f(\cdot)$, as $h \to 0$ and $\kappa \to 0$. Then, the approximations and the full memory segments are close for small κ and h, and the drift rate at iterate n of the chain is approximated as follows:

Drift rate under the approximating memory segments =

$$\begin{aligned} &\int_{-\bar{\theta}}^{0-} \int_{U^h} b(\bar{\xi}_{a,n}^{h,\kappa}, \alpha, \theta) \bar{r}_{a,n}^{h,\kappa,\prime}(d\alpha,\theta)\mu_c(d\theta) + b(\bar{\xi}_{a,n}^{h,\kappa}, u_n^{h,\kappa}, 0)\mu_c(\{0\}) \\ &\quad \approx \int_{-\bar{\theta}}^{0-} \int_{U^h} b(\bar{\xi}_n^{h,\kappa}, \alpha, \theta) \bar{r}_n^{h,\kappa,\prime}(d\alpha,\theta)\mu_c(d\theta) + b(\bar{\xi}_n^{h,\kappa}, u_n^{h,\kappa}, 0)\mu_c(\{0\}). \end{aligned} \tag{1.36}$$

Condition (1.35) is quite strong because it concerns the behavior at each iterate. Consider the following weaker condition, which allows us to consider averages of the differences between the full control memory segment and its approximations over a finite time interval. For bounded and continuous $f(\cdot)$, replace (1.35) by the assumption that

$$E \left| \int_t^{t+\Delta} ds \int_{-\bar{\theta}}^{0} \int_{U^h} f(s,\alpha,\theta) \left[r^{h,\kappa,\prime}(d\alpha, s+\theta) - \tilde{r}_a^{h,\kappa,\prime}(d\alpha, s, \theta) \right] \mu_c(d\theta) \right| \to 0 \tag{1.37}$$

as $h \to 0$, uniformly in the control and in t for each $\Delta > 0$. Using this, (1.34), and the timescale equivalences in Theorem 3.1 will allow us to asymptotically approximate the drift term in (1.33) by

$$\int_{-\bar{\theta}}^{0} \left[\int_0^t \int_{U^h} b_h(\bar{\xi}^{h,\kappa}(q_\tau^{h,\kappa}(s)), \alpha, \theta) r_\tau^{h,\kappa,\prime}(d\alpha, s+\theta) \right] \mu_c(d\theta) ds. \tag{1.38}$$

Let $V^{h,\kappa}(\hat{x},\hat{u})$ denote the optimal cost function for the model modified as above, using approximating memory segments $\bar{\xi}_{a,n}^{h,\kappa}$ and $\bar{r}_{a,n}^{h,\kappa,\prime}$. Then we have the following result.

Theorem 1.4. *Assume the conditions of Theorems* 1.1, 1.2, *or* 1.3, *but with the use of memory segment approximations* $\bar{\xi}^{h,\kappa}_{a,n}$ *and* $\bar{r}^{h,\kappa,\prime}_{a,n}$ *satisfying* (1.34) *and* (1.37), *resp. Then* $V^{h,\kappa}(\bar{\xi}^h_0, \bar{u}^h_0) \to V(\bar{x}(0), \bar{u}(0))$ *as* $h \to 0$ *and then* $\kappa \to 0$.

7.2 Computational Procedures

7.2.1 Delay in the Path Only: State Representations and the Bellman Equation

Theorem 1.1 gave sufficient conditions for a numerical approximation to the optimal control problem for system (1.5) and cost function (1.6) to converge to the optimal value as the approximation parameter h goes to zero. But it does not give any hint as to how the approximation might be constructed so that the numerical procedure is actually reasonable from a computational perspective. Suppose that the process ξ^h_n is locally consistent and the transition probabilities satisfy (1.11). Because the transition probabilities in (1.11) depend on $\hat{\xi}$, a key problem is that the state space must include the information that is needed to define $\hat{\xi}$, and this might require considerable memory. The effective use of dynamic programming methods requires that the system (i.e., the memory) state be embedded into a finite-state Markov chain. The size and structure of this chain determines the numerical feasibility of the algorithm, and this is the subject of the rest of this section. The next section and Chapter 8 show some advantages of the implicit approximation method as well as of methods motivated by it. Keep in mind that the reflection directions depend only on the reflecting point, as the reflection directions do not depend on delayed values and are not controlled.

A first and crude Markov chain representation. We will begin the discussion of representations and approximations of the path memory segment with a rather crude form. Until further notice, continue to use the interpolation $\xi^h(\cdot)$ defined above (1.7a). Let us start with the memory state at step n being $\bar{\xi}^h_n$, defined in (1.7b), which we recall is a piecewise-constant function with $\bar{\xi}^h_n(\theta) = \xi^h(t^h_n + \theta)$, for $\theta \in [-\bar{\theta}, 0]$. All of its values must be in G_h, except possibly the most recent one, $\bar{\xi}^h_n(0) = \xi^h_n$, which can take values in either G_h or ∂G^+_h.

The $\bar{\xi}^h_n$ can be represented in terms of a finite-state Markov process as follows. Let $\overline{\Delta}^h = \inf_{\alpha,\hat{\xi}} \Delta t^h(\hat{\xi}, \alpha)$, where $\alpha \in U^h$ and $\hat{\xi} \in D(G_h; [-\bar{\theta}, 0])$. Suppose (w.l.o.g.) that $\bar{\theta}/\overline{\Delta}^h = K^h$ is an integer. The interpolated time interval $[t^h_n - \bar{\theta}, t^h_n]$ is covered by at most K^h intervals of length $\bar{\Delta}^h$. The reflection states do not appear in the construction of $\bar{\xi}^h_n(\theta)$, for $\theta < 0$, but it is possible that $\bar{\xi}^h_n(0) \in \partial G^+_h$. Suppose that $\xi^h_n \in G_h$. Let $\xi^h_{n,i}, i > 0$, denote the ith non-reflection state before step n, and $\Delta t^h_{n,i}$ the associated interpolation interval. Then we can represent $\bar{\xi}^h_n$ in terms of $\{(\xi^h_{n,K^h}, \Delta t^h_{n,K^h}), \ldots, (\xi^h_{n,1}, \Delta t^h_{n,1}), \xi^h_n\}$.

If $\xi_n^h \notin G_h$, so that it is a reflecting state, then to compute the transition probability to the next state the values of the path before step n are not needed and the above vector is still a complete description of the needed memory.

This new representation clearly evolves as a $(2K^h + 1)$-dimensional controlled Markov chain, although it will usually be much too complicated to be of any practical use for computation. If the interpolation interval $\Delta t^h(\hat{\xi}, \alpha)$ is not constant, then the construction of the $\bar{\xi}_n^h$ requires that we keep a record of the values of both the $\xi_i^h, \Delta t_i^h$, for the indices i that contribute to $\bar{\xi}_n^h$. The use of constant interpolation intervals simplifies this problem. Consider the special case where $\Delta t^h(\hat{\xi}, \alpha)$ is a constant. This would be the case if $\sigma(\cdot)$ were a constant and an approximation analogous to that in the example in Section 6.4 were used. Then the vector $\{(\xi_{n,K^h}^h, \ldots, \xi_{n,1}^h, \xi_n^h\}$ evolves as a Markov process and $\bar{\xi}_n^h$ is a piecewise-constant and right-continuous (except possibly at $\theta = 0$) interpolation of these values, with $\bar{\xi}_n^h(0) = \xi_n^h$. We can identify $\bar{\xi}_n^h$ with this vector without ambiguity.

Transforming to a constant interpolation interval. If $\Delta t^h(\hat{\xi}, \alpha)$ is not constant, then (6.2.7) showed how to transform the transition probabilities to yield a chain with a constant interpolation interval for the no-delay case, and we now write the analogous equations for the delay case. Let $\bar{p}^h(\cdot)$ denote the transition probabilities for the constant interpolation interval case and use the form (1.11) for $p^h(\hat{\xi}, \tilde{x}|\alpha)$. Suppose (w.l.o.g.) that a state does not transit to itself in that $p^h(\hat{\xi}, \hat{\xi}(0)|\alpha) = 0$. To get the transition probabilities $\bar{p}^h(\cdot)$ for the delay case with the constant interpolation interval $\bar{\Delta}^h$, use the analog of (6.2.7):

$$\begin{aligned} \bar{p}^h(\hat{\xi}, \tilde{x}|\alpha) &= p^h(\hat{\xi}, \tilde{x}|\alpha)\left(1 - \bar{p}^h(\hat{\xi}, \hat{\xi}(0)|\alpha)\right) \\ \bar{p}^h(\hat{\xi}, \hat{\xi}(0)|\alpha) &= 1 - \frac{\bar{\Delta}^h}{\Delta t^h(\hat{\xi}, \alpha)}. \end{aligned} \tag{2.1}$$

A one-dimensional example with a constant interpolation interval. Let $\Delta t^h(\hat{\xi}, \alpha) = \bar{\Delta}^h$, so that the interpolation interval is constant. Detailed examination of the memory vector suggests various ways of simplifying the state space. To simplify the presentation, until further notice we let $x(t)$ be one-dimensional with $G = [0, B]$, where $B > 0$ is assumed to be an integral multiple of the approximation parameter h. We assume that nonreflection states move only to their nearest neighbors. Then $G_h = \{0, h, \ldots, B\}$ and the reflection states are $\{-h, B + h\}$.

For $\hat{\xi}(0) \in G_h$, the Bellman equation for the process defined by this chain with cost (1.13) can be written as

$$V^h(\hat{\xi}) = \inf_{\alpha \in U^h}\left[e^{-\beta\bar{\Delta}^h}\sum_{\pm} p^h(\hat{\xi}, \hat{\xi}(0) \pm h|\alpha)V^h(\hat{y}^{\pm}) + k(\hat{\xi}, \alpha)\bar{\Delta}^h\right]. \tag{2.2a}$$

The terms $\hat{y}^{\pm}$ denote the functions on $[-\bar{\theta}, 0]$ that represent the memory segment at the next step, where the state of the chain is $\xi_1^h = \hat{\xi}(0) \pm h$. The values are obtained as follows:

$$\hat{y}^{\pm}(\theta) = \hat{\xi}(\theta + \bar{\Delta}^h), \ -\bar{\theta} \le \theta < -\bar{\Delta}^h,$$

$$\hat{y}^{\pm}(\theta) = \hat{\xi}(0), \ -\hat{\Delta}^h \le \theta < 0, \quad \hat{y}^{\pm}(0) = \hat{\xi}(0) \pm h.$$

If $\hat{\xi}(0)$ is a reflecting state, then there is no shift and only the value $\hat{\xi}(0)$ changes. It becomes ξ_1^h. In particular, if $\hat{\xi}(0) = -h$, then $\Delta t^h(\hat{\xi}, \alpha) = 0$ and

$$V^h(\hat{\xi}) = V^h(\hat{\xi}^+) + q_1 h, \tag{2.2b}$$

where $\hat{\xi}^+(\theta)$ equals $\hat{\xi}(\theta)$, except at $\theta = 0$ where $\hat{\xi}^+(0) = 0$. If $\hat{\xi}(0) = B + h$, then $\Delta t^h(\hat{\xi}, \alpha) = 0$ and

$$V^h(\hat{\xi}) = V^h(\hat{\xi}^-) + q_2 h, \tag{2.2c}$$

where $\hat{\xi}^-(\theta) = \hat{\xi}(\theta)$, except for $\theta = 0$, where $\hat{\xi}^-(0) = B$. Owing to the contraction due to the discounting, there is a unique solution to (2.2).

More simply, as noted above we can represent $\bar{\xi}_n^h$ unambiguously as

$$\bar{\xi}_n^h = (\xi_{n,K^h}^h, \ldots, \xi_{n,1}^h, \xi_n^h).$$

If $\xi_n^h \in G_h$, then we can represent $\bar{\xi}_{n+1}^h$ unambiguously as

$$\bar{\xi}_{n+1}^h = (\xi_{n,K^h-1}^h, \ldots, \xi_{n,1}^h, \xi_n^h, \xi_{n+1}^h).$$

If $\xi_n^h = -h$, then we can represent $\bar{\xi}_{n+1}^h$ unambiguously as

$$\bar{\xi}_{n+1}^h = (\xi_{n,K^h}^h, \cdots, \xi_{n,1}^h, 0),$$

and analogously if $\xi_n^h = B + h$. With this representation, the maximum number of possible values can be very large, up to $(B/h + 1)^{K^h}(B/h + 3)$ where, typically, $K^h = O(1/h^2)$.

Simplifying the state representation by using differences. The representation that is used for the memory segment in the above one-dimensional example requires a state space of enormous size. This can be reduced by using the standard data compression method of using only the current ξ_n^h and the differences between successive values. This gives the representation

$$\bar{\xi}_n^h = (c_{n,K^h}^h, \cdots, c_{n,1}^h, \xi_n^h),$$

where

$$\begin{aligned} c_{n,1}^h &= \xi_{n,1}^h - \xi_n^h \\ c_{n,i}^h &= \xi_{n,i}^h - \xi_{n,i-1}^h, \ \text{for } 1 < i \le K^h. \end{aligned} \tag{2.3}$$

If the path can move only its nearest neighbors, then the $c_{n,i}^h$ take at most two values, and the number of values in the state space is reduced to $2^{K^h}(B/h+3)$. The two values and the reconstruction of the $\xi_{n,i}^h$ from them are easily determined by an iterative procedure. For example, if $\xi_n^h = -h$, then $\xi_{n,1}^h = 0$. If $\xi_n^h = 0$, then $\xi_{n,1}^h \in \{0, h\}$. If ξ_n^h is not a reflecting or boundary value then $\xi_{n,1}^h = \xi_{n-1}^h = \xi_n^h \pm h$. If $\xi_{n,i}^h = 0$, then $\xi_{n,i-1}^h \in \{0, h\}$. If $\xi_{n,i}^h$ is not a boundary value (it cannot be a reflecting state), then $\xi_{n,i-1}^h = \xi_{n,i}^h \pm h$, and so forth.

If $\Delta t^h(\hat{\xi}, \alpha)$ is not constant, so that we need to use (2.1) to transform the transition probabilities, we then have the possibility of transitions from a state to itself, since $\bar{p}^h(\hat{\xi}, \hat{\xi}(0)|\alpha)$ might not now be zero. Because $\xi_{n+1}^h - \xi_n^h \in \{-h, 0, h\}$, each of the $c_{n,i}^h$ can take as many as three values and we have at most $(B/h+3)3^{K^h}$ points in the state space. Theorem 1.1 holds. Keep in mind that the memory state at time $n+1$ must be computable from the memory state at time n and the new value ξ_{n+1}^h. The use of differences reduces the memory requirements, but at the price of increased computation. It would be worthwhile to evaluate other data coding and compression schemes, even those with a small loss of information.

The approaches in Section 3 and in the next chapter use fewer intervals to cover $[-\bar{\theta}, 0]$ and have the promise of being more efficient in terms of memory requirements as they use approximations to the path over interpolation intervals that are larger than $\bar{\Delta}^h$.

7.2.2 Delay in Both Path and Control

Now suppose that both the control and the path are delayed, with the maximum delay being $\bar{\theta}$ for each. The memory requirements can be greatly increased. In this subsection, we suppose that $\Delta t_n^h = \bar{\Delta}^h$, a constant, and give a representation of the memory segment of the control process that is an analog of the representation that was used for the path in the previous subsection. The general case will be dealt with in the next chapter.

For illustrative purposes, let us continue to work with a one-dimensional example and the notation of the previous subsection. Let $u_{n,i}^h$ denote the control action that was used in the ith no-reflection step before step n. Let $p^h(\hat{\xi}, \hat{u}; \tilde{x}|\alpha)$ denote the probability $P\{\xi_1^h = \tilde{x} | \bar{\xi}_0^h = \hat{\xi}, \bar{u}_0^h = \hat{u}, u_0^h = \alpha\}$. Analogously to what was done in the previous subsection, the memory variables can be embedded into a Markov process, with values at time n being

$$\left\{(\xi_{n,K^h}^h, u_{n,K^h}^h), \ldots, (\xi_{n,1}^h, u_{n,1}^h), \xi_n^h\right\}.$$

The analog of (2.2a) with cost function (1.29) is, for $\hat{\xi}(0) \in G_h$,

$$V^h(\hat{\xi}, \hat{u}) = \inf_{\alpha \in U^h} \left[e^{-\beta \bar{\Delta}^h} \sum_{\pm} p^h(\hat{\xi}, \hat{u}; \hat{\xi}(0) \pm h|\alpha) V^h(\hat{y}^{\pm}, \hat{u}_\alpha) + \bar{k}(\hat{\xi}, \hat{u}, \alpha)\bar{\Delta}^h \right], \tag{2.4}$$

where $\hat{y}^{\pm}$ denotes the new "path memory sections" defined below (2.2a). The new "control memory segment" depends on the current choice of control, namely α. The interpolated form is $\hat{u}_\alpha$, defined by

$$\hat{u}_\alpha(\theta) = \hat{u}(\theta + \bar{\Delta}^h), \; -\bar{\theta} \le \theta < -\bar{\Delta}^h,$$
$$\hat{u}_\alpha(\theta) = \alpha, \; -\bar{\Delta}^h \le \theta < 0.$$

It can be unambiguously represented in the form

$$\hat{u}_\alpha = \left(u^h_{n,K^h-1}, \ldots, u^h_{n,1}, \alpha\right).$$

The reflecting states are treated as for the no-delay case. Because of the contraction due to the discounting, there is a unique solution to (2.4).

We can use the more efficient representation (2.3) for the path variable. However, the total memory requirements with this approach would be large, unless U^h itself can be approximated by only a few values. Suppose that $U = U^h$ consists of only the two points $\{0, 1\}$. Then the number of points needed to represent the control memory segment is 2^{K^h}, comparable to what was needed for the one-dimensional problem of the previous subsection where only the path was delayed. If only the control were delayed, then this crude representation for the control memory would be more acceptable.

The dynamics depend on delayed values of the control, but not the current value. In this case, $p^h(\hat{\xi}, \hat{u}; \tilde{x}|\alpha)$ does not depend on the current control choice α, and (2.4) simplifies to

$$V^h(\hat{\xi}, \hat{u}) = \inf_{\alpha \in U^h} \left[e^{-\beta \bar{\Delta}^h} \sum_{\pm} p^h(\hat{\xi}, \hat{u}; \hat{\xi}(0) \pm h) V^h(\hat{y}^{\pm}, \hat{u}_\alpha) + \bar{k}(\hat{\xi}, \hat{u}, \alpha)\bar{\Delta}^h \right]. \tag{2.5}$$

7.2.3 A Comment on Higher-Dimensional Problems

The discussion in the previous subsection concentrated on one-dimensional models. The representations of the memory all extend to higher-dimensional problems, but the required memory grows exponentially in the dimension. When the path only is delayed, there are representations that are analogous to (2.3). Consider a two-dimensional problem in a box $[0, B_1] \times [0, B_2]$, with the same path delay in each coordinate, no control delay, and discretization level h in each coordinate. The ξ^h_n in (2.3) is replaced by vector containing the current two-dimensional value of the chain. The difference $c_i = \xi^h_{n,i} - \xi^h_{n,i-1}$ is now a two-dimensional vector. The values can be computed iteratively, as for the one-dimensional case, but the somewhat boring details will not be presented here.

7.3 The Implicit Numerical Approximation: Path Delayed

The implicit method of constructing the approximating chain that was introduced in Section 6.5 can play an important role in reducing the memory requirements and state space size. It also serves as the basis of a variety of other useful approximations with memory requirements that are less than what was needed in Section 2.[7] The equations (6.5.6) provided a simple way of getting the transition probabilities and interpolation interval for the implicit approximation method directly from those for the explicit approximation method for the no-delay problem. The approach is the same for the problem with delays. In this section, we concentrate on the model where only the path is delayed. Further developments are in the next chapter.

Let $\delta > 0$ be the discretization interval for the time variable, with $h^2/\delta \to 0$ as $h \to 0, \delta \to 0$. As in Section 6.5, let $\xi_n^{h,\delta}$ denote the state process for the spatial component, $\phi_n^{h,\delta}$ that for the time variable, and define $\zeta_n^{h,\delta} = (\phi_n^{h,\delta}, \xi_n^{h,\delta})$. To get the transition probabilities, one starts with the delay form of (6.5.6), where the $p^h(\cdot)$ are defined as in Section 1.

7.3.1 Local Consistency and the Memory Segment

Transition probabilities. In this section, let $\bar{\xi}_{r,n}^{h,\delta}$ denote the path memory segment that is used at iterate n for the chain. It will replace the $\bar{\xi}_n^h$ that was used in Sections 1 and 2 and will be defined precisely in (3.8) after defining the transition probabilities and interpolations. As for the method of Sections 1 and 2, it is a function on $[-\bar{\theta}, 0]$ with the value at $\theta = 0$ being $\bar{\xi}_{r,n}^{h,\delta}(0) = \xi_n^{h,\delta}$. The canonical value of $\bar{\xi}_{r,n}^{h,\delta}$ is again denoted by $\hat{\xi}$. The subscript r is used owing to the relationship with the random delay approximation of (4.2.7). With the implicit approximation method, there are several possibilities for the interpolation that defines the memory segment, and the choice affects the computational complexity.

Let $p^{h,\delta}(\hat{\xi}, i\delta; \tilde{x}, i\delta|\alpha)$ denote the probability that $\xi_{n+1}^{h,\delta} = \tilde{x}$ and that $\phi_{n+1}^{h,\delta} = i\delta$, given all past data and $\bar{\xi}_{r,n}^{h,\delta} = \hat{\xi}$, $\phi_n^{h,\delta} = i\delta$, $u_n^{h,\delta} = \alpha$ (i.e., the time variable is not advancing). Let $p^{h,\delta}(\hat{\xi}, i\delta; \hat{\xi}(0), i\delta + \delta|\alpha)$ denote the probability that $\xi_{n+1}^{h,\delta} = \xi_n^{h,\delta}$ and $\phi_{n+1}^{h,\delta} = i\delta + \delta$, given all past data and the values $u_n^{h,\delta} = \alpha$, and $\bar{\xi}_{r,n}^{h,\delta} = \hat{\xi}$, $\phi_n^{h,\delta} = i\delta$, with $\hat{\xi}(0) = \xi_n^{h,\delta}$ (i.e., the time variable is advancing, and the spatial state does not change). These probabilities depend on the past only via the value of the current path memory segment $\hat{\xi}$.

[7] Since we do not know the rate of convergence as a function of the parameters of the various approximations, this assertion is not quantifiable at the present time, except by computations and simulations for selected problems.

Now, adapting the procedure that led to (6.5.6) to the delay case yields the transition probabilities and interpolation intervals $\Delta t^{h,\delta}(\hat{\xi},\alpha)$ for the $\zeta_n^{h,\delta} = (\phi_n^{h,\delta}, \xi_n^{h,\delta})$ process in terms of those for the ξ_n^h process as:

$$\begin{aligned} p^{h,\delta}\left(\hat{\xi}, i\delta; \tilde{x}, i\delta \middle| \alpha\right) &= p^h\left(\hat{\xi}, \tilde{x}\middle|\alpha\right)\left(1 - p^{h,\delta}\left(\hat{\xi}, i\delta; \hat{\xi}(0), i\delta+\delta\middle|\alpha\right)\right) \\ p^{h,\delta}\left(\hat{\xi}, i\delta; \hat{\xi}(0), i\delta+\delta\middle|\alpha\right) &= \frac{\Delta t^h(\hat{\xi},\alpha)}{\Delta t^h(\hat{\xi},\alpha)+\delta}, \end{aligned} \tag{3.1}$$

$$\Delta t^{h,\delta}(\hat{\xi},\alpha) = \frac{\delta \Delta t^h(\hat{\xi},\alpha)}{\Delta t^h(\hat{\xi},\alpha)+\delta}. \tag{3.2}$$

Redefine

$$\Delta t_n^{h,\delta} = \Delta t^{h,\delta}(\bar{\xi}_n^h, u_n^h), \quad t_n^{h,\delta} = \sum_{i=0}^{n-1} \Delta t_i^{h,\delta}. \tag{3.3}$$

An alternative form of the implicit process. An alternative construction allows both the spatial and time variable to change simultaneously. Then the transition probabilities for the spatial component is just (1.11), (1.12), the conditional probability that time advances at step n is just $\Delta t^h(\hat{\xi},\alpha)/\delta$, and the interpolation interval is $\Delta t^h(\hat{\xi},\alpha)$. This procedure is equivalent to reindexing the process determined by (3.1) by omitting the indices at which the time variable advances. The corresponding spatial path is that of the explicit procedure. This variation will be useful in Chapter 8.

Local consistency and dynamical representations. Define $\Delta\xi_n^{h,\delta} = \xi_{n+1}^{h,\delta} - \xi_n^{h,\delta}$ and the martingale differences

$$\beta_n^{h,\delta} = \left[\Delta\xi_n^{h,\delta} - E_n^{h,\delta}\Delta\xi_n^{h,\delta}\right] I_{\{\xi_n^{h,\delta}\in G_h\}},$$

$$\beta_{0,n}^{h,\delta} = (\phi_{n+1}^{h,\delta} - \phi_n^{h,\delta}) - E_n^{h,\delta}(\phi_{n+1}^{h,\delta} - \phi_n^{h,\delta}),$$

where $E_n^{h,\delta}$ is the expectation conditioned on the data to step n. Let $E_{\hat{\xi},n}^{h,\delta,\alpha}$ denote the expectation conditioned on the data to step n with $u_n^{h,\delta} = \alpha$ and $\bar{\xi}_{r,n}^{h,\delta} = \hat{\xi}$. Then, for $\xi_n^{h,\delta} \in G_h$, the definitions (3.1), (3.2), and (1.8) yield the analog of (6.5.7):

$$\begin{aligned} E_{\hat{\xi},n}^{h,\delta,\alpha}\Delta\xi_n^{h,\delta} &= b_h(\hat{\xi},\alpha)\Delta t^{h,\delta}(\hat{\xi},\alpha) = b(\hat{\xi},\alpha)\Delta t^{h,\delta}(\hat{\xi},\alpha) + o(\Delta t^{h,\delta}(\hat{\xi},\alpha)), \\ E_{\hat{\xi},n}^{h,\delta,\alpha}\beta_n^{h,\delta}[\beta_n^{h,\delta}]' &= a_h(\hat{\xi})\Delta t^{h,\delta}(\hat{\xi},\alpha) = a(\hat{\xi})\Delta t^{h,\delta}(\hat{\xi},\alpha) + o(\Delta t^{h,\delta}(\hat{\xi},\alpha)), \\ E_{\hat{\xi},n}^{h,\delta,\alpha}\left[\phi_{n+1}^{h,\delta} - \phi_n^{h,\delta}\right] &= \Delta t_n^{h,\delta}. \end{aligned} \tag{3.4}$$

The reflecting states are dealt with as in Section 1. The use of the process $\zeta_n^{h,\delta}$ leads to some intriguing possibilities for efficient representation of the memory data for the delay problem. Note that either the spatial variable $\xi_n^{h,\delta}$ changes or the time variable $\phi_n^{h,\delta}$ advances at each iteration, but not both. There are several choices for the timescale of the continuous-time interpolations. We will start by using the $\Delta t_n^{h,\delta}$ defined in (3.3) as the interpolation intervals, and construct $\xi^{h,\delta}(\cdot)$. Then we will define an interpolation with which it will be convenient to define the memory segment $\bar{\xi}_{r,n}^{h,\delta}$.

Let $\Delta z_n^{h,\delta}$ denote the reflection term at step n, with components $\Delta y_{i,n}^{h,\delta}$. Recall the definition of the time $d^{h,\delta}(\cdot)$ given in (6.5.23). Let $\xi^{h,\delta}(\cdot)$ and $\phi^{h,\delta}(\cdot)$ denote the continuous-time interpolations of the $\{\xi_n^{h,\delta}\}$ and $\{\phi_n^{h,\delta}\}$, resp., with the intervals $\{\Delta t_n^{h,\delta}\}$ when the path memory segments $\{\bar{\xi}_{r,n}^{h,\delta}\}$ are used. We always define $\phi_0^{h,\delta} = 0$. Then we can write

$$\xi_{n+1}^{h,\delta} = \xi_n^{h,\delta} + b_h(\bar{\xi}_{r,n}^{h,\delta}, u_n^{h,\delta})\Delta t_n^{h,\delta} + \beta_n^{h,\delta} + \Delta z_n^{h,\delta}, \tag{3.5}$$

$$\xi^{h,\delta}(t) = \xi_0^{h,\delta} + \sum_{i=0}^{d^{h,\delta}(t)-1} b_h(\bar{\xi}_{r,i}^{h,\delta}, u_i^{h,\delta})\Delta t_i^{h,\delta} + \sum_{i=0}^{d^{h,\delta}(t)-1} \beta_i^{h,\delta} + \sum_{i=0}^{d^{h,\delta}(t)-1} \Delta z_i^{h,\delta}, \tag{3.6}$$

$$\phi_{n+1}^{h,\delta} = \phi_n^{h,\delta} + \Delta t_n^{h,\delta} + \beta_{0,n}^{h,\delta}. \tag{3.7}$$

Interpolations using $\phi^{h,\delta}(\cdot)$ as the timescale. Definition of the memory segment $\bar{\xi}_{r,n}^{h,\delta}$. In analogy to the definition (1.7a), define $\bar{\xi}_n^{h,\delta} = \{\xi^{h,\delta}(t_n^{h,\delta} + \theta), \theta \in [-\bar{\theta}, 0]\}$. If $\bar{\xi}_{r,n}^{h,\delta}$ were simply a segment of the interpolated process $\xi^{h,\delta}(\cdot)$, say $\bar{\xi}_n^{h,\delta}$, then the issues concerning the number of required values of the memory variable that arose in Section 2 would arise here in the same way, and there would be no advantage in the use of the implicit approximation procedure. Consider the alternative where the time variables $\phi_n^{h,\delta}$ determine interpolated time, in that real (i.e., interpolated) time advances (by an amount δ) only when the time variable is incremented and it does not advance otherwise. This will be an analog of the "random" Approximation 4 defined by (4.2.7).

To make this precise, consider $\xi_n^{h,\delta}$ at only the times that $\phi_n^{h,\delta}$ changes. Suppose that $\bar{\theta}/\delta = Q_\delta$ is an integer. Recall the definition (6.5.13) where $v_0^{h,\delta} = 0$, and, for $n > 0$,

$$v_n^{h,\delta} = \inf\{i > v_{n-1}^{h,\delta} : \phi_i^{h,\delta} - \phi_{i-1}^{h,\delta} = \delta\}.$$

The path memory segment denoted by $\bar{\xi}_{r,n}^{h,\delta}$ is defined to be the function on $[\bar{\theta}, 0]$, with the following values: For any l and n satisfying $v_l^{h,\delta} \le n < v_{l+1}^{h,\delta}$, set

$$\bar{\xi}^{h,\delta}_{r,n}(0) = \xi^{h,\delta}_n,$$

$$\bar{\xi}^{h,\delta}_{r,n}(\theta) = \begin{cases} \xi^{h,\delta}_{v^{h,\delta}_l}, & \theta \in [-\delta, 0), \\ \vdots & \\ \xi^{h,\delta}_{v^{h,\delta}_{l-Q_\delta+1}}, & \theta \in [-\bar{\theta}, -\bar{\theta}+\delta). \end{cases} \tag{3.8}$$

Figure 3.1 illustrates the construction of $\xi^{h,\delta}(\cdot)$ and $\bar{\xi}^{h,\delta}_{r,n}(\cdot)$ for $\bar{\theta}/\delta = 3$ and $v^{h,\delta}_l \le n < v^{h,\delta}_{l+1}$, and where we define $\sigma^{h,\delta}_l = t^{h,\delta}_{v^{h,\delta}_l}$.

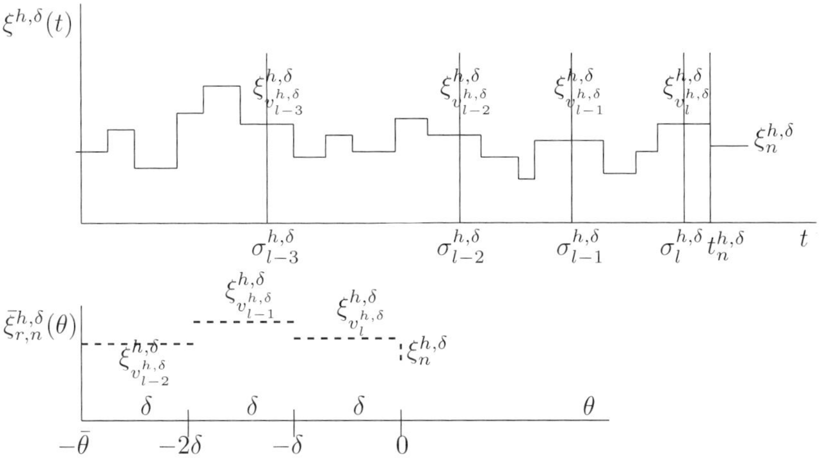

Figure 3.1. Illustration of $\bar{\xi}^{h,\delta}_{r,n}(\theta)$, for $v^{h,\delta}_l \le n < v^{h,\delta}_{l+1}$, $\bar{\theta}/\delta = 3$.

Recall that, in this section, $\hat{\xi}$ denotes the canonical value of the memory $\bar{\xi}^{h,\delta}_{r,n}$. It can be represented as the piecewise-constant right-continuous interpolation with interval δ of its values

$$\big(\hat{\xi}(-\bar{\theta}), \ldots, \hat{\xi}(-\delta), \hat{\xi}(0)\big)$$

with a discontinuity at $\theta = 0$, and we can unambiguously call the above set $\hat{\xi}$.

The possible transitions are as follows. If the time variable advances at the current step, then we have the shift

$$\big(\hat{\xi}(-\bar{\theta}), \ldots, \hat{\xi}(-\delta), \hat{\xi}(0)\big) \to \big(\hat{\xi}(-\bar{\theta}+\delta), \ldots, \hat{\xi}(-\delta), \hat{\xi}(0), \hat{\xi}(0)\big). \tag{3.9a}$$

This implies that $\hat{\xi}(0) \in G_h$, as otherwise there must be a reflection at the current step and the time variable could not advance. Let $\hat{\xi}(0) \in G_h$ and suppose that the time variable does not advance. Then

$$\hat{\xi} = \big(\hat{\xi}(-\bar{\theta}), \ldots, \hat{\xi}(-\delta), \hat{\xi}(0)\big) \to \big(\hat{\xi}(-\bar{\theta}), \ldots, \hat{\xi}(-\delta), \xi_1\big), \tag{3.9b}$$

where, conditioned on the time variable not advancing and the use of control value α, the probability that $\xi_1 = \tilde{x}$ is $p^h(\hat{\xi}, \tilde{x}|\alpha)$. Suppose that $\hat{\xi}(0) \notin G_h$, so that it is a reflecting point. Then

$$\hat{\xi} = (\hat{\xi}(-\bar{\theta}), \ldots, \hat{\xi}(-\delta), \hat{\xi}(0)) \to (\hat{\xi}(-\bar{\theta}), \ldots, \hat{\xi}(-\delta), \xi_1), \tag{3.9c}$$

where $\xi_1 \in G_h$ is the state that the reflecting state $\hat{\xi}(0)$ moves to, with the transition probabilities satisfying (6.2.2).

Size of the state space. For the one-dimensional problem discussed at the end of Section 2, the maximum size of the state space that is required with the use of $\bar{\xi}^{h,\delta}_{r,n}$ for the path memory segment is

$$(B/h + 1)^{\bar{\theta}/\delta}(B/h + 3) \tag{3.10}$$

compared with $(B/h+1)^{K^h}(B/h+3)$ there, where commonly $K^h = O(1/h^2)$. This saving is partly due to the fact that, for the implicit approximation procedure, the memory consists of the samples at iterates separated by many steps, and not the set of values or differences in the values for each of those individual steps. One could approximate the values of the $\hat{\xi}(-i\delta), i > 0$, further by discretizing to a coarser set of values. Further reductions in the size of the state space will be dealt with in the next chapter, where we discuss the advantages of using differences of the values in lieu of the values themselves, and also develop alternative constructions that are motivated by the implicit approximation procedure and are likely to be advantageous.

Note on the interpolation interval. An important additional point to note is that the implicit approximation procedure does not require the use of a constant interpolation time interval. It allows us to use the original time intervals $\Delta t^{h,\delta}_n \approx \Delta t^h_n$, and not the minimal value $\overline{\Delta}^h$. This is computationally advantageous when the values $\Delta t^h(\hat{\xi}, \alpha)$ vary a great deal, as for example when the upper bound on the control is large or when $a(\cdot)$ is not constant. In the example of Section 6.4, $\Delta t^h(x) = h^2/\sigma^2(x)$, and if $\sigma^2(\cdot)$ varies a great deal, the transformation to a constant interval might entail a considerable increase in the dimension of the memory segment $\bar{\xi}^h_n$ that was used in Section 2. The implicit approximation procedure does not have this disadvantage.

The effective maximum delay. The approximation procedure that we have just illustrated has replaced the true maximum delay by a random delay. The actual effective maximum delay for the example in the figure is $t^{h,\delta}_n - \sigma^{h,\delta}_{l-2}$. In general, for $\sigma^{h,\delta}_l \le t^{h,\delta}_n < \sigma^{h,\delta}_{l+1}$, the maximum delay is $t^{h,\delta}_n - \sigma^{h,\delta}_{l-Q_\delta+1}$. As $\delta \to 0$, the delays converge to their values for the original model (1.5). Let δ be fixed. It is shown in Theorem 4.1 that, as $h \to 0$, the interpolated times between increases in the time variable $\phi^{h,\delta}(\cdot)$ are exponentially distributed with mean δ. The interval between a random time and the most recent time before it that $\phi^{h,\delta}(\cdot)$ increased is also (asymptotically) exponentially distributed with mean δ, and these intervals are asymptotically mutually independent. Thus, as $h \to 0$, the maximum delay is the sum of Q_δ exponentially distributed and mutually independent random variables, each with mean δ. Hence it has an Erlang distribution of order Q_δ, and with total mean $\bar{\theta}$.

7.3.2 The Cost Function and Bellman Equation

With the use of the process $\zeta_n^{h,\delta}$, with $\bar{\xi}_{r,0}^{h,\delta} = \hat{\xi}$ and the control sequence $u^{h,\delta} = \{u_n^{h,\delta}, n < \infty\}$ used, an approximation to the discounted cost function (3.4.3) is

$$W^{h,\delta}(\hat{\xi}, u^{h,\delta}) = E_{\hat{\xi}}^{h,\delta,u^{h,\delta}} \sum_{n=0}^{\infty} e^{-\beta\phi_n^{h,\delta}} \left[k(\bar{\xi}_{r,n}^{h,\delta}, u_n^{h,\delta})\delta I_{\{\phi_{n+1}^{h,\delta} \neq \phi_n^{h,\delta}\}} + q'\Delta y_n^{h,\delta}\right]. \tag{3.11}$$

By using the last line of (3.4) and taking a conditional expectation, the term $\delta I_{\{\phi_{n+1}^{h,\delta} \neq \phi_n^{h,\delta}\}}$ can be replaced by $\Delta t_n^{h,\delta}$. It will be shown in Theorem 3.2 that (3.11) is well defined and is asymptotically equal to

$$E_{\hat{\xi}}^{h,\delta,u^{h,\delta}} \sum_{n=0}^{\infty} e^{-\beta t_n^{h,\delta}} \left[k(\bar{\xi}_{r,n}^{h,\delta}, u_n^{h,\delta})\Delta t_n^{h,\delta} + q'\Delta y_n^{h,\delta}\right]. \tag{3.12}$$

With the form (3.12), the effective canonical cost rate when the memory segment is $\hat{\xi}$ and control value α is used is just $k(\hat{\xi}, \alpha)$ times δ times the probability that the time variable advances, and the product is $k(\hat{\xi}, \alpha)\Delta t^{h,\delta}(\hat{\xi}, \alpha)$.

The Bellman equation. The Bellman equation can be based on either (3.11) or (3.12). They might yield different results but will be asymptotically equal by Theorem 3.2. For (3.11) and $\hat{\xi}(0) = \xi_0^{h,\delta} \in G_h$, the Bellman equation is[8]

$$\begin{aligned} V^{h,\delta}(\hat{\xi}) = \inf_{\alpha \in U^h} \Bigg[& \sum_{\tilde{x}} p^{h,\delta}\left(\hat{\xi}, \phi; \tilde{x}, \phi \middle| \alpha\right) V^{h,\delta}\left(\hat{\xi}(-\bar{\theta}), \ldots, \hat{\xi}(-\delta), \tilde{x}\right) \\ & + e^{-\beta\delta} p^{h,\delta}\left(\hat{\xi}, \phi; \hat{\xi}(0), \phi + \delta\right) V^{h,\delta}\left(\hat{\xi}(-\bar{\theta} + \delta), \ldots, \hat{\xi}(-\delta), \hat{\xi}(0), \hat{\xi}(0)\right) \\ & + k(\hat{\xi}, \alpha)\Delta t^{h,\delta}(\hat{\xi}, \alpha) \Bigg], \end{aligned} \tag{3.13}$$

where $V^{h,\delta}(\hat{\xi})$ is the optimal value. The analog for (3.12) can be written as (using a more succinct notation)

---

[8] The time variable ϕ does not appear in the state as the dynamical terms are time-independent.

$$V^{h,\delta}(\hat{\xi}) = \inf_{\alpha \in U^h} E_{\hat{\xi}}^{h,\delta,\alpha} \left[e^{-\beta \Delta t^{h,\delta}(\hat{\xi},\alpha)} V^{h,\delta}\left(\bar{\xi}_{r,1}^{h,\delta}\right) + k(\hat{\xi},\alpha)\Delta t^{h,\delta}(\hat{\xi},\alpha) \right], \tag{3.14}$$

where $\bar{\xi}_{r,1}^{h,\delta}$ is the successor memory segment to $\hat{\xi}$ under control value α. If $\hat{\xi}(0) \notin G_h$, then for either (3.11) or (3.12)

$$V^{h,\delta}(\hat{\xi}) = E_{\hat{\xi}}^{h,\delta,\alpha} \left[V^{h,\delta}(\bar{\xi}_{r,1}^{h,\delta}) + q' \Delta y_0^{h,\delta} \right], \tag{3.15}$$

where $\Delta y_0^{h,\delta}$ is the vector of the components of the reflection term from state $\hat{\xi}(0)$. These equations make it clear that the full state at iterate n is $\bar{\xi}_{r,n}^{h,\delta}$, namely, the current values of the spatial variable $\xi_n^{h,\delta}$, together with its value at the last $Q_\delta = \bar{\theta}/\delta$ times that the time variable advances.

7.3.3 The Use of Averaging in Constructing the Path Memory Approximation

One might be tempted to use an average of the path values over the intervals in lieu of the samples $\xi_{v_i^{h,\delta}}^{h,\delta}$ in (3.8). This can be done, but it entails a considerable increase in the memory requirements. One possibility is as follows. Let $v_l^{h,\delta} \le n < v_{l+1}^{h,\delta}$. Define

$$\xi_{av,l,n}^{h,\delta} = \frac{\sum_{i=v_l^{h,\delta}+1}^{n} \xi_i^{h,\delta} \Delta t_i^{h,\delta}}{\sum_{i=v_l^{h,\delta}+1}^{n} \Delta t_i^{h,\delta}}, \quad \xi_{av,l}^{h,\delta} = \frac{\sum_{i=v_l^{h,\delta}+1}^{v_{l+1}^{h,\delta}-1} \xi_i^{h,\delta} \Delta t_i^{h,\delta}}{\sum_{i=v_l^{h,\delta}+1}^{v_{l+1}^{h,\delta}-1} \Delta t_i^{h,\delta}}.$$

Then replace the $\xi_{v_l^{h,\delta}}^{h,\delta}$ in (3.8) by $\xi_{av,l,n}^{h,\delta}$, the path average over the interval $[v_l^{h,\delta}+1, n]$. Replace $\xi_{v_{l-i}^{h,\delta}}^{h,\delta}$ by $\xi_{av,l-i}^{h,\delta}$. The ratio can be computed recursively on each interval: At the beginning of the lth cycle, set $\xi_{av,l,v_l^{h,\delta}+1}^{h,\delta} = \xi_{v_l^{h,\delta}+1}^{h,\delta}$, and for $n > v_l^{h,\delta}+1$, use

$$\xi_{av,l,n+1}^{h,\delta} = \frac{\xi_{n+1}^{h,\delta} \Delta t_{n+1}^{h,\delta}}{\sum_{i=v_l^{h,\delta}+1}^{n+1} \Delta t_i^{h,\delta}} + \frac{\xi_{av,l,n}^{h,\delta}}{1 + \Delta t_{n+1}^{h,\delta} / \sum_{i=v_l^{h,\delta}+1}^{n} \Delta t_i^{h,\delta}}.$$

The computation is simpler if the interpolation interval is constant. In general, one needs to keep track of the running sums of the weighted path variables and the accumulated time, which introduces two new variables, one whose dimension is that of G. The set of such values will have to be discretized. For example, discretize the possible values and update the approximations by randomization if the new values fall between the allowable discrete points. The randomization method could be analogous to what is to be done in Section 8.4 for the periodic-Erlang approximation for the control variables. This will, in any case, yield a value that is a close to a convex combination of a subset of values within the interval, so it might be worth considering. Similar considerations apply to the approximations that are used for the path in Chapter 8, but the issue will not be pursued further.

A simpler procedure is to use a linear interpolation of the values in (3.8), which would not entail any increase in the required memory.

7.3.4 Timescales

The Interpolation $\psi^{h,\delta}(\cdot)$ and its timescale. The discrete-parameter process $\{\xi_n^{h,\delta}\}$ with memory segments $\{\bar{\xi}_{r,n}^{h,\delta}\}$ (or the variations discussed in the next chapter) are used for the numerical computations. The proofs of convergence in Section 8.5 will be based on a continuous-time process $\psi^{h,\delta}(\cdot)$ that is analogous to those defined by (1.14), (6.3.10), and (6.5.12), analogously to what was done in [58, Chapters 10 and 11]. Next, recalling the method of defining (1.14), let us define the interpolation $\psi_n^{h,\delta}(\cdot)$. Let $\nu_n, n < \infty$, be mutually independent and identically and exponentially distributed with unit mean (as above (6.3.4)), and independent of $\{\zeta_n^{h,\delta}, u_n^h\}$. Then set $\Delta\tau_n^{h,\delta} = \nu_n \Delta t_n^{h,\delta}$ and $\tau_n^{h,\delta} = \sum_{i=0}^{n-1} \Delta\tau_n^{h,\delta}$. Recall the definition of $d_\tau^{h,\delta}(s)$ from (6.5.23) and let $r_\tau^{h,\delta}(\cdot)$ denote the relaxed control representation of the interpolation (intervals $\Delta\tau_n^{h,\delta}$) of the control process. Analogously to what was done in getting (1.14), define the interpolation $\bar{\xi}_r^{h,\delta}(\cdot)$ (with intervals $\{\Delta t_n^{h,\delta}\}$) of the memory segment by $\bar{\xi}_r^{h,\delta}(s) = \bar{\xi}_{r,n}^{h,\delta}$, for $t_n^{h,\delta} \le s < t_{n+1}^{h,\delta}$, and set $q_\tau^{h,\delta}(s) = t_{d_\tau^{h,\delta}(s)}^{h,\delta}$. With these definitions, $\bar{\xi}_{r,d_\tau^{h,\delta}(s)}^{h,\delta} = \bar{\xi}_r^{h,\delta}(q_\tau^{h,\delta}(s))$. Let $\psi^{h,\delta}(\cdot)$ denote the interpolation of the sequence $\xi_n^{h,\delta}$ using the random intervals $\Delta\tau_n^{h,\delta}$. Then, analogously to (1.14),

$$\psi^{h,\delta}(t) = \xi_0^{h,\delta} + \int_0^t \int_{U^h} b_h(\bar{\xi}_{r,d_\tau^{h,\delta}(s)}^{h,\delta}, \alpha) r_\tau^{h,\delta}(d\alpha\, ds) + B_\tau^{h,\delta}(t) + z_\tau^{h,\delta}(t), \quad (3.16)$$

where the drift term can be written as

$$\int_0^t \int_{U^h} b_h(\bar{\xi}_r^{h,\delta}(q_\tau^{h,\delta}(s)), \alpha) r_\tau^{h,\delta}(d\alpha\, ds),$$

and the quadratic variation of the martingale $B_\tau^{h,\delta}(\cdot)$ is

$$\int_0^t a_h(\bar{\xi}_r^{h,\delta}(q_\tau^{h,\delta}(s)))ds.$$

Asymptotic equivalence of the timescales. It follows from the proof of Theorem 6.5.1 that the timescales used in the $\xi^{h,\delta}(\cdot)$ and the $\psi^{h,\delta}(\cdot)$ processes coincide asymptotically. That is, $q^{h,\delta}(s) - s \to 0$, $\phi^{h,\delta}(s) - s \to 0$ and $q_\tau^{h,\delta}(s) -$

$s \to 0$. The following theorem reasserts this result in the context of the current chapter.

Theorem 3.1. *Assume local consistency,* (A3.1.1), (A3.1.2), (A3.2.1), (A3.2.2) *and* (A3.4.3), *with system* (1.5) *and memory segment* (3.8). *Let* $\phi^{h,\delta}(\cdot)$ *denote the interpolation of the* $\phi_n^{h,\delta}$ *with the intervals* $\Delta t_n^{h,\delta}$, *and suppose that* h/δ *is bounded as* $h \to 0$ *and* $\delta \to 0$. *Then Theorem* 6.5.1 *holds and for each* $T < \infty$,

$$\lim_{h,\delta\to 0} \sup_{\hat{\xi},u^{h,\delta}} E_{\hat{\xi}}^{h,\delta,u^{h,\delta}} \sup_{-\bar{\theta}\le\theta\le 0, t\le T} \left| \psi^{h,\delta}(t+\theta) - \bar{\xi}^{h,\delta}_{d_\tau^{h,\delta}(t)}(\theta) \right| = 0. \tag{3.17}$$

If the memory segments $\bar{\xi}_n^h$ *are used, as in Section* 2, *then the index* δ *is redundant and we have*

$$\lim_{h\to 0} \sup_{\hat{\xi},u^h} E_{\hat{\xi}}^{h,u^h} \sup_{-\bar{\theta}\le\theta\le 0, t\le T} \left| \psi^h(t+\theta) - \bar{\xi}^h_{d_\tau^h(t)}(\theta) \right| = 0. \tag{3.18}$$

An alternative construction of the implicit procedure. Time and spatial variables changing simultaneously. Recall the comments on the alternative construction of an implicit procedure below (3.3), where we allowed the possibility that both the path and time variables change simultaneously. With the memory segment taking any of the forms that were discussed, the resulting processes and costs are asymptotically equivalent to those for the implicit procedure.

7.3.5 Convergence Theorems

The next theorem asserts that the cost functions (3.11) and (3.12) are well defined and asymptotically equal.

Theorem 3.2. *Assume local consistency,* (A3.1.1), (A3.1.2), (A3.2.1), (A3.2.2), *and* (A3.4.3), *and the model* (1.5) *with memory segment* (3.8). *Then* (3.11) *is asymptotically equal to* (3.12) *uniformly in the control and in the initial condition* $\hat{\xi}$, *where the function* $\hat{\xi}$ *is piecewise-constant, with intervals* δ *and with values in* G_h.

Proof. To show that the sum involving $k(\cdot)$ in (3.11) is well defined, first note that it can be bounded by a constant times the expectation of $\int_0^\infty e^{-\beta\phi^{h,\delta}(s)}ds$. By Theorem 3.1 or Theorem 6.5.1, for each $K > 0$ there is an $\epsilon_1 > 0$, which does not depend on the controls, initial conditions, or T, such that for small enough h, δ,

$$P\left\{\phi^{h,\delta}(T+K) - \phi^{h,\delta}(T) \ge \epsilon_1 \middle| \text{data to } T\right\} > \epsilon_1, \quad \text{w.p.1.}$$

Hence, for each $K > 0$ there is $\epsilon_2 > 0$, not depending on the controls, initial conditions, or T, such that for small enough h, δ,

$$E\left[e^{-\beta(\phi^{h,\delta}(T+K)-\phi^{h,\delta}(T))}\Big|\text{data to } T\right\} \leq e^{-\epsilon_2} \quad \text{w.p.1.}$$

This implies that the "tail" of the sum (3.11) can be neglected and we need only consider the sum $\sum_{i=0}^{N^{h,\delta}(t)}$ where $N^{h,\delta}(t) = \min\{n : t_n^{h,\delta} \geq t\}$ for arbitrary t. But, by Theorem 3.1 or Theorem 6.5.1, for such a sum the asymptotic values are the same if $\phi_i^{h,\delta}$ is replaced by $t_i^{h,\delta}$ for $i \leq N^{h,\delta}(t)$. Hence the terms involving $k(\cdot)$ in (3.11) and (3.12) are asymptotically equal. The above estimates and Lemma 6.3.1 yield the same result for the terms involving $\Delta y_n^{h,\delta}$. ■

The convergence theorem. As in Theorem 1.1, approximate the initial condition $\bar{x}(0)$ by $\bar{\xi}_0^{h,\delta}$ (in the sense of uniform convergence as $h \to 0, \delta \to 0$), and let it be constant on the Q_δ intervals $[-\bar{\theta}, -\bar{\theta}+\delta), \ldots, [-\delta, 0)$, with all values being in G_h. Because by Theorem 3.2 we can use (3.12) for the cost function when proving convergence, the proof of the next theorem is nearly identical to that of Theorem 1.1, which is to be given in Section 8.5.

Theorem 3.3. *Assume local consistency,* (A3.1.1), (A3.1.2), (A3.2.1)–(A3.2.3), *and* (A3.4.3), *with system* (1.5) *and cost function* (3.4.3). *The memory segment for the numerical approximation is* (3.8). *Let* $p^{h,\delta}(\cdot)$ *be derived via* (3.1)–(3.3) *from the transition probabilities* $p^h(\cdot)$ *that are locally consistent (in the sense of* (1.8)). *Let* $\bar{\xi}_0^{h,\delta}$ *approximate the continuous initial condition* $\hat{x}$ *as in Theorem* 1.1. *Let* h/δ *be bounded. With either* (3.11) *or* (3.12) *used,* $V^{h,\delta}(\bar{\xi}_0^{h,\delta}) \to V(\hat{x})$ *as* $h \to 0, \delta \to 0$. *The analogous result holds for the analog of the cost functional* (1.28) *if* (A3.4.1) *and* (A3.4.2) *are assumed and the conditions on the reflection directions are dropped.*

7.4 The Implicit Approximation Procedure and the Random Delay Model

The intervals between time advances, δ fixed. Consider the implicit approximation procedure of Section 3, with the value of δ fixed and only $h \to 0$. The following theorem shows that the sequence of times between shifts in the time variable converges to a sequence of i.i.d. random variables, each of which is exponentially distributed with mean δ. Define $\hat{\sigma}_l^{h,\delta} = \tau_{v_l^{h,\delta}}^{h,\delta}$, where $v_l^{h,\delta}$ was defined above (3.8). Recall the definition of $\sigma_l^{h,\delta} = t_{v_l^{h,\delta}}^{h,\delta}$ below (3.8).

In the theorem, we ignore the time-shift steps in the indexing. This does not change the distribution of the quantities of interest. The resulting path is that for the explicit procedure if the same controls are used.

Theorem 4.1. *Assume the model of Section 3 and that $\Delta t^h(\hat{\xi},\alpha) = O(h^2)$, with the assumptions of Theorem 3.3, but with δ fixed. As $h \to 0$, $\phi^{h,\delta}(\cdot)$ converges to a Poisson process with rate $1/\delta$ and jump size δ, and this process on $[t,\infty)$ is independent of the other weak-sense limits on $[0,t]$. The conditional distribution of $\sigma_{l+1}^{h,\delta} - \sigma_l^{h,\delta}$, given the data to time $\sigma_l^{h,\delta}$, converges to an exponentially distributed random variable with mean δ, and the conditional mean value converges to δ, all uniformly in the data and l. Now let $\delta = O(h)$ and replace $\sigma_{l+1}^{h,\delta} - \sigma_l^{h,\delta}$ by $[\sigma_{l+1}^{h,\delta} - \sigma_l^{h,\delta}]/\delta$. Then the results of the previous sentence hold, but with mean unity. The analogous results hold if the $\hat{\sigma}_l^{h,\delta}$ are used in lieu of the $\sigma_l^{h,\delta}$.*

Proof. Fix $\delta > 0$. Let $R(\cdot)$ be a Poisson process with rate $1/\delta$ and jump size δ. Approximate $\phi^{h,\delta}(\cdot)$ as follows. For each n, if $R(\cdot)$ has multiple jumps on $[t_n^{h,\delta}, t_{n+1}^{h,\delta})$, then ignore any jump beyond the first, and assign the remaining jump (if any) to time $t_n^{h,\delta}$. The difference between this process and both $\phi^{h,\delta}(\cdot)$ and $R(\cdot)$ converges weakly to zero as $h \to 0$. This yields the first two assertions of the theorem. The assertion concerning the convergence of the conditional mean follows from this and the uniform integrability of $\{\sigma_{l+1}^{h,\delta} - \sigma_l^{h,\delta};\, h,l,\delta\}$. If $\delta = O(h)$, then the result for the $[\sigma_{l+1}^{h,\delta} - \sigma_l^{h,\delta}]/\delta$ follows by a rescaling of time and amplitude. A similar argument is used if the $\hat{\sigma}_l^{h,\delta}$ are used in lieu of the $\sigma_l^{h,\delta}$. ■

Convergence to the random delay model if δ is fixed. If $\delta > 0$ is fixed and only $h \to 0$, then the limit is the optimal value for the Approximation 4 of Section 4.2. This assertion follows from Theorem 4.1 and the proof of Theorem 3.3 (see Section 8.5) and is stated in the following theorem.

Theorem 4.2. *Let the initial condition for (1.1) be $\bar{x}(0)$, assumed to be continuous and G-valued. Let the G-valued piecewise constant $\bar{x}^\delta(0)$ (intervals δ) converge to $\bar{x}(0)$ uniformly on $[-\bar{\theta},0]$. Assume the conditions of Theorem 3.3, but with the memory segment defined by (4.2.7), the random case. Hence the model is (4.1.9b), with $\bar{x}_a = \bar{x}_r^\delta$, for which we suppose that there is a weak-sense unique solution for each control and initial condition $\bar{x}^\delta(0)$. Let $V^\delta(\hat{\xi})$ denote the optimal cost for this model. Let $\bar{\xi}_0^{h,\delta}$ approximate $\bar{x}^\delta(0)$ (in the sense of uniform convergence as $h \to 0$), with values in G_h, and use (3.1) and (3.2) for the transition probabilities and interpolation intervals. Then, with $\delta > 0$ fixed, $V^{h,\delta}(\bar{\xi}_0^{h,\delta}) \to V^\delta(\bar{x}^\delta(0))$. As $\delta \to 0$, $V^\delta(\bar{x}^\delta(0)) \to V(\bar{x}(0))$. The same results hold for the analog of the cost function (1.28) for the implicit procedure.*

Consider the analog of Theorem 1.4 *for the implicit procedure. Then, under the analogs of its conditions for the implicit method and the path memory segment* $\bar{\xi}^{h,\delta}_{r,n}$ *used in lieu of* $\bar{\xi}^{h}_{n}$, *its conclusions hold.*

Comment. The implicit approximation algorithm illustrates one way of reducing the memory requirement over that needed for the procedure of Sections 1 or 2. In addition, one does not need the interval $\Delta t^h(\hat{\xi}, \alpha)$ to be constant, which is a considerable advantage when $\sigma(\cdot)$ is either small or is not a constant. The motivation for the implicit approximation procedure was the desire for a simpler representation of the path memory segment for the approximating process. However, the randomness of the effective delays with this procedure might be too large unless δ is small. Approximations that aim at compromises between the explicit procedure of Section 1 and the implicit approximation procedure will be discussed in the next chapter. Reliable numerical comparisons are still lacking, however.

8

Path and Control Delayed: Continued

8.0 Outline of the Chapter

It is clear that one needs to start with good approximations to the original model in order to deal effectively with the memory problem, and this led to the implicit procedure in Section 7.3 and to general conditions on approximations in Theorem 7.1.4. The development is continued in this chapter. Chapter 4 developed a variety of approximations for the original model. The "random delay" form (4.2.7) led to the implicit approximation of Section 7.3. Other interesting approximations were the periodic (4.2.6) and the periodic-Erlang in (4.2.8). The periodic model in Chapter 4 was developed as a first step to a finite-dimensional memory for the case where the path only was delayed. In that model, for each t the path memory segment $\bar{x}_p^\delta(t)$ that is used in the dynamics is a function on $[-\bar{\theta}, 0]$ with values $\bar{x}_p^\delta(t, \theta)$. It is piecewise-constant in θ, but with a discontinuity at $\theta = 0$, where it takes the value $x(t)$. In each of the time intervals $[l\delta, l\delta + \delta), l = 0, 1 \ldots$, only the value at $\theta = 0$ changed, as it is $x(t)$. At times $l\delta, l = 1, 2 \ldots$, the entire segment shifted left, dropping the "oldest" part. An issue was the measurement of the passage of time between the "shifts" of the memory vector. Because this variable takes values in the continuum $[0, \delta)$, it needed to be discretized in some way. One way was via the periodic-Erlang method, where time advanced at random moments, and which was motivated by the anticipated demands of the numerical problem. It was seen in Section 4.3 that both the periodic and the periodic-Erlang approximation to the original model can be quite good. But, as noted previously, given the great sensitivity of many problems with delays, one always needs to exercise care in the use of any particular approximation, although the numerical problem will be very hard for any model with great sensitivity.

This chapter adapts these approximations to the numerical problem. The periodic model is introduced in Section 1, when only the path is delayed. Although the method does reduce the memory requirements for the path memory segment, it still has the problem of tracking the time since the last

H.J. Kushner, *Numerical Methods for Controlled Stochastic Delay Systems*,
doi: 10.1007/978-0-8176-4621-9_8,

shift. A step to remedying this is developed in Section 2, where an analog of the periodic-Erlang method is introduced. This handles the measurement of the passage of time when only the path is delayed, and it is seen that there can be considerable savings in memory. The periodic approximation that was used in Chapter 4 was based on the particular discretization of the delay interval $[-\bar{\theta}, 0]$ that was illustrated in Figure 4.2.3, where the last interval on the right has length $\delta/2$, and the others have length δ. As noted below (4.2.6) and (4.2.8), the choice of $\delta/2$ for the rightmost interval was made because the effective delay over the interval varied between zero and δ due to the periodicity, and its mean value is $\delta/2$. Other choices that maintain the correct mean could be used as well. For ease of reference to the models and results of Chapter 4, we keep this form.

With these methods, one needs to keep track of the path values at the shift times as well as the current path value. Memory can be saved by working with the differences of the path values at the successive shift times, and this is discussed in Section 3.

Owing to the continuity properties of the solution path, there are many useful ways to approximate the path memory segment. Whatever the means of doing it, the approximation becomes a state variable for the problem, and it must be able to be embedded into a finite-state Markov chain so that dynamic programming methods can be used. This fundamental fact limits our freedom and was a consideration in the choice of the methods that are discussed.

The control process does not necessarily have the continuity properties of the path. In fact, little is usually known about it *a priori.* Because of this, when the control is delayed, one cannot approximate the control memory segment by simply sampling the control and interpolating, unless all the control values are used as in Subsection 7.2.2. However, the relaxed control representation is continuous in time and can provide a useful approach to approximation. This idea is developed in Section 4 for the periodic and periodic-Erlang models. The methods that are presented should be taken as suggestive and tentative. Much more work is needed. Section 5 contains proofs of the theorems of Chapter 7 and of Sections 1, 2, and 4 of this chapter. Section 6 concerns the singular control problem. If the singular control is not delayed in the dynamics, then the numerical approximations are essentially a simple combination of what was done in Chapter 6 and in this chapter. When the singular control is delayed in the dynamics, then one needs an additional approximation step. Section 7 contains some remarks on neutral equations. It is seen that the approximations developed previously can be carried over. Section 8 concerns the ergodic cost problem.

8.1 Periodic Approximations to the Delay: Path Delayed

For simplicity in the development, we start with a chain approximation to the periodic approximation of the original model that was given in Section 4.2, where the maximum delay was periodic in $[\bar{\theta} - \delta/2, \bar{\theta} + \delta/2]$, and extend it in subsequent sections. It is to be based on the explicit procedure of Section 7.1. Define Q_δ^+ by $\bar{\theta} = Q_\delta^+ \delta + \delta/2$ and suppose that it is an integer. We start by supposing that $\Delta t^h(\hat{\xi}, \alpha) = \bar{\Delta}^h$, a constant. This assumption will be commented on later and dropped in the next section. Define $\bar{L}^{h,\delta} = \delta/\bar{\Delta}^h$ and suppose that it is an integer. Because the processes involved in the approximating chain will depend on the two parameters h and δ, they will be indexed by both. So we use the terminology $\xi_n^{h,\delta}, u_n^{h,\delta}$, and so forth. Hopefully, there will be no confusion with the implicit approximation procedure of Section 7.3.

The memory segment for the approximating chain is based on the construction in (4.2.6) and Figure 4.2.3. Its value at step n is denoted by $\bar{\xi}_{p,n}^{h,\delta}$, and is defined as follows. Define $t_n^{h,\delta} = \sum_{i=0}^{n-1} \Delta t^h(\bar{\xi}_{p,i}^{h,\delta}, u_i^h)$ and $\nu_l^{h,\delta} = l\delta/\bar{\Delta}^h$, the number of (nonreflection, if any) iterates of the chain that are required for an interpolated time interval of length $l\delta$. Let $l\delta \leq t_n^{h,\delta} < l\delta + \delta$. Then define the memory segment by:

$$\bar{\xi}_{p,n}^{h,\delta}(0) = \xi_n^{h,\delta},$$

$$\bar{\xi}_{p,n}^{h,\delta}(\theta) = \begin{cases} \xi^{h,\delta}(l\delta) = \xi_{\nu_l^{h,\delta}}^{h,\delta}, & \theta \in [-\delta/2, 0), \\ \xi^{h,\delta}(l\delta - \delta) = \xi_{\nu_{l-1}^{h,\delta}}^{h,\delta}, & \theta \in [-\delta/2 - \delta, -\delta/2), \\ \vdots & \\ \xi^{h,\delta}(l\delta - Q_\delta^+\delta) = \xi_{\nu_{l-Q_\delta^+}^{h,\delta}}^{h,\delta}, & \theta \in [-\bar{\theta}, -\bar{\theta} + \delta). \end{cases} \tag{1.1}$$

The construction is illustrated in Figure 1.1 and is only one of the many asymptotically consistent possibilities for constructing an approximation with the desired period.

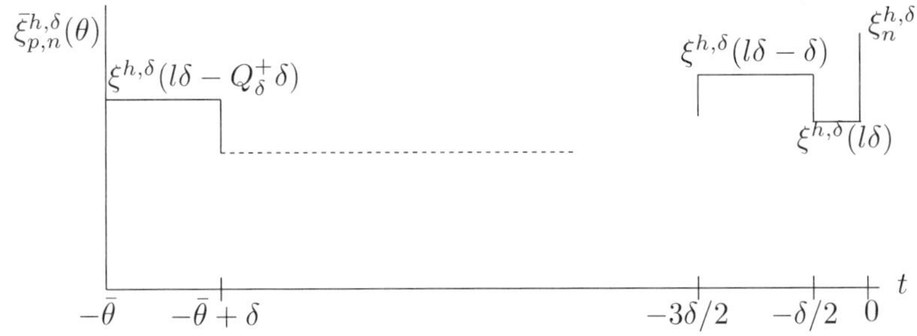

Figure 1.1. Illustration of $\bar{\xi}_{p,n}^{h,\delta}$, periodic delay model, $l\delta \leq t_n^{h,\delta} < l\delta + \delta$.

For $l\delta \leq t_n^{h,\delta} < l\delta + \delta$, the path memory segment can be unambiguously identified with the vector $\left(\xi_{\nu_{l-Q_\delta^+}^{h,\delta}}^{h,\delta}, \ldots, \xi_{\nu_l^{h,\delta}}^{h,\delta}, \xi_n^{h,\delta} \right)$.

Notes on computation. For the random memory segment model of (7.3.8), we needed to keep track only of the values of the state at the current iterate

n, namely, $\xi_n^{h,\delta}$, and at the previous Q_δ iterates at which the time variable $\phi_i^{h,\delta}$ increased. Owing to the random way that the time variable advanced, there was no need to keep track of the elapsed time since the last time shift as it provided no information on the time of the next increase in the time variable. For the model of this section, at iterate n, where $l\delta \le t_n^{h,\delta} < l\delta + \delta$, we need to keep track of $n - l\delta/\bar{\Delta}^h$, the number of (nonreflection, for the reflecting boundary model) iterates that have passed since the last time shift, the last time that $t_n^{h,\delta}$ equalled δ times an integer. This new variable takes $\bar{L}^{h,\delta}$ values. This increases the size of the state space for the dynamic programming problem, but it reduces the randomness in the algorithm.

Let $L_n^{h,\delta} = n(\text{mod } \bar{L}^{h,\delta})$ denote the number of iterates since the last shift. For $l\delta \le t_n^{h,\delta} < l\delta + \delta$, $\xi_n^h \in G_h$, and $L_n^{h,\delta} < \bar{L}^{h,\delta} - 1$, the state transitions are

$$\left(\xi^{h,\delta}_{\nu^{h,\delta}_{l-Q^+_\delta}}, \ldots, \xi^{h,\delta}_{\nu^{h,\delta}_l}, \xi^{h,\delta}_n\right) \to \left(\xi^{h,\delta}_{\nu^{h,\delta}_{l-Q^+_\delta}}, \ldots, \xi^{h,\delta}_{\nu^{h,\delta}_l}, \xi^{h,\delta}_{n+1}\right); \quad L^{h,\delta}_{n+1} = L^{h,\delta}_n + 1. \tag{1.2a}$$

For $L_n^{h,\delta} = \bar{L}^{h,\delta} - 1$, $t_{n+1}^{h,\delta} = l\delta + \delta$ and the transitions are

$$\left(\xi^{h,\delta}_{\nu^{h,\delta}_{l-Q^+_\delta}}, \ldots, \xi^{h,\delta}_{\nu^{h,\delta}_l}, \xi^{h,\delta}_n\right) \to \left(\xi^{h,\delta}_{\nu^{h,\delta}_{l-Q^+_\delta+1}}, \ldots, \xi^{h,\delta}_{\nu^{h,\delta}_l}, \xi^{h,\delta}_{n+1}, \xi^{h,\delta}_{n+1}\right); \quad L^{h,\delta}_{n+1} = 0. \tag{1.2b}$$

Nonconstant $\Delta t^h(\hat{\xi}, \alpha)$. Suppose that the values of $\Delta t^h(\hat{\xi}, \alpha)$ can be approximated as integral multiples of some number $\bar{\Delta}^h$, with $\delta/\bar{\Delta}^h = \bar{L}^{h,\delta}$ being an integer. Then the same procedure can be used, the only change being that the value of $L_n^{h,\delta}$ might increase by more than one at some steps. It is set to zero when it reaches or exceeds $\bar{L}^{h,\delta}$. If its assumptions are adjusted to the model of this section, then Theorem 7.3.3 holds, where the limit as $h \to 0$ (and δ is fixed) is the model with memory segment (4.2.6). If, for the chosen value of $\bar{\Delta}^h$, there are values of $\hat{\xi}$ and α where $\Delta t^h(\hat{\xi}, \alpha)$ is not an integral multiple of it, then one can update $L_n^{h,\delta}$ there by a randomization procedure, so that the mean increase is $\Delta t^h(\hat{\xi}, \alpha)/\bar{\Delta}^{h,\delta}$. We omit the details, as it is partly covered by the procedure in the next section, which requires less memory.

8.2 A Periodic-Erlang Model

The increase in the required memory for the periodic memory segment form (1.1) (over what is needed for the random memory segment model of Section 7.3) is due to the need to keep track of the time that has elapsed since the last shift. If the interpolation interval is constant and $\bar{L}^{h,\delta}$ is very large, then the memory requirements might be quite large. If the interpolation interval is not constant and/or we don't wish to transform the problem into the constant interval form, then we can approximate the time variable part of the memory state by a randomization procedure that will be described shortly.

In Section 7.3, the transition probability (7.3.1) was used to determine the process of shifts of the time variable that was used to construct the interpolated path memory segment. The probability that time advanced by δ at any step was the second line of (7.3.1), which is approximated by $\Delta t^h(\hat{\xi}, \alpha)/\delta$ if this quantity is small. For the process that was constructed with the transition probabilities (7.3.1), either the spatial or time component changed at each step, but not both. In the paragraph below (7.3.3), the modification where both the spatial state and the time variable could change simultaneously was noted, in which case the conditional probability that time advances is $\Delta t^h(\hat{\xi}, \alpha)/\delta$, always assumed to be no larger than unity.. This variation will be adapted for use in this section.

The memory approximation procedure. Divide the interval $[0, \delta]$ into subintervals of length δ_0, where $\delta/\delta_0 = \bar{L}^{\delta_0,\delta}$ is an integer. In the first part of Section 1, where the explicit procedure was used, the time interval was the constant $\bar{\Delta}^h$, hence interpolated time advanced by $\bar{\Delta}^h$ at each step. In this section, we use an adaptation of the method of Section 7.3, but with δ_0 replacing δ and allow the possibility that the spatial and temporal states change simultaneously. After each $\bar{L}^{\delta_0,\delta}$ of the δ_0-shifts, time is advanced by δ and the memory segment structure shifts. By Theorem 7.4.1, for δ and δ_0 fixed, as $h \to 0$ the time between the main δ-shifts becomes Erlang of order $\bar{L}^{\delta_0,\delta}$ and total mean δ. Because the chain, control, and time processes depend on the three quantities h, δ and δ_0, we index them by this triple and use the notation $\xi_n^{h,\delta.\delta_0}$, and so forth, and

$$\Delta t_n^{h,\delta_0,\delta} = \Delta t^h(\bar{\xi}_n^{h,\delta_0,\delta}, u_n^{h,\delta_0,\delta}).$$

More precisely, we use the following procedure, guided by the periodic-Erlang model of Approximation 5 in (4.2.8). Use the explicit approximation procedure for the evolution of the $\xi_n^{h,\delta_0,\delta}$. Set $L_0^{h,\delta_0,\delta} = 0$ and let $L_n^{h,\delta_0,\delta} \in \{0, 1, \ldots, \bar{L}^{\delta_0,\delta} - 1\}$ denote the number of δ_0-shifts (mod $\bar{L}^{\delta_0,\delta}$) up to iterate n. Let $\sup_{\hat{\xi},\alpha} \Delta t^h(\hat{\xi}, \alpha)/\delta_0 \leq 1$. For $L_n^{h,\delta_0,\delta} < \bar{L}^{\delta_0,\delta} - 1$, we have $L_{n+1}^{h,\delta_0,\delta} = L_n^{h,\delta_0,\delta} + 1$ with conditional probability $\Delta t_n^{h,\delta_0,\delta}/\delta_0$, and $L_{n+1}^{h,\delta_0,\delta} = L_n^{h,\delta_0,\delta}$ otherwise. For $L_n^{h,\delta_0,\delta} = \bar{L}^{\delta_0,\delta} - 1$, $L_{n+1}^{h,\delta_0,\delta} = 0$ with conditional probability $\Delta t_n^{h,\delta_0,\delta}/\delta_0$, and $L_{n+1}^{h,\delta_0,\delta} = L_n^{h,\delta_0,\delta}$ otherwise. Define $m_0^{h,\delta_0,\delta} = 0$ and, for $i > 0$, define

$$m_i^{h,\delta_0,\delta} = \min\left\{n > m_{i-1}^{h,\delta_0,\delta} : L_n^{h,\delta_0,\delta} = 0, L_{n-1}^{h,\delta_0,\delta} = \bar{L}^{\delta_0,\delta} - 1\right\}.$$

Let $m_l^{h,\delta_0,\delta} \leq n < m_{l+1}^{h,\delta_0,\delta}$. If $m_l^{h,\delta_0,\delta}$ is used as a subscript, then we might write it simply as m_l. The path memory segment $\bar{\xi}_{e,n}^{h,\delta_0,\delta}$ that is analogous to (4.2.8) is defined by:

$$\bar{\xi}_{e,n}^{h,\delta_0,\delta}(0) = \xi_n^{h,\delta_0,\delta},$$

$$\bar{\xi}_{e,n}^{h,\delta_0,\delta}(\theta) = \begin{cases} \xi_{m_l}^{h,\delta_0,\delta}, & \theta \in [-\delta/2, 0), \\ \xi_{m_{l-1}}^{h,\delta_0,\delta}, & \theta \in [-\delta/2 - \delta, -\delta/2), \\ \vdots & \\ \xi_{m_{l-Q_\delta^+}}^{h,\delta_0,\delta}, & \theta \in [-\bar{\theta}, -\bar{\theta} + \delta). \end{cases} \tag{2.1}$$

This construction is illustrated in Figure 2.1.

Figure 2.1. Illustration of $\bar{\xi}_{e,n}^{h,\delta_0,\delta}$, Erlang delay model, $m_l^{h,\delta_0,\delta} \le n < m_{l+1}^{h,\delta_0,\delta}$.

For $m_l^{h,\delta_0,\delta} \le n < m_{l+1}^{h,\delta_0,\delta}$, we can unambiguously identify the path memory segment with the vector of values in (2.1):

$$\tilde{X}_{e,n}^{h,\delta_0\delta} = \left(\xi_{m_{l-Q_\delta^+}}^{h,\delta_0,\delta}, \ldots, \xi_{m_l}^{h,\delta_0,\delta}, \xi_n^{h,\delta_0,\delta}\right).$$

For $L_n^{h,\delta_0,\delta} < \bar{L}^{\delta_0,\delta} - 1$, the transitions of the path memory variables are

$$\tilde{X}_{e,n}^{h,\delta_0\delta} \to \tilde{X}_{e,n+1}^{h,\delta_0\delta} = \left(\xi_{m_{l-Q_\delta^+}}^{h,\delta_0,\delta}, \ldots, \xi_{m_l}^{h,\delta_0,\delta}, \xi_{n+1}^{h,\delta_0,\delta}\right), \tag{2.2a}$$

and

$$L_n^{h,\delta_0,\delta} \to L_{n+1}^{h,\delta_0,\delta} = \begin{cases} L_n^{h,\delta_0,\delta} + 1, & \text{w.p. } \Delta t_n^{h,\delta_0,\delta}/\delta_0, \\ L_n^{h,\delta_0,\delta}, & \text{w. p. } 1 - \Delta t_n^{h,\delta_0,\delta}/\delta_0. \end{cases} \tag{2.2b}$$

For $L_n^{h,\delta_0,\delta} = L^{\delta_0,\delta} - 1$,

$$\left.\begin{aligned} \tilde{X}_{e,n}^{h,\delta_0\delta} &\to \tilde{X}_{e,n+1}^{h,\delta_0\delta} = \left(\xi_{m_{l-Q_\delta^+}}^{h,\delta_0,\delta}, \ldots, \xi_{m_l}^{h,\delta_0,\delta}, \xi_{n+1}^{h,\delta_0,\delta}\right) \\ L_{n+1}^{h,\delta_0,\delta} &= L_n^{h,\delta_0,\delta} \end{aligned}\right\} \text{ w.p. } 1 - \frac{\Delta t_n^{h,\delta_0,\delta}}{\delta_0}, \tag{2.3a}$$

and

$$\left.\begin{aligned} &\tilde{X}_{e,n}^{h,\delta_0\delta} \to \tilde{X}_{e,n+1}^{h,\delta_0\delta} = \left(\xi_{m_{l-Q_\delta^+ +1}}^{h,\delta_0,\delta}, \ldots, \xi_{m_l}^{h,\delta_0,\delta}, \xi_{n+1}^{h,\delta_0,\delta}, \xi_{n+1}^{h,\delta_0,\delta}\right) \\ &L_{n+1}^{h,\delta_0,\delta} = 0 \end{aligned}\right\} \text{ w.p. } \frac{\Delta t_n^{h,\delta_0,\delta}}{\delta_0}. \tag{2.3b}$$

We call $L_n^{h,\delta_0,\delta}$ the Erlang state, and the times that it goes from the value $\bar{L}^{\delta_0,\delta} - 1$ to the value zero are called the shift or δ-shift times.

Comments. The periodic-Erlang model is a compromise between the random delay model of Section 7.3 and the periodic delay model of the last section. For any fixed δ, the quality of the approximation is better than for the random model, but it requires more memory. The random delay model would require a smaller value of δ for an equivalent accuracy. But keep in mind that the dimension of the state space for the random model is $Q_\delta + 1$, implying an exponential increase in memory as $\delta \to 0$ (see Section 7.2 and the next section on this point). The quality for the periodic-Erlang model would not be as good as that for the periodic model of the previous subsection, but the memory requirements are less, perhaps much less, and as $\delta_0 \to 0$, the two models converge to each other. Simulations indicate that there will often be considerable savings with little loss of accuracy with the periodic-Erlang model. Obvious analogs of asymptotic timescale equivalence results of Theorems 6.5.1 and 7.3.1 hold, and the details are left to the reader.

The next theorem follows from Theorem 7.1.4. The cost function is either (7.3.12) or (7.1.28) with the current memory segment $\bar{\xi}_{e,n}^{h,\delta_0,\delta}$ used. In particular, the analog of (7.3.12) with control $u = \{u_n^{h,\delta_0,\delta}\}$ used is

$$E_{\hat{\xi}}^{h,\delta_0,\delta,u} \sum_{n=0}^{\infty} e^{-\beta t_n^{h,\delta_0,\delta}} \left[k(\bar{\xi}_{e,n}^{h,\delta_0,\delta}, u_n^{h,\delta_0,\delta})\Delta t_n^{h,\delta_0,\delta} + q'\Delta y_n^{h,\delta_0,\delta}\right]. \tag{2.4}$$

Theorem 2.1. *Assume the conditions of Theorem 7.3.3, but with memory segment for the original model defined by (4.2.8), the periodic-Erlang case. Hence the original model is (4.1.9b), with $\bar{x}_a = \bar{x}_e^{\delta_0,\delta}$, for which we suppose that there is a weak-sense unique solution for each control and initial condition. Assume the form of this section for the numerical procedure with $V^{h,\delta_0,\delta}(\hat{\xi})$ denoting the optimal cost for the numerical model, $V^{\delta_0,\delta}(\hat{\xi})$ for that of (4.2.8), and $V^\delta(\hat{\xi})$ when the periodic memory segment (4.2.6) is used. Let the piecewise-constant $\bar{x}^\delta(0)$ approximate $\bar{x}(0)$, and let $\bar{\xi}_0^{h,\delta_0,\delta} \to \bar{\xi}_0^{\delta_0,\delta} \to \bar{x}^\delta(0)$ in the sup norm as $h \to 0$ and then $\delta_0 \to 0$, where all the functions are G_h-valued. Then as $h \to 0$ and then $\delta_0 \to 0$, $V^{h,\delta_0,\delta}(\bar{\xi}_0^{h,\delta_0,\delta}) \to V^{\delta_0,\delta}(\bar{\xi}_0^{\delta_0,\delta}) \to V^\delta(\bar{x}^\delta(0))$.*

8.3 The Number of Points in the State Space: Path Only Delayed

The value of δ. Consider solving a parabolic PDE on a finite time interval via finite differences with spatial interval h and (for the implicit approximation procedure) time interval δ, and with the classical estimates of rate of convergence holding. The rate of convergence is $O(h^2) + O(\delta^2)$ for the implicit approximation procedure, vs. $O(h^2) + O(\text{max time increment})$ for the explicit approximation procedure [91, Chapter 6]. For the explicit approximation procedure, the value of the time increment is $O(h^2)$. Thus, for $\delta = O(h)$, the rates of convergence would be of the same order. There is no proof that such estimates hold for the control problem of concern here. But numerical data for the no-delay problems suggests that one should use $\delta = O(h)$. In this chapter, the value of δ is either constant or of this order.

8.3.1 The Implicit and Periodic-Erlang Approximation Methods: Reduced Memory

Reduced memory requirements. We will work with the random delay approximation method of Section 7.3, but identical estimates hold for the periodic-Erlang model of Section 2. To illustrate the memory issues, consider the one-dimensional model mentioned in the example in the paragraph below (7.3.9c). Recall the definition $Q_\delta = \bar{\theta}/\delta$, an integer. The vector $\hat{\xi} = (\hat{\xi}(-\bar{\theta}), \hat{\xi}(-\bar{\theta}+\delta), \cdots, \hat{\xi}(-\delta), \hat{\xi}(0))$, which represents the canonical value of the memory segment $\bar{\xi}^{h,\delta}_{r,n}$ for the approximating chain, as in the left-hand sides of (7.3.9), can be represented in terms of the vector of differences

$$\hat{D} = \left(\hat{\xi}(-\bar{\theta}) - \hat{\xi}(-\bar{\theta}+\delta), \ldots, \hat{\xi}(-\delta) - \hat{\xi}(0), \hat{\xi}(0)\right) \equiv (D(Q_\delta), \ldots, D(0)). \tag{3.1}$$

If $\hat{\xi}(0)$ is a reflection point, then it moves immediately to the closest point in G_h. Otherwise, for the example of concern, with this representation and if the time variable does not advance, the transitions are to one of the two values

$$(D(Q_\delta), \ldots, D(2), D(1) \mp h, D(0) \pm h).$$

If the time variable advances, then the transition is to

$$(D(Q_\delta - 1), \ldots, D(1), 0, D(0)).$$

The variable $D(0)$ takes $B/h + 3$ possible values. Because there are a potentially unbounded number of steps between successive increases of the time variable, the differences $D(i), i \geq 2$, can take values in the set $G_h - G_h$, which is the set of points $\{B, B-h, \ldots, -B\}$. Hence there are $2B/h+1$ possible values. The value of $D(1)$ is in the set $G_h - G_h^+ = \{B+h, B, B-h, \ldots, -B-h\}$, as $\hat{\xi}(0)$ takes values in $G_h \cup \partial G_h^+$ and $\hat{\xi}(-\delta)$ takes values in G_h. This set has $(2B/h+3)$ points. The maximum required number of points is therefore

$$(B/h+3)(2B/h+3)(2B/h+1)^{\bar{\theta}/\delta-1}. \tag{3.2}$$

This is smaller than that for the explicit approximation procedure for the same value of h, but larger than (7.3.10), which is that for the representation (7.3.9). The advantage of the representation (3.1) in terms of differences is that the sizes of the components can be truncated, as we will now see.

An approximation result. We will use the following version of the approximation method of Theorem 7.1.4. Suppose that we approximate the path memory segment for the numerical process as follows. Let κ be the approximation parameter and let the associated chain be denoted by $\{\xi_n^{h,\delta,\kappa}\}$, with controls $\{u_n^{h,\delta,\kappa}\}$. In the development below, κ will be a truncation parameter. The approximation to the path memory segment at step n is a nonanticipative function $\bar{\xi}_{a,n}^{h,\delta,\kappa} = \{\bar{\xi}_{a,n}^{h,\delta,\kappa}(\theta), \theta \in [-\bar{\theta}, 0]\}$, which can be embedded into a finite-state Markov chain, and with the value $\bar{\xi}_{a,n}^{h,\delta,\kappa}(0) = \xi_n^{h,\delta,\kappa}$. Define $\Delta t_n^{h,\delta,\kappa} = \Delta t^{h,\delta}(\bar{\xi}_{a,n}^{h,\delta,\kappa}, u_n^h)$. The continuous-time interpolation analogous to (7.3.16) is

$$\begin{aligned}\psi^{h,\delta,\kappa}(t) = \xi_0^{h,\delta} &+ \int_0^t \int_{U^h} b_h(\bar{\xi}_a^{h,\delta,\kappa}(q_\tau^{h,\delta,\kappa}(s)), \alpha) r_\tau^{h,\delta,\kappa}(d\alpha\, ds) \\ &+ B_\tau^{h,\delta,\kappa}(t) + z_\tau^{h,\delta,\kappa}(t).\end{aligned} \tag{3.3}$$

The exact form of $\bar{\xi}_{a,n}^{h,\delta,\kappa}$ will be defined below. The quadratic variation of the martingale $B_\tau^{h,\delta,\kappa}(\cdot)$ is

$$\int_0^t a_h(\bar{\xi}_a^{h,\delta,\kappa}(q_\tau^{h,\delta,\kappa}(s)))ds.$$

The process $\bar{\xi}_a^{h,\delta,\kappa}(\cdot)$ (resp., $\xi^{h,\delta,\kappa}(\cdot)$) is the interpolation of $\{\bar{\xi}_{a,n}^{h,\delta,\kappa}\}$ (resp., of $\{\xi_n^{h,\delta,\kappa}\}$) with intervals $\{\Delta t_n^{h,\delta,\kappa}\}$. The processes $r_\tau^{h,\delta,\kappa}(d\alpha\, ds)$, and $q_\tau^{h,\delta,\kappa}(s)$ are defined analogously to those without the κ, but under the condition that the modified path memory segment $\bar{\xi}_{a,n}^{h,\delta,\kappa}$ is used in the dynamics and cost rate at step n of the chain. See the discussion above (7.1.14) concerning the role of $q_\tau^{h,\delta}(\cdot)$.

Suppose that

$$\lim_{\delta\to 0} \lim_{\kappa\to 0,\, h\to 0} \sup_{t\le T, \text{control}} P\left\{ \sup_{-\bar{\theta}\le\theta\le 0} \left|\xi^{h,\delta,\kappa}(t+\theta) - \bar{\xi}_a^{h,\delta,\kappa}(t,\theta)\right| \ge \epsilon \right\} = 0, \tag{3.4}$$

for each $T<\infty$ and $\epsilon>0$. Then it follows from Theorem 7.1.4, under appropriate conditions on the boundaries and the cost rate and dynamical functions, that as $h\to 0$, $\kappa\to 0$, and then $\delta\to 0$, the limit of the processes (3.3) and the associated costs are processes and costs for the original model (7.1.5) where the path memory segment at time t is $\bar{x}(t)$.

Now fix δ. Consider the variant where as $\kappa \to 0$, the sup over any arbitrary time interval of the absolute value of the difference between the path memory segment process $\bar{\xi}_a^{h,\delta,\kappa}(\cdot)$ and that based on the approximation (7.3.8) goes to zero in probability as $\kappa \to 0$ and $h \to 0$. Then, as $h \to 0$ and $\kappa \to 0$, the optimal processes and costs will converge to those where the path memory segment at time t is $\bar{x}_r^\delta(t)$, defined by (4.2.7).

Truncated state variables. We now exploit these approximation results to reduce the memory requirement by truncating the range of the values of the $D(i)$. The basic idea behind the approximation is that over the number of steps that are required for the time variable to advance, with a high probability (for small h), the sample number of values taken by the $D(i)$ will be much less than $2B/h$ due to cancellations of positive and negative steps. Thus one can truncate the range of values of the $D(i), i \geq 1$, leaving $N_1 < 2B/h$ allowed values, where the probability that any of the $D(i), i \leq Q_\delta$, over the approximating delay interval is not one of the N_1 allowed values is smaller than some predetermined number. We will only indicate the possibilities. Much further work is needed to get the best truncations.

One possible approach is to use the martingale properties of the $B_\tau^{h,\delta,\kappa}(\cdot)$ process to get bounds on the path excursions. A cleaner approach is to use the properties of weak convergence, as will now be done. Let $\sigma_n^{h,\delta,\kappa}$ be defined as $\sigma_n^{h,\delta}$ was above Figure 7.3.1, but where the spatial chain is $\{\xi_n^{h,\delta,\kappa}\}$. Define

$$\gamma_{l+1}^{h,\delta,\kappa} = \min\left\{t > \sigma_l^{h,\delta,\kappa} : \left|\xi^{h,\delta,\kappa}(t) - \xi^{h,\delta,\kappa}(\sigma_l^{h,\delta,\kappa})\right| \geq \kappa\right\} \wedge \sigma_{l+1}^{h,\delta,\kappa}.$$

The idea is to use $\xi^{h,\delta,\kappa}(\gamma_{l+1}^{h,\delta,\kappa}) - \xi^{h,\delta,\kappa}(\sigma_l^{h,\delta,\kappa})$ in lieu of $\xi^{h,\delta}(\sigma_{l+1}^{h,\delta}) - \xi^{h,\delta}(\sigma_l^{h,\delta})$ for the components of the representation (3.1) of (7.3.9) in terms of differences.

An estimate based on a Wiener process approximation. Fix δ, with only $h \to 0$. As $h \to 0$, the process $B_\tau^{h,\delta}(\cdot)$ in (7.3.16) converges weakly to the stochastic integral $\int_0^t \sigma(\bar{x}_r^\delta(s))dw(s)$. Hence the distribution of the first time than $|B_\tau^{h,\delta}(\cdot)|$ exceeds κ will converge weakly to that of the stochastic integral. Start by supposing that the process is one-dimensional and that $\sigma(\hat{x}) = 1$, so that the limit is a Wiener process with unit variance.

The density of the time of escape of a Wiener process starting at time zero from the interval $[-\kappa, \kappa]$ is bounded by [42, Chapter 2, remark 8.3, adjusted for the double-sided barrier][1]

$$e^{-\kappa^2/2t} \frac{4\kappa}{\sqrt{2\pi}t^{3/2}}. \tag{3.5}$$

[1] This is a crude upper bound. We have added the density of the escape time from $(-\infty, \kappa)$ to that for the escape time from $(-\kappa, \infty)$.

The probability that the truncated process will not equal the untruncated process over a time interval $[0, \tau_\delta]$ is asymptotically bounded by

$$\frac{4\kappa}{\sqrt{2\pi}} \int_0^{\tau_\delta} e^{-\kappa^2/2t} \frac{1}{t^{3/2}} dt = \frac{8}{\sqrt{2\pi}} \int_{\kappa/\sqrt{\tau_\delta}}^{\infty} e^{-v^2/2} dv \leq 8e^{-(\kappa^2/2\tau_\delta)} \left[\frac{\sqrt{\tau_\delta}}{\kappa}\right], \quad (3.6)$$

where the right-hand inequality is taken from [24, Chapter 7, (1.8)]. Because in our application τ_δ is asymptotically a random variable that is exponentially distributed with mean δ, the expectation of the right side of (3.6) is

$$8Ee^{-(\kappa^2/2\tau_\delta)} \left[\frac{\sqrt{\tau_\delta}}{\kappa}\right] = 8 \int_0^{\infty} e^{-(\kappa^2/2u)} \left[\frac{\sqrt{u}}{\kappa}\right] \frac{1}{\delta} e^{-u/\delta} du, \quad (3.7)$$

which can be written as

$$A(\kappa, \delta) \equiv 8 \int_0^{\infty} \frac{\sqrt{u\delta}}{\kappa} e^{-\kappa^2/2\delta u} e^{-u} du. \quad (3.8)$$

Because there are Q_δ intervals in the construction of each memory segment, the probability that there is a truncation in the construction of the path memory segment at an arbitrary time is asymptotically bounded by

$$1 - (1 - A(\kappa, \delta))^{Q_\delta} \leq Q_\delta A(\kappa, \delta). \quad (3.9)$$

Let us use the form $\kappa = kB$, where $0 < k < 1$. Suppose for example that $B = 10, \bar{\theta} = .1, \delta = .01, \kappa = .8$. Then the left side of (3.9) is 0.0012 and the right side is 0.0187. If $\kappa = .6$, then the left side is 0.0225. A bound on the required number of points is (3.2), where B is replaced by κ in the two right-hand factors.

If $\sigma^2(\cdot)$ is not constant, then use a time change argument and the largest value of the variance to get an upper bound. The estimate is unchanged if the drift term $b(\cdot)$ is included. We have supposed that δ is constant while $h \to 0$. But the results will continue to hold if $\delta \to 0$ sufficiently slowly as $h \to 0$.

An alternative procedure for memory reduction. The use of cruder approximations can also be helpful to reduce the memory requirements. For example, for the above special case, suppose that the differences are to be approximated by values in $S = \{0, \pm 2h, \pm 4h, \ldots, \pm B\}$. If a difference falls between a pair of any such points, then assign by a randomization. One could also approximate such intermediate values by alternating the assignments, first to the closest higher point, then to the closest lower point, and so forth. Smaller sets than S can be used as well, with appropriate assignment policies.

8.4 Control and Path Delayed

Subsection 7.2.2 contained some comments on the case where the control was also delayed. It was assumed in that section that the interpolation interval

$\Delta t^h(\cdot)$ was a constant $\bar{\Delta}^h$. At step n, the control memory segment involved the entire history of the control over interpolated time $[t_n^h - \bar{\theta}, t_n^h]$. The required memory can be excessive if $\bar{\Delta}^h$ is very small or the control takes more than two or three values. In this section, we will develop analogs of the approximation methods of Section 1, where the delays are periodic, and of Section 7.3 and of Section 2 of this chapter, where the time intervals defining the piecewise-constant memory segments were random. These methods required less memory and allowed the use of arbitrary $\Delta t^h(\cdot)$, and these advantages will also hold for the control memory segment.

The continuous-time interpolations of the control process u_n^h have no regularity properties that are analogous to the (asymptotic) continuity of the interpolations of the ξ_n^h. Because of this, it is natural to work with the relaxed control representations. It is supposed that local consistency (7.1.23) holds. This will be the case if the transition probabilities are defined by (7.1.24), where the functions $N^h(\cdot)$, $D^h(\cdot)$ yield local consistency for the nondelay problem. The system is (3.2.4) and we start with the cost function (3.4.4). The cost function for the numerical problem is (7.1.26). Also, for ease of visualizing the various constructions, suppose that $U = U^h$ contains only a finite number of points.

8.4.1 A Periodic Approximating Memory Segment

We start with an analog of the periodic memory segment of Section 1, as it is relatively simple and will provide a good introduction to the other forms. Recall the definition $(Q_\delta^+ + 1/2)\delta = \bar{\theta}$ and suppose, for notational simplicity, that $\mu_c([-\delta/2, 0)) = 0$; that is, there are no delays in the control memory with values on the half open interval $[-\delta/2, 0)$. Until the last paragraph of this subsection, it is assumed that, for $\hat{\xi}(0) \in G_h$, $\Delta t^h(\hat{\xi}, \hat{u}, \alpha) = \bar{\Delta}^h$, a constant. This assumption will be dropped in the last paragraph and in subsequent subsections. If we have a reflecting boundary and $\xi_n^h \notin G_h$, then there is no control at step n. To simplify the notation, we will ignore such steps in computing the control memory segment below. Recall the definitions $\bar{L}^{h,\delta} = \delta/\bar{\Delta}^h$, $\nu_l^{h,\delta} = l\delta/\bar{\Delta}^h$, and $L_n^{h,\delta} = n(\mathrm{mod}(\bar{L}^{h,\delta}))$. Thus for $\nu_l^{h,\delta} \le n < \nu_{l+1}^{h,\delta}$ we have $L_n^{h,\delta} = n - \nu_l^{h,\delta}$. To construct the approximation to the control memory segment, we use the method and notation of Section 1 and divide the interval $[-\bar{\theta}, 0]$ into $Q_\delta^+ + 1$ segments, as depicted in Figure 1.1. The control memory segment will be represented as a relaxed control whose derivative $\bar{r}_{p,n}^{h,\delta,\prime}(d\alpha, \theta)$ at step n of the chain is constant in θ on each of the θ-intervals $[-\bar{\theta}, -\bar{\theta} + \delta), \ldots, [-3\delta/2, -\delta/2), [-\delta/2, 0)$ on the horizontal axis in Figure 1.1. For $l\delta \le t_n^{h,\delta} < l\delta + \delta$, we use the notation $r_{p,l,n}^{h,\delta,\prime}(d\alpha)$ for the value of $\bar{r}_{p,n}^{h,\delta,\prime}(d\alpha, \theta)$ for $\theta \in [-\delta/2, 0)$, $r_{p,l-1}^{h,\delta,\prime}(d\alpha)$ for the value of $\bar{r}_{p,n}^{h,\delta,\prime}(d\alpha, \theta)$ for $\theta \in [-3\delta/2, -\delta/2)$, and so forth. These control terms will now be defined.

Until the last paragraph of this subsection, it will be assumed that n and l are such that $\nu_l^{h,\delta} \le n < \nu_{l+1}^{h,\delta}$. Define

$$r^{h,\delta,\prime}_{p,l-i}(d\alpha) = \sum_{j=\nu^{h,\delta}_{l-i}}^{\nu^{h,\delta}_{l-i+1}-1} I_{\{u^{h,\delta}_j=\alpha\}} \Big/ \bar{L}^{h,\delta}, \quad i = 1, \ldots, Q^+_\delta. \tag{4.1}$$

This is the average of the relaxed control representations of the actual control values that are used on the interpolated time interval $[l\delta - i\delta, l\delta - i\delta + \delta)$, or, equivalently, on the set $[v^{h,\delta}_{l-i}, v^{h,\delta}_{l-i+1})$ of iterates. Keep in mind that, at iterate n, the value of $\xi^{h,\delta}_n$ is known, but the value of $u^{h,\delta}_n$ is to be computed. Define (where $0/0 = 0$)

$$r^{h,\delta,\prime}_{p,l,n}(d\alpha) = \sum_{j=\nu^{h,\delta}_l}^{n-1} I_{\{u^{h,\delta}_j=\alpha\}} \Big/ L^{h,\delta}_n. \tag{4.2}$$

Summarizing, the values of $\bar{r}^{h,\delta,\prime}_{p,n}(\theta, d\alpha)$ are given by

$$\bar{r}^{h,\delta,\prime}_{p,n}(d\alpha, \theta) = \begin{cases} r^{h,\delta,\prime}_{p,l,n}(d\alpha), & \theta \in [-\delta/2, 0), \\ r^{h,\delta,\prime}_{p,l-1}(d\alpha), & \theta \in [-\delta/2 - \delta, -\delta/2), \\ \vdots & \\ r^{h,\delta,\prime}_{p,l-Q^+_\delta}(d\alpha), \, \theta \in [-\bar{\theta}, -\bar{\theta} + \delta). & \end{cases} \tag{4.3}$$

This construction is illustrated in Figure 4.1, where $\nu^{h,\delta}_l \leq n < \nu^{h,\delta}_{l+1}$. In the figure, the top line gives the derivative of the relaxed control in each of the subintervals, and the lowest line indicates the ranges of the iterates that were used to compute the values.

The expression (4.2) is the average of the relaxed control representations of the controls computed over the set of iterates $[\nu^{h,\delta}_l, n - 1]$. With this approximation, on each set $[\nu^{h,\delta}_l, \nu^{h,\delta}_{l+1})$ of iterates we need to track $n(\text{mod}(\bar{L}^{h,\delta}))$ and the numerator of (4.2). The first of these quantities takes $\bar{L}^{h,\delta}$ possible values $\{0, 1, \ldots, \bar{L}^{h,\delta} - 1\}$, which can be large if h is small.

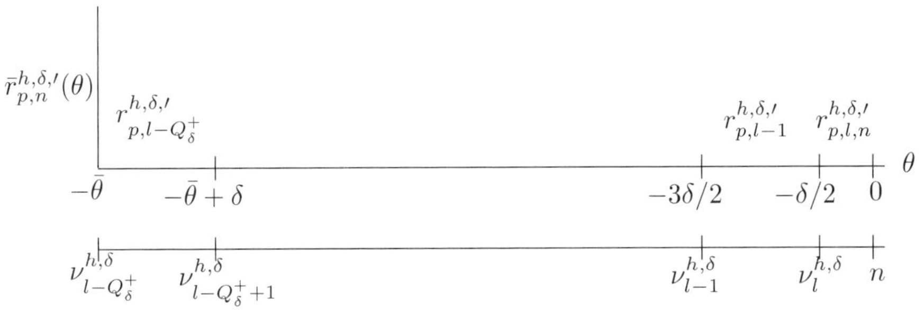

Figure 4.1. Illustration of the control memory segment.

Since U^h has only a finite number of points, the total number of possible values for the numerator of (4.2) is finite, but can be very large. The representation does not use significantly less memory than the method of Subsection 7.1.3, but it provides a basis for the further approximations that will be given below.

The transitions of the control memory segment. Analogously to (1.2a), for $L_n^{h,\delta} < \bar{L}^{h,\delta} - 1$ the control memory-state and its transition can be unambiguously represented by the form

$$\begin{aligned} \bar{r}_{p,n}^{h,\delta,\prime}(d\alpha) &= \left(r_{p,l-Q_\delta^+}^{h,\delta,\prime}(d\alpha), \ldots, r_{p,l-1}^{h,\delta,\prime}(d\alpha), r_{p,l,n}^{h,\delta,\prime}(d\alpha)\right) \\ &\longrightarrow \bar{r}_{p,n+1}^{h,\delta}(d\alpha) = \left(r_{p,l-Q_\delta^+}^{h,\delta,\prime}(d\alpha), \ldots, r_{p,l-1}^{h,\delta,\prime}(d\alpha), r_{p,l.n+1}^{h,\delta,\prime}(d\alpha)\right), \end{aligned} \tag{4.4}$$

which involves only an updating of the index in (4.2) from n to $(n+1)$.

Now suppose that $L_n^{h,\delta} = \bar{L}^{h,\delta} - 1$. Then $L_{n+1}^{h,\delta} = 0$, $t_{n+1}^{h,\delta} = l\delta + \delta$ and $\nu_{l+1}^{h,\delta} = n + 1$. The numerator of (4.2) becomes

$$\sum_{j=\nu_l^{h,\delta}}^{n-1} I_{\{u_j^{h,\delta}=\alpha\}} + I_{\{u_n^{h,\delta}=\alpha\}} = \sum_{j=\nu_l^{h,\delta}}^{\nu_{l+1}^{h,\delta}-1} I_{\{u_j^{h,\delta}=\alpha\}}.$$

Dividing this by $\bar{L}^{h,\delta}$ yields $r_{p,l}^{h,\delta,\prime}(d\alpha)$, we begin the next cycle, and the control memory state transits as the following analog of (1.2b):

$$\begin{aligned} \bar{r}_{p,n}^{h,\delta,\prime}(d\alpha) &= \left(r_{p,l-Q_\delta^+}^{h,\delta,\prime}(d\alpha), \ldots, r_{p,l-1}^{h,\prime}(d\alpha), r_{p,l,n}^{h,\delta,\prime}(d\alpha)\right) \\ &\longrightarrow \bar{r}_{p,n+1}^{h,\delta,\prime}(d\alpha) = \left(r_{p,l-Q_\delta^+ +1}^{h,\delta,\prime}(d\alpha), \ldots, r_{p,l-1}^{h,\delta,\prime}(d\alpha), r_{p,l}^{h,\delta,\prime}(d\alpha), 0\right). \end{aligned} \tag{4.5}$$

The last value is zero, as $t_{n+1}^{h,\delta} = l\delta + \delta$ and no control values for the interval of interpolated time $[l\delta + \delta, l\delta + 2\delta)$ have yet been computed.

A representation of the drift term. Let $u^{h,\delta}(\cdot)$ denote the continuous-time interpolation of the controls $\{u_n^{h,\delta}\}$, with relaxed control representation $r_p^{h,\delta}(\cdot)$, and $\bar{r}_p^{h,\delta,\prime}(\cdot)$ the continuous-time interpolation of the memory segments $\{\bar{r}_{p,n}^{h,\delta,\prime}\}$, all with intervals $\bar{\Delta}^h$. Then, for $t \in [l\delta, l\delta + \delta)$ and $\mu_c([-\delta/2, 0)) = 0$, and the periodic memory segment $\bar{\xi}_{p,n}^{h,\delta}$ used for the path, we can represent the drift term at time t in the form

$$\begin{aligned} \bar{b}^{h,\delta}(\bar{\xi}_p^{h,\delta}(t), \bar{r}_p^{h,\delta,\prime}(t), u^{h,\delta}(t)) &= \mu_c(\{0\}) \int_{U^h} b(\bar{\xi}_p^{h,\delta}(t), \alpha, 0) r_p^{h,\delta,\prime}(d\alpha, t) \\ &+ \sum_{i=1}^{Q_\delta^+} \int_{-i\delta-\delta/2}^{(-i\delta+\delta-\delta/2)-} \mu_c(d\theta) \int_{U^h} b(\bar{\xi}_p^{h,\delta}(t), \alpha, \theta) r_{p,l-i}^{h,\delta,\prime}(d\alpha). \end{aligned} \tag{4.6}$$

The analog of (4.1)–(4.3) for arbitrary $\Delta t^h(\cdot)$. Let us now write the analog of (4.1)–(4.3) when the interpolation intervals $\Delta t^h(\cdot)$ are not constant. The expressions are hard to use for the memory segments in a numerical procedure but will be approximated by the method in the next subsection. Because the number of iterates between the successive δ-shifts is now random, a weighting over the appropriate random intervals will replace (4.1) and (4.2). Redefine $\Delta t_n^{h,\delta} = \Delta t^h(\bar{\xi}_{p,n}^{h,\delta}, \bar{r}_{p,n}^{h,\delta,\prime}, u_n^h)$ and $\tilde{\nu}_l^{h,\delta} = \min\{n : t_n^{h,\delta} \geq l\delta\}$.

For $i = 1, \ldots, Q_\delta^+$, redefine

$$r_{p,l-i}^{h,\delta,\prime}(d\alpha) = \sum_{j=\tilde{\nu}_{l-i}^{h,\delta}}^{\tilde{\nu}_{l-i+1}^{h,\delta}-1} I_{\{u_j^{h,\delta}=\alpha\}} \Delta t_j^{h,\delta} \Big/ \sum_{j=\tilde{\nu}_{l-i}^{h,\delta}}^{\tilde{\nu}_{l-i+1}^{h,\delta}-1} \Delta t_j^{h,\delta}. \tag{4.7}$$

For $\tilde{\nu}_l^{h,\delta} \leq n < \tilde{\nu}_{l+1}^{h,\delta}$, redefine

$$r_{p,l,n}^{h,\delta,\prime}(d\alpha) = \sum_{j=\tilde{\nu}_l^{h,\delta}}^{n-1} I_{\{u_j^{h,\delta}=\alpha\}} \Delta t_j^{h,\delta} \Big/ \sum_{j=\tilde{\nu}_l^{h,\delta}}^{n-1} \Delta t_j^{h,\delta} = \frac{N_{l,n}^{h,\delta}(d\alpha)}{D_{l.n}^{h,\delta}}. \tag{4.8}$$

With these definitions, for $n = \tilde{\nu}_{l+1}^{h,\delta}$,

$$r_{p,l}^{h,\delta,\prime}(d\alpha) = \frac{N_{l,n}^{h,\delta}(d\alpha)}{D_{l,n}^{h,\delta}}. \tag{4.9}$$

With the definition $\bar{\sigma}_l^{h,\delta} = t_{\tilde{\nu}_l^{h,\delta}}^{h,\delta}$, (4.7) can be written as

$$\frac{\displaystyle\int_{\bar{\sigma}_{l-i}^{h,\delta}}^{\bar{\sigma}_{l-i+1}^{h,\delta}} r_p^{h,\delta,\prime}(d\alpha, s)ds}{\bar{\sigma}_{l-i+1}^{h,\delta} - \bar{\sigma}_{l-i}^{h,\delta}}, \tag{4.10}$$

and (4.8) can be written as

$$\frac{\displaystyle\int_{\bar{\sigma}_l^{h,\delta}}^{t_n^{h,\delta}} r_p^{h,\delta,\prime}(d\alpha, s)ds}{t_n^{h,\delta} - \bar{\sigma}_l^{h,\delta}}, \tag{4.11}$$

where we continue to let $r_p^{h,\delta,\prime}(d\alpha, t)$ denote the derivative of the relaxed control corresponding to the interpolation of $\{u_n^{h,\delta}\}$ with intervals $\{\Delta t_n^{h,\delta}\}$.

8.4.2 A Periodic-Erlang Approximation

When $\Delta t^h(\hat{\xi}, \hat{u}, \alpha)$ is not constant, in principle the expressions (4.7)–(4.11) could be used in (4.3) for the approximation to the control memory segment.

Unfortunately, they are no simpler than (4.1) and (4.2) and generally would require more memory. In this subsection, the ideas of the periodic-Erlang approach of Section 2 and Approximation 5a of Section 4.4 will be adapted to reduce these memory requirements. The method that is to be proposed is well motivated and promising, but it is only one among many possibilities for dealing with the memory requirements.

It will no longer be necessary that $\Delta t^h(\cdot)$ be constant. Recall the definitions of δ_0, $\bar{L}^{\delta_0,\delta}$, $L_n^{h,\delta_0,\delta}$, and $m_l^{h,\delta_0,\delta}$ from Section 2. As in the previous subsection, the approximation to the control memory segment will be in terms of the derivative of a relaxed control, and it will be piecewise-constant. In the previous subsection, the intervals of constancy in θ of the relaxed control derivative $\bar{r}_{p,n}^{h,\delta,\prime}(d\alpha,\theta)$ were of length δ, except the top one, whose length was $\delta/2$. In Section 2, the θ-subintervals where the memory segment $\bar{\xi}_{p,n}^{h,\delta_0,\delta}(\theta)$ was constant were of these same lengths, and the values that were taken were determined by a sampling of the interpolated chain at intervals that were determined (at least asymptotically, as $h \to 0$) by an Erlang distribution. The variable that measured the passage of time since the last δ-shift took $\bar{L}^{\delta_0,\delta}$ values. This is opposed to the much larger value $\bar{L}^{h,\delta} = \delta/\bar{\Delta}^h$ (as in the first part of the previous subsection), where, if we need to transform the transition probabilities to get a constant interpolation interval, we would have $\bar{\Delta}^h = \min_{\hat{\xi},\hat{u},\alpha} \Delta t^h(\hat{\xi},\hat{u},\alpha)$.

There are several issues of concern. The first is that of Section 2, in simplifying the measurement of the passage of time since the last δ-shift. The next issue is similar, it concerns the problem that (4.2) and (4.8) are updated at each step of the chain, which might also require an enormous memory. These problems will be dealt with by an adaptation of the method of Section 2 and the form of the periodic-Erlang approximation (4.4.7). We will use the subscript *ee*, as in (4.4.7), to denote the approximation.

The approximation method. At step n of the chain, the approximation to the path memory segment will be $\bar{\xi}_{e,n}^{h,\delta_0,\delta}$ defined in (2.1), and the approximation $\bar{r}_{ee,n}^{h,\delta_0,\delta,\prime}$ to the control memory segment will be defined below. Redefine $\Delta t_n^{h,\delta_0,\delta} = \Delta t^h(\bar{\xi}_{e,n}^{h,\delta_0,\delta}, \bar{r}_{ee,n}^{h,\delta_0,\delta,\prime}, u_n^{h,\delta_0,\delta})$, and $t_n^{h,\delta_0,\delta} = \sum_{i=0}^{n-1} \Delta t_i^{h,\delta_0,\delta}$. Define $\bar{\sigma}_l^{h,\delta_0,\delta} = t_{m_l^{h,\delta_0,\delta}}^{h,\delta_0,\delta}$, where the renewal time $m_l^{h,\delta_0,\delta}$ was defined above (2.1), but use the current interpolation intervals.

Let $m_l^{h,\delta_0,\delta} \le n < m_{l+1}^{h,\delta_0,\delta}$ or, equivalently, $\bar{\sigma}_l^{h,\delta_0,\delta} \le t_n^{h,\delta_0,\delta} < \bar{\sigma}_{l+1}^{h,\delta_0,\delta}$. When using $m_l^{h,\delta_0,\delta}$ as an index of summation, we will write it simply as m_l. We will approximate (4.8) by the form

$$\tilde{r}_{ee,l.n}^{h,\delta,\delta_0,\prime}(d\alpha) = \frac{\tilde{N}_{l,n}^{h,\delta_0,\delta}(d\alpha)}{\tilde{D}_{l,n}^{h,\delta_0,\delta}}, \tag{4.12}$$

where the functions $\tilde{N}$ and $\tilde{D}$ will now be defined. For convenience rewrite (2.2b) and the second lines of (2.3a) and (2.3b) as follows. For $L_n^{h,\delta_0,\delta} <$

$\bar{L}^{\delta_0,\delta} - 1$,[2]

$$L_n^{h,\delta_0,\delta} \to L_{n+1}^{h,\delta_0,\delta} = \begin{cases} L_n^{h,\delta_0,\delta}, & \text{w.p. } (1 - \Delta t_n^{h,\delta_0,\delta}/\delta_0), \\ L_n^{h,\delta_0,\delta} + 1, & \text{w.p. } \Delta t_n^{h,\delta_0,\delta}/\delta_0. \end{cases} \tag{4.13}$$

If $L_n^{h,\delta_0,\delta} = \bar{L}^{\delta_0,\delta} - 1$, then

$$L_n^{h,\delta_0,\delta} \to L_{n+1}^{h,\delta_0,\delta} = \begin{cases} L_n^{h,\delta_0,\delta}, & \text{w.p. } (1 - \Delta t_n^{h,\delta_0,\delta}/\delta_0), \\ 0, & \text{w.p. } \Delta t_n^{h,\delta_0,\delta}/\delta_0. \end{cases} \tag{4.14}$$

The two lines of (4.13) correspond to, respectively,

$$\tilde{N}_{l,n}^{h,\delta_0,\delta}(d\alpha) \to \tilde{N}_{l,n+1}^{h,\delta_0,\delta}(d\alpha) = \begin{cases} \tilde{N}_{l,n}^{h,\delta_0,\delta}(d\alpha), \\ \tilde{N}_{l,n}^{h,\delta_0,\delta}(d\alpha) + I_{\{u_n^{h,\delta_0,\delta}=\alpha\}}. \end{cases} \tag{4.15}$$

In the event of the first line of (4.14), $\tilde{N}_{l,n+1}^{h,\delta_0,\delta}(d\alpha) = \tilde{N}_{l,n}^{h,\delta_0,\delta}(d\alpha)$. In the event of the second line, $\tilde{N}_{l,n+1}^{h,\delta_0,\delta}(\alpha) = 0$, the lth cycle is completed, and we define

$$\tilde{r}_{ee,l}^{h,\delta,\delta_0,\prime}(d\alpha) = \frac{\tilde{N}_{l,n}^{h,\delta_0,\delta}(d\alpha) + I_{\{u_n^{h,\delta_0,\delta}=\alpha\}}}{\bar{L}^{\delta_0,\delta}}. \tag{4.16}$$

Let $I_n^{h,\delta_0,\delta}$ denote the indicator function of the event $L_n^{h,\delta_0,\delta} \neq L_{n+1}^{h,\delta_0,\delta}$; i.e., that the Erlang state changes at step n. Then the ratio (4.16) can be written as

$$\tilde{r}_{ee,l}^{h,\delta,\delta_0,\prime}(d\alpha) = \frac{\sum_{i=m_l}^{m_{l+1}-1} I_{\{u_i^{h,\delta_0,\delta}=\alpha\}} I_i^{h,\delta_0,\delta}}{\sum_{i=m_l}^{m_{l+1}-1} I_i^{h,\delta_0,\delta}}. \tag{4.17a}$$

The expression (4.12) will be

$$\tilde{r}_{ee,l,n}^{h,\delta,\delta_0,\prime}(d\alpha) = \frac{\sum_{i=m_l}^{n-1} I_{\{u_i^{h,\delta_0,\delta}=\alpha\}} I_i^{h,\delta_0,\delta}}{\sum_{i=m_l}^{n-1} I_i^{h,\delta_0,\delta}}. \tag{4.17b}$$

The denominator of the last expression defines $\tilde{D}_{l,n}^{h,\delta_0,\delta}$.

The control memory segment of the approximating chain. Summarizing the above discussion, for $m_l^{h,\delta_0,\delta} \le n < m_{l+1}^{h,\delta_0,\delta}$, the values of the control memory segment for the periodic-Erlang approximation are given by the process $\bar{r}_{ee,n}^{h,\delta_0,\delta,\prime}(d\alpha,\theta)$ defined by

[2] The probabilities in (4.13) and (4.14) are the conditional probabilities, given the path and control data to and including iterate n.

$$\bar{r}^{h,\delta_0,\delta,\prime}_{ee,n}(d\alpha,\theta) = \begin{cases} \tilde{r}^{h,\delta_0,\delta,\prime}_{ee,l,n}(d\alpha), & \theta \in [-\delta/2, 0), \\ \tilde{r}^{h,\delta_0,\delta,\prime}_{ee,l-1}(d\alpha), & \theta \in [-\delta/2-\delta, -\delta/2), \\ \vdots \\ \tilde{r}^{h,\delta_0,\delta,\prime}_{ee,l-Q^+_\delta}(d\alpha), & \theta \in [-\bar{\theta}, -\bar{\theta}+\delta). \end{cases} \tag{4.18}$$

For $\theta = 0$, we define, as usual, $\bar{r}^{h,\delta_0,\delta,\prime}_{ee,n}(d\alpha, 0) = I_{\{u^{h,\delta_0,\delta}_n = \alpha\}}$.

Evaluating the randomization errors in (4.17). By (4.13) and (4.14), $\Delta t^{h,\delta_0,\delta}_n/\delta_0$ is the conditional expectation of $I^{h,\delta_0,\delta}_n$ given the data to and including iterate n. Hence (4.17a) can be written as

$$\frac{\left[\sum_{i=m_l}^{m_{l+1}-1} I_{\{u^{h,\delta_0,\delta}_i=\alpha\}} \Delta t^{h,\delta_0,\delta}_i/\delta_0 + \rho^{h,\delta_0,\delta}_{l,N}\right]}{\left[\sum_{i=m_l}^{m_{l+1}-1} \Delta t^{h,\delta_0,\delta}_i/\delta_0 + \rho^{h,\delta_0,\delta}_{l,D}\right]},$$

which equals

$$= \frac{\left[\sum_{i=m_l}^{m_{l+1}-1} I_{\{u^{h,\delta_0,\delta}_i=\alpha\}} \Delta t^{h,\delta_0,\delta}_i + \delta_0\rho^{h,\delta_0,\delta}_{l,N}\right]}{\left[\sum_{i=m_l}^{m_{l+1}-1} \Delta t^{h,\delta_0,\delta}_i + \delta_0\rho^{h,\delta_0,\delta}_{l,D}\right]}, \tag{4.19}$$

where the ρ-terms represent the randomization noise. In particular,

$$\delta_0\rho^{h,\delta_0,\delta}_{l,D} = \sum_{i=m_l}^{m_{l+1}-1} \delta_0 \left[I^{h,\delta_0,\delta}_i - \Delta t^{h,\delta_0,\delta}_i/\delta_0\right],$$

whose summands are martingale differences with conditional variance

$$\delta_0 \Delta t^{h,\delta_0,\delta}_i (1 - \Delta t^{h,\delta_0,\delta}_i/\delta_0).$$

The conditional (given the data to the start of the lth cycle) variance of the randomization error in the denominator of (4.19) is bounded by the conditional expectation (given the same data) of $\delta_0[\bar{\sigma}^{h,\delta_0,\delta}_{l+1} - \bar{\sigma}^{h,\delta_0,\delta}_l]$. The sum in the denominator of (4.19) is $[\bar{\sigma}^{h,\delta_0,\delta}_{l+1} - \bar{\sigma}^{h,\delta_0,\delta}_l]$. Thus as $h \to 0$ and then $\delta_0 \to 0$, the effect of the randomization error in (4.17a) goes to zero. The analogous result holds for (4.17b). These facts will be essential to the proof of Theorem 4.1.

Comment on the memory requirements. Suppose that $U = U^h = \{0, 1\}$. Then we need only keep track of the values for $\alpha = 1$. The denominator in (4.17b) is the value of the Erlang state at iterate $n-1$, and the numerator is the number of times that the control has taken the value unity at the steps up to and including $(n-1)$ where the Erlang state has advanced since it last had the value zero. Thus, (4.17b) is the fraction of times in the current cycle up to and including iterate $(n-1)$ that the control has taken the value unity at the times when the Erlang state advanced.

The expression (4.17a) takes $\bar{L}^{\delta_0,\delta}+1$ values, as the number of times that the control can take the value α per cycle can be any of $0,1,\ldots,\bar{L}^{\delta_0,\delta}$. The denominator of (4.17b) is the current Erlang state and takes the values $(0,1,\ldots,\bar{L}^{\delta_0,\delta}-1)$. If the value is k, then the numerator can take any of the values $(0,1,\ldots,k)$. Thus it requires $\bar{L}^{\delta_0,\delta}(\bar{L}^{\delta_0,\delta}+1)/2$ points to record the evolution of (4.17b). Summarizing, with the representation (4.17), the function that represents the entire control memory segment takes $[\bar{L}^{\delta_0,\delta}+1]^{Q_\delta^+}[\bar{L}^{\delta_0,\delta}][\bar{L}^{\delta_0,\delta}+1]/2$ values. This quantity does not depend on the value of h. For the direct method of Subsection 7.2.2, the total number of required points would be $2^{\bar{\theta}/\bar{\Delta}^h}$, which would be substantially larger as, typically, $\bar{\Delta}^h=O(h^2)$. The required number can be reduced further if we discretize the $\tilde{r}^{h,\delta_0,\delta,\prime}_{ee,l-i}(d\alpha)$ more coarsely.

The evolution of the control memory segment. In the event of (4.13) or the first line of (4.14), we can represent the transitions of the control memory vector as follows:

$$\begin{aligned}\bar{r}^{h,\delta_0,\delta,\prime}_{ee,n}(d\alpha) &= \left(\tilde{r}^{h,\delta_0,\delta,\prime}_{ee,l-Q_\delta^+}(d\alpha),\ldots,\tilde{r}^{h,\delta_0,\delta,\prime}_{ee,l-1}(d\alpha),\tilde{r}^{h,\delta_0,\delta,\prime}_{ee,l,n}(d\alpha)\right)\\ \longrightarrow \bar{r}^{h,\delta_0,\delta}_{ee,n+1}(d\alpha) &= \left(\tilde{r}^{h,\delta_0,\delta,\prime}_{ee,l-Q_\delta^+}(d\alpha),\ldots,\tilde{r}^{h,\delta_0,\delta,\prime}_{ee,l-1}(d\alpha),\tilde{r}^{h,\delta_0,\delta,\prime}_{ee,l,n+1}(d\alpha)\right),\end{aligned} \tag{4.20}$$

In the event of the second line of (4.14),

$$\begin{aligned}\bar{r}^{h,\delta_0,\delta}_{ee,n}(d\alpha) &\longrightarrow \bar{r}^{h,\delta_0,\delta}_{ee,n+1}(d\alpha)\\ &= \left(\tilde{r}^{h,\delta_0,\delta,\prime}_{ee,l-Q_\delta^++1}(d\alpha),\ldots,\tilde{r}^{h,\delta_0,\delta,\prime}_{ee,l-1}(d\alpha),\tilde{r}^{h,\delta_0,\delta,\prime}_{ee,l}(d\alpha),0\right).\end{aligned} \tag{4.21}$$

A continuous time interpolation. Definitions. Let $r^{h,\delta_0\delta}(\cdot)$ denote the continuous time interpolation (with intervals $\{\Delta t_n^{h,\delta_0\delta}\}$) of the relaxed control representation of the controls $\{u_n^{h,\delta_0,\delta}\}$. Let $r_\tau^{h,\delta_0\delta}(\cdot)$ denote the interpolation with intervals $\{\Delta\tau_n^{h,\delta_0\delta}\}$. In the current context, the κ in the form (7.1.31) becomes the pair (δ_0,δ), and the subscript "a" is ee for the control memory segments. For future use, let us define the following alternative notation for the control memory segments. The *full* memory segment at iterate n, based on the actual *controls* that are used, is the function of α,t, and $\theta\in[-\bar{\theta},0)$ defined by

$$\bar{r}_n^{h,\delta_0,\delta,\prime}(d\alpha,\theta)=r^{h,\delta_0,\delta,\prime}(d\alpha,t_n^{h,\delta_0,\delta}+\theta).$$

Define the functions of α,t, and $\theta\in[-\bar{\theta},0]$ that are based on the actual *control memory segments* that are used:

$$\left.\begin{aligned} \tilde{r}^{h,\delta_0,\delta,\prime}_{ee}(d\alpha,t,\theta) &= \bar{r}^{h,\delta_0,\delta,\prime}_{ee,n}(d\alpha,\theta), \quad \text{for } \theta\in[-\bar{\theta},0) \\ \tilde{r}^{h,\delta_0,\delta,\prime}_{ee}(d\alpha,t,0) &= I_{\{u^{h,\delta_0,\delta}_n=\alpha\}} \end{aligned}\right\} t\in[t^{h,\delta_0,\delta}_n, t^{h,\delta_0,\delta}_{n+1}), \tag{4.22a}$$

$$\left.\begin{aligned} \tilde{r}^{h,\delta_0,\delta,\prime}_{ee,\tau}(d\alpha,t,\theta) &= \bar{r}^{h,\delta_0,\delta,\prime}_{ee,n}(d\alpha,\theta), \quad \text{for } \theta\in[-\bar{\theta},0) \\ \tilde{r}^{h,\delta_0,\delta,\prime}_{ee,\tau}(d\alpha,t,0) &= I_{\{u^{h,\delta_0,\delta}_n=\alpha\}} \end{aligned}\right\} t\in[\tau^{h,\delta_0,\delta}_n, \tau^{h,\delta_0,\delta}_{n+1}). \tag{4.22b}$$

For future use, for each $\theta \in [-\bar{\theta},0]$, let $\tilde{r}^{h,\delta_0,\delta}_{ee,\tau}(d\alpha,t,\theta)$ denote the relaxed control process whose derivative at time t is $\tilde{r}^{h,\delta_0,\delta,\prime}_{ee,\tau}(d\alpha,t,\theta)$. It is piecewise-constant in θ (on the intervals $[-d/2,0), [-3\delta/2,-\delta/2),\ldots$).

Recall the definitions

$$\begin{aligned} &\bar{\xi}^{h,\delta_0,\delta}_n \text{ is the path segment } \{\xi^{h,\delta_0,\delta}(t^{h,\delta_0,\delta}_n+\theta), \theta\in[-\bar{\theta},0]\}, \\ &\bar{\xi}^{h,\delta_0,\delta}_e(\cdot) \text{ is the interpolation of } \{\bar{\xi}^{h,\delta_0,\delta}_{e,n}\} \text{ with intervals } \{\Delta t^{h,\delta_0,\delta}_n\}. \end{aligned} \tag{4.23}$$

A continuous time interpolation. With the path memory segment (2.1) and control memory segment (4.18), the interpolated path process (with intervals $\{\Delta\tau^{h,\delta_0,\delta}_n\}$) can be written as

$$\begin{aligned} \psi^{h,\delta_0,\delta}(t) &= \xi^h_0 \\ &+ \int_{-\bar{\theta}}^0 \left[\int_0^t \int_{U^h} b_h(\bar{\xi}^{h,\delta_0,\delta}_e(q^{h,\delta_0,\delta}_\tau(s)),\alpha,\theta)\tilde{r}^{h,\delta_0,\delta,\prime}_{ee,\tau}(d\alpha,s,\theta)ds\right]\mu_c(d\theta) \\ &+ B^{h,\delta_0,\delta}_\tau(t) + z^{h,\delta_0,\delta}_\tau(t), \end{aligned} \tag{4.24}$$

where the martingale $B^{h,\delta_0,\delta}_\tau(\cdot)$ has quadratic variation process

$$\int_0^t a_h(\bar{\xi}^{h,\delta_0,\delta}_e(q^{h,\delta_0,\delta}_\tau(s)))ds,$$

and, as below (6.5.12), it can be represented as (modulo an asymptotically negligible error)

$$B^{h,\delta_0,\delta}_\tau(t) = \int_0^t \sigma(\bar{\xi}^{h,\delta_0,\delta}_e(q^{h,\delta_0,\delta}_\tau(s)))dw^{h,\delta_0,\delta}(s), \tag{4.25}$$

where the martingale $w^{h,\delta_0,\delta}(\cdot)$ has quadratic variation It and converges weakly to a standard Wiener process as either $h\to 0$, $(h,\delta_0)\to 0$, or $(h,\delta_0,\delta)\to 0$.

The convergence theorem. Let $V^\delta(\hat{x},\hat{u})$ denote the optimal cost for system (7.1.2) and cost (7.1.4), but with the periodic memory segments, where that for the control is given in (4.4.3), and that for the path is given in (4.2.6). Let $V^{h,\delta}(\hat{x},\hat{u})$ denote the analogous value for the numerical approximation, where the approximation for the path and control memory segments are (1.1) and

(4.3), resp. Let $V^{\delta_0,\delta}(\hat{x},\hat{u})$ denote the optimal cost when the control and path memory segments are the periodic-Erlang forms (4.2.8) and (4.4.7), resp. Let $V^{h,\delta_0,\delta}(\hat{x},\hat{u})$ denote the optimal cost for the numerical approximation when the path memory segment is that of (2.1) and that for the control is that of this section in (4.18).

Theorem 4.1. *Assume local consistency,* (A3.1.2), (A3.1.3), (A3.2.1)–(A3.2.3), *and* (A3.4.3), *with system* (7.1.2) *and cost function* (7.1.4). *For the chain, let the memory segment for the path be as in* (2.1) *and that for the control as in* (4.18), *with cost function* (2.4). *Let $h = o(\delta_0)$ and let $\bar{\xi}_0^{h,\delta_0,\delta} \in D(G_h; [-\bar{\theta}, 0])$ be piecewise-constant, and converge to $\bar{x}^\delta(0)$, as $h \to 0$ and $\delta_0 \to 0$, where $\bar{x}^\delta(0)$ is piecewise-constant as for the path memory segments for the periodic approximation in* (4.2.6). *Similarly, let $\bar{u}_0^{h,\delta_0,\delta} \in D(U^h; [-\bar{\theta}, 0))$ be piecewise-constant, have values $\bar{u}_0^{h,\delta_0,\delta}(\theta) \in U^h$, and converge in $D(U; [-\bar{\theta}, 0))$ to $\bar{u}^\delta(0)$ as $h \to 0$ and $\delta_0 \to 0$. Let δ be fixed. Then as $h \to 0$ and $\delta_0 \to 0$,*

$$V^{h,\delta_0,\delta}(\bar{\xi}_0^{h,\delta_0,\delta}, \bar{u}_0^{h,\delta_0,\delta}) \to V^\delta(\bar{x}^\delta(0), \bar{u}^\delta(0)). \tag{4.26}$$

The analogous result holds for the analog of the cost functional (7.1.28) *if* (A3.4.1) *and* (A3.4.2) *are assumed and the conditions on the reflection directions are dropped.*

8.5 Proofs of Convergence

8.5.1 Proofs of Theorems from Chapter 7

Proof of Theorem 7.1.3. Until further notice, suppose that only the path is delayed. Thus the cost function for the chain is (7.1.28). Let $u^h = \{u_n^h, n < N_G^h\}$ be the optimal control sequence for the process $\{\xi_n^h\}$. The control stops at time N_G^h. But for convenience in the notation of the proof we suppose that some arbitrary admissible control is used for $n \geq N_G^h$. Let $r_\tau^h(\cdot)$ denote the relaxed control representation of the interpolation of $\{u_n^h\}$ with intervals $\{\Delta\tau_n^h\}$. In all cases, $\xi^h(\cdot)$ and $\psi^h(\cdot)$ are constructed from the approximating chain ξ_n^h, via the appropriate interpolation.

The main new issue in the delay case (over that for the no-delay case in [58]) is that the memory-segment process $\bar{\xi}^h(t)$ appears in the functions $b(\cdot), \sigma(\cdot), k(\cdot)$. For $s \geq 0$, recall the definitions $d_\tau^h(s) = \max\{n : \tau_n^h \leq s\}$ and $q_\tau^h(s) = t_{d_\tau^h(s)}^h$ from (6.5.23), and recall the representation (7.1.14):

$$\psi^h(t) = \xi_0^h + \int_0^t \int_{U^h} b_h(\bar{\xi}^h(q_\tau^h(s)), \alpha) r_\tau^h(d\alpha\, ds) + B_\tau^h(t) + z_\tau^h(t), \tag{5.1}$$

where $B_\tau^h(\cdot)$ is a martingale with quadratic variation process

$$\int_0^t a_h(\bar{\xi}^h(q_\tau^h(s)))ds.$$

Because we are starting with the case where the process stops on hitting the boundary, there is no reflection term, but we include the term $z_\tau^h(\cdot)$ in (5.1) for future use.

There is martingale $w^h(\cdot)$ with quadratic variation process It, where I is the identity matrix if $x(t)$ is vector-valued, and it is just unity for the scalar case,[3] such that [58, Section 10.4.1][4]

$$B_\tau^h(t) = \int_0^t \sigma_h(\bar{\xi}^h(q_\tau^h(s)), u_\tau^h(s))dw^h(s) = \int_0^t \sigma(\bar{\xi}^h(q_\tau^h(s)))dw^h(s) + \epsilon^h(t), \tag{5.2}$$

where the martingale error term $\epsilon^h(\cdot)$ converges weakly and in mean square (on any bounded time interval) to the zero process as $h \to 0$. The discontinuities of $w^h(t)$ go to zero as $h \to 0$, and it converges to a standard Wiener process. The proofs of these assertions are the same as for the no-delay case in [58, Section 10.4]. Theorem 7.3.1 applies and shows that the timescales for $\xi^h(\cdot)$ and $\psi^h(\cdot)$ are asymptotically equal and that $\sup_{s \le T} |q_\tau^h(s) - s| \to 0$ for any $T < \infty$.

The sequence of processes $\{\psi^h(\cdot), \xi^h(\cdot), r_\tau^h(\cdot), B_\tau^h(\cdot), w^h(\cdot), q_\tau^h(\cdot)\}$ is tight and all weak-sense limits are continuous. Take a weakly convergent subsequence, which (abusing notation) we also index by h, and with limit denoted by $(x(\cdot), \xi(\cdot), r(\cdot), B(\cdot), w(\cdot), q(\cdot))$. Assume the Skorokhod representation (Theorem 2.1.6) so that the limits can be assumed to be w.p.1. By the timescale equivalence of Theorem 7.3.1, $x(\cdot) = \xi(\cdot)$. Thus the memory segment process $\bar{\xi}^h(\cdot)$ converges to $\bar{x}(\cdot)$, which is continuous in time. By the continuity assumption (A3.1.1) and the weak convergence, we can write

$$\int_0^t \int_{U^h} b_h(\bar{\xi}^h(q_\tau^h(s)), \alpha) r_\tau^h(d\alpha\, ds) \to \int_0^t \int_U b(\bar{x}(s), \alpha) r(d\alpha\, ds). \tag{5.3}$$

To prove that

$$\int_0^t \sigma(\bar{\xi}^h(q_\tau^h(s)))dw^h(s) \to \int_0^t \sigma(\bar{x}(s))dw(s) \tag{5.4}$$

we need to work with a discretized form as in the proof of Theorem 3.5.1. For any function $f(\cdot)$ of a real variable and $\kappa > 0$, define $f_\kappa(s) = f(n\kappa)$ for $n\kappa \le s < n\kappa + \kappa$. By the martingale and quadratic variation properties of $w^h(\cdot)$, the mean square value of the martingale process

$$\begin{aligned} &\int_0^t \sigma(\bar{\xi}^h(q_\tau^h(s)))dw^h(s) - \int_0^t \sigma(\bar{\xi}_\kappa^h(q_\tau^h(s)))dw^h(s) \\ &\qquad = \int_0^t \left[\sigma(\bar{\xi}^h(q_\tau^h(s))) - \sigma(\bar{\xi}_\kappa^h(q_\tau^h(s)))\right] dw^h(s) \end{aligned} \tag{5.5}$$

[3] The probability space might have to be augmented by adding an "independent" Wiener process to construct $w^h(\cdot)$ if $a_h(\hat{\xi})$ is degenerate at any point.

[4] In using the reference [58], note that our $B_\tau^h(\cdot)$ is called $M^h(\cdot)$ there.

is just the integral of the mean value of the square of the term in brackets in the right-hand integrand, and the $\limsup_{h\to 0}$ goes to zero as $\kappa \to 0$, due to the convergence of $\xi^h(\cdot)$ and $q^h_\tau(\cdot)$ to continuous processes. It follows that the supremum over any finite time interval of the expression (5.5) goes to zero in mean square as $h \to 0$ and then $\kappa \to 0$.

For fixed $\kappa > 0$, the integral in the right side of the first line of (5.5) can be written as a sum, and the weak convergence implies that its limit as $h \to 0$ is $\int_0^t \sigma(\bar{x}_\kappa(s))dw(s)$. These arguments imply that, for small κ, the left side of (5.4) is arbitrarily close to $\int_0^t \sigma(\bar{x}_\kappa(s))dw(s)$ as $h \to 0$. The completion of the proof of (5.4) involves taking the limit as $\kappa \to 0$ in this last expression. To justify this last step, we must show that $x(\cdot)$ is nonanticipative with respect to $w(\cdot)$.

The nonanticipativity of $(x(\cdot), y(\cdot), r(\cdot))$ with respect to the Wiener process $w(\cdot)$ is proved by using the method of Theorem 3.5.1. The analog of (3.5.2) for the current case is

$$Eh\big(\psi^h(s_i), w^h(s_i), \langle r^h_\tau, \phi_j\rangle(s_i), i \le I, j \le J\big)\big(w^h(t+T) - w^h(t)\big) = 0. \tag{5.6}$$

By the weak convergence and uniform integrability of $\{w^h(t), h > 0\}$ for each t,

$$Eh\left(x(s_i), w(s_i), \langle r, \phi_j\rangle(s_i), i \le I, j \le J\right)\left(w(t+T) - w(t)\right) = 0. \tag{5.7}$$

The nonanticipativity is a consequence of this expression and the arbitrariness of the functions and times in (5.7), analogously to the situation dealt with in Theorem 3.5.1. With the nonanticipativity proved, let $\kappa \to 0$ in $\int_0^t \sigma(\bar{x}_\kappa(s))dw(s)$ to get (5.4). Thus we have proved that the limit satisfies (7.1.5), but without the $z(\cdot)$ term, for some relaxed control $r(\cdot)$.

By the convergence $\psi^h(\cdot) \to x(\cdot)$ and (A3.4.2), $\xi^h_{N^h_G} \to x(\tau)$, where τ is the first time that $x(\cdot)$ touches ∂G. By the definitions of $W^h(\cdot)$ and $V^h(\cdot)$ we have $V^h(\bar{\xi}^h_0) = W^h(\bar{\xi}^h_0, u^h)$. By the weak convergence, the convergence of the initial condition $\bar{\xi}^h_0$ to $\bar{x}(0)$, and the continuity properties of $k(\cdot)$ and $g_0(\cdot)$ in (A3.4.1), it follows that $W^h(\bar{\xi}^h_0, u^h) \to W(\bar{x}(0), r)$. By the minimality of $V(\bar{x}(0))$, we must have $\liminf_h V^h(\bar{\xi}^h_0) \ge V(\bar{x}(0))$.

To complete the proof of the convergence $V^h(\bar{\xi}^h_0) \to V(\bar{x}(0))$, we need only prove that

$$\limsup_h V^h(\bar{\xi}^h_0) \le V(\bar{x}(0)). \tag{5.8}$$

For the no-delay case this was proved in [58, Chapters 10,11], based on [58, Theorem 10.3.1], and the form of this theorem for the delay case was given in Theorem 3.5.4. The method can be readily adapted to our needs. First, Theorem 3.5.4 will be restated in slightly simpler form for the model (3.1.1) and cost function (3.4.1). [By Theorem 3.5.4, there is an analogous construction when the control is delayed and/or if the boundary is reflecting.] By Theorem 3.5.4, for each $\epsilon > 0$ there is an ϵ-optimal control $u^\epsilon(\cdot)$ of the following type.

There are finite sets $U_\epsilon \subset U$ (and we can suppose w.l.o.g. that $U_\epsilon \subset U^h$) and a $\kappa > 0$ such that $u^\epsilon(\cdot)$ is U_ϵ-valued, right-continuous, and constant on intervals $[n\kappa, n\kappa + \kappa)$, $n = 0, 1, \ldots$. Let $w(\cdot)$ denote the driving Wiener process. The control can be represented as follows, for $\alpha \in U_\epsilon$:

$$\begin{aligned} &P\left\{u^\epsilon(n\kappa) = \alpha \middle| \text{data to time } n\kappa\right\} \\ &\quad = P\left\{u^\epsilon(n\kappa) = \alpha \middle| \bar{x}(0), w(i\kappa), i \le n; u^\epsilon(i\kappa), i < n\right\}, \end{aligned} \tag{5.9}$$

where the probability on the right side is continuous in the initial condition and in the samples $w(i\kappa), i < n$, for each value of the set of controls in the conditioning, and there are only a finite number of such control values for each n. This construction is only for use in the convergence proof. It has no value for numerical purposes.

The next step is to apply this control to the ξ_n^h process. The adaptation will be denoted by $\{u_n^{h,\epsilon}\}$, and the timing (to be described below) will be determined by the scale of the corresponding continuous-time interpolation $\psi^h(\cdot)$. The continuous-time interpolation of $\{u_n^{h,\epsilon}\}$ with (intervals $\{\Delta\tau_n^h\}$) will be called $u^{h,\epsilon}(\cdot)$. This adaptation is done by following [58, Theorem 10.5.2], with a slightly altered notation. For $n = 0, 1, \ldots$, define $\eta_n^h = \min\{\tau_k^h : \tau_k^h \ge n\kappa\}$, the first jump of $\psi^h(\cdot)$ at or after interpolated time $n\kappa$. Thus $\eta_0^h = 0$. The control $u^{h,\epsilon}(\cdot)$ will be constant on the intervals $[\eta_n^h, \eta_{n+1}^h)$, with values determined by the adaptation of (5.9):

$$\begin{aligned} &P\left\{u^{h,\epsilon}(\eta_n^h) = \alpha \middle| \text{data to time } \eta_n^h\right\} \\ &\quad = P\left\{u^\epsilon(\eta_n^h) = \alpha \middle| \bar{\xi}^h(0), w^{h,\epsilon}(\eta_i^h), i \le n : u^\epsilon(\eta_i^h), i < n\right\}, \end{aligned} \tag{5.10}$$

where $w^{h,\epsilon}(\cdot)$ is the analog of the process $w^h(\cdot)$ in (5.2) that corresponds to the use of this control. Now, applying this control to $\psi^h(\cdot)$ and using a weak convergence argument and the continuity of the law (5.9) in the initial condition and the samples of the $w(\cdot)$, implies that $W^h(\bar{\xi}^h(0), u^{h,\epsilon}) \to W(\bar{x}(0), u^\epsilon)$. Because $W(\bar{x}(0), u^\epsilon) \le V(\bar{x}(0)) + \epsilon$ and $V^h(\bar{\xi}^h(0)) \le W^h(\bar{\xi}^h(0), u^{h,\epsilon})$ and ϵ is arbitrary, (5.8) follows. This completes the proof of Theorem 7.1.3 when only the path is delayed.

Delay in the control. Now let there be delays in the control and use the system model (7.1.2). By the weak convergence, the bracketed term in (7.1.25) converges to

$$\int_0^t \int_U b(\bar{x}(s), \alpha, v) r(d\alpha, ds + v)$$

for all $v \in [-\bar{\theta}, 0]$, and the analogous convergence holds for the bracketed term in (7.1.27). The rest of the details are as for the case where only the state is delayed. ■

Theorems 7.1.1 and 7.1.2. Reflecting boundaries. Now replace the absorbing boundary by a reflecting boundary. The main new issue in the proof concerns the asymptotic properties of the reflection process. Under (A3.2.1) and (A3.2.2), it is shown in Theorem 3.5.5 (see also [56, Theorem 3.6.1] and [58, Theorem 11.1.2]) that $\{z^h(\cdot)\}$ is tight with all weak-sense limits being continuous. This implies that $\{\psi^h(\cdot), \xi^h(\cdot)\}$ is tight and that the weak-sense limit processes are continuous. Let h index a weakly convergent subsequence of $\{\psi^h(\cdot), \xi^h(\cdot), r^h_\tau(\cdot), B^h_\tau(\cdot), w^h(\cdot), q^h_\tau(\cdot), z^h(\cdot)\}$. Then the proof that the weak-sense limit $(x(\cdot), x(\cdot), r(\cdot), w(\cdot), z(\cdot))$ of $\{\psi^h(\cdot), \xi^h(\cdot), r^h_\tau(\cdot), w^h(\cdot), z^h(\cdot)\}$ satisfies (7.1.2) is the same as what was done above for the absorbing boundary case, except for the addition of the term $z(\cdot)$. The proof that the limit $z(\cdot)$ is a reflecting process for the limit $x(\cdot)$ is shown in Theorem 3.5.5 (see also [58, Theorem 11.1.2]).

Now consider the convergence of the cost functionals and recall (A3.4.3). Define $z(\cdot) = \sum_i d_i y_i(\cdot)$. Under the first part of (A3.4.3), the processes $y^h_i(\cdot)$ converge to $y_i(\cdot)$. If the reflection directions on the faces adjoining some edge or corner are linearly dependent, then some of the components $y_i(\cdot)$ of the limit $z(\cdot)$ might not be uniquely defined. In this case, the second part of (A3.4.3) supposes that the coefficients q_i corresponding to any such linearly dependent set are identical. Then the sum of the associated $y^h_i(\cdot)$ would converge to the sum of the limit values. The proof is completed using these observations and the uniform integrability implied by the estimates in Lemma 6.3.1. ■

Proof of Theorem 7.1.4. The proof is essentially a consequence of the proofs of the previous theorems and the properties (7.1.34) and either (7.1.35), (7.1.36), or (7.1.37). We will use (7.1.35). The arguments under (7.1.36) and (7.1.37) are similar. Recall the definition of $\tilde{r}^{h,\kappa,\prime}_{a,\tau}(\cdot)$ in (7.1.31b).

Let $r^{h,\kappa}_\tau(\cdot)$ denote the relaxed control representation of the interpolation (intervals $\{\Delta\tau^{h,\kappa}_n\}$) of the actual controls $\{u^{h,\kappa}_n\}$ that are used. Write equation (7.1.33) as

$$\psi^{h,\kappa}(t) = \xi^h_0 + \int_{-\bar{\theta}}^0 \left[\int_0^t \int_{U^h} b_h(\bar{\xi}^{h,\kappa}(q^{h,\kappa}_\tau(s)), \alpha, \theta) r^{h,\kappa,\prime}_\tau(d\alpha, s+\theta)ds\right] \mu_c(d\theta) \\ + B^{h,\kappa}_\tau(t) + z^{h,\kappa}_\tau(t) + \epsilon^{h,\kappa}(t), \tag{5.11}$$

where the error term $\epsilon^{h,\kappa}(\cdot)$ accounts for the use of the $\bar{\xi}^{h,\kappa}(t)$ and $r^{h,\kappa,\prime}_\tau(d\alpha, s+\theta)$ in lieu of the $\bar{\xi}^{h,\kappa}_a(t)$ and $\tilde{r}^{h,\kappa,\prime}_{a,\tau}(d\alpha, s, \theta)$ processes that are actually used for the memory segments in the dynamics. Analogously to the expressions below (7.1.14), we can write

$$B^{h,\kappa}_\tau(t) = \int_0^t \sigma(\bar{\xi}^{h,\kappa}(q^{h,\kappa}_\tau(s)))dw^{h,\kappa}(s) + \epsilon^{h,\kappa}_1(t), \tag{5.12}$$

where the error term is due to the use of $\sigma(\cdot)$ and $\bar{\xi}^{h,\kappa}(\cdot)$ in lieu of $\sigma_h(\cdot)$ and $\bar{\xi}^{h,\kappa}_a(\cdot)$, resp. By (7.1.34) and (7.1.35), the asymptotic continuity of $\bar{\xi}^{h,\kappa}(q^{h,\kappa}_\tau(\cdot))$, the timescale equivalence shown in Theorem 7.3.1, and the fact that $\sigma_h(\hat{x}) - \sigma(\hat{x}) \to 0$ uniformly in $\hat{x}$, the sups of the error terms in (5.11) and

(5.12) over any finite time interval go to zero in mean square as $h \to 0$ and then $\kappa \to 0$. It then follows that the limits of $\psi^{h,\kappa}(\cdot)$ and $V^{h,\kappa}(\cdot)$ as $h \to 0$ and then $\kappa \to 0$ are the same as if the memory segment processes were $\bar{\xi}^{h,\kappa}(\cdot)$ and $\tilde{r}^{h,\kappa,\prime}(\cdot)$. From this point on, the proof is just that of Theorems 7.1.1–7.1.3. ■

Proof of Theorem 7.3.3. Owing to the asymptotic equivalences of the timescales shown in Theorem 7.3.1,

$$\lim_{h\to 0,\delta\to 0} \sup_{\text{control}} \sup_n E \sup_{-\bar{\theta}\le\theta\le 0} \left|\bar{\xi}^{h,\delta}_{r,n}(\theta) - \bar{\xi}^{h,\delta}_{n}(\theta)\right] = 0.$$

Because the control is not delayed, the proof follows from Theorem 7.1.4 with δ, r replacing κ, a.

Comment on Theorem 7.4.2. Theorem 7.4.1 implies that the sequence of intervals between shifts converges to a sequence of i.i.d. random variables, each of which is exponentially distributed with mean δ. Using this fact, the details are similar to those of Theorems 7.1.1–7.1.3 and are omitted.

8.5.2 Proof of Theorem 4.1

We will discuss only the problem with a reflecting boundary. Let $h \to 0, \delta_0 \to 0$, with $h = o(\delta_0)$ and δ fixed. Recall the definition of the (piecewise-constant in θ) relaxed control $\tilde{r}^{h,\delta_0,\delta}_{ee,\tau}(d\alpha, t, \theta)$ that was given below (4.22) and that of $r^{h,\delta_0\delta}_{\tau}(\cdot)$ that was given above (4.22). It follows from Theorem 7.4.1 that the intervals between the δ-shifts converge to the constant δ. The set, indexed by (h, δ_0),

$$\Big\{\psi^{h,\delta_0,\delta}(\cdot), \xi^{h,\delta_0,\delta}(\cdot), r^{h,\delta_0,\delta}_{\tau}(\cdot), w^{h,\delta_0,\delta}(\cdot), y^{h,\delta_0,\delta}(\cdot);$$
$$\tilde{r}^{h,\delta_0,\delta}_{ee,\tau}(\cdot,\theta), \theta \in [-\bar{\theta}, 0]; \bar{\sigma}^{h,\delta_0,\delta}_{l}, l < \infty\Big\}$$

is tight. Let

$$\left(x^{\delta}(\cdot), x^{\delta}(\cdot), r^{\delta}(\cdot), w^{\delta}(\cdot), y^{\delta}(\cdot); \tilde{r}^{\delta}_{p}(\cdot,\theta), \theta \in [-\bar{\theta}, 0]; l\delta, l < \infty\right)$$

denote the limit of a weakly convergent subsequence (also indexed by h, δ_0 for notational convenience). Suppose that the Skorokhod representation is used, so that we can suppose that the convergence is w.p.1 in the appropriate topologies. Then $q^{h,\delta_0,\delta}_{\tau}(t) \to t$ and $\bar{\xi}^{h,\delta_0,\delta}_{e}(t) \to \bar{x}^{\delta}_{p}(t)$, where $\bar{x}^{\delta}_{p}(t)$ is now obtained from $x^{\delta}(\cdot)$ as it was obtained from $x(\cdot)$ in (4.2.6).

By the w.p.1 convergence, the difference between the drift term in (4.24) and

$$\int_{-\bar{\theta}}^{0}\left[\int_{0}^{t}\int_{U^h} b(\bar{x}^{\delta}_{p}(s), \alpha, \theta)\tilde{r}^{h,\delta_0,\delta,\prime}_{ee,\tau}(d\alpha, s, \theta)ds\right]\mu_c(d\theta) \tag{5.13}$$

goes to zero as $h \to 0$. Let $\bar{\sigma}^{h,\delta_0,\delta}_{l} \le t^{h,\delta_0,\delta}_{n} < \bar{\sigma}^{h,\delta_0,\delta}_{l+1}$. Then neglecting the randomization errors in the representation (4.19) of (4.17a), (4.19) can be written as

$$\frac{\displaystyle\int_{\bar{\sigma}_l^{h,\delta_0,\delta}}^{\bar{\sigma}_{l+1}^{h,\delta_0,\delta}} r^{h,\delta_0,\delta,\prime}(d\alpha,s)ds}{\bar{\sigma}_{l+1}^{h,\delta_0,\delta}-\bar{\sigma}_l^{h,\delta_0,\delta}},$$

whose limit (as $h \to 0, \delta_0 \to 0$) is

$$\frac{r^\delta(d\alpha, l\delta+\delta) - r^\delta(d\alpha, l\delta)}{\delta}. \tag{5.14}$$

Similarly, neglecting the randomization errors in (4.17b), it can be written as

$$\frac{\displaystyle\int_{\bar{\sigma}_l^{h,\delta_0,\delta}}^{t_n^{h,\delta_0,\delta}} r^{h,\delta_0,\delta,\prime}(d\alpha,s)ds}{t_n^{h,\delta_0,\delta}-\bar{\sigma}_l^{h,\delta_0,\delta}},$$

whose limit (as $h \to 0, \delta_0 \to 0$) is

$$\frac{r^\delta(d\alpha, t) - r^\delta(d\alpha, l\delta)}{t - l\delta}. \tag{5.15}$$

The comments below (4.19) concerning the randomization errors imply that they can be neglected as $h \to 0, \delta_0 \to 0$.

Recalling the form (4.4.3), where the periodic approximation $\tilde{r}_p^\delta(\cdot)$ to the control memory segment was obtained in terms of the relaxed control representation of the actual control $u(\cdot)$ that was used, we see that the relationship between our limits $r^\delta(\cdot)$ and $\tilde{r}_p^\delta(\cdot)$ is just that between the terms $r(\cdot)$ and $\tilde{r}_p^\delta(\cdot)$ in (4.4.3). Thus the limit of the control memory segment is just the periodic approximation. By the above convergence of the $\tilde{r}_{ee,\tau}^{h,\delta_0,\delta}(\cdot)$, the asymptotic difference between (5.13) and

$$\int_{-\bar{\theta}}^0 \left[\int_0^t \int_U b(\bar{x}_p^\delta(s),\alpha,\theta)\tilde{r}_p^{\delta,\prime}(d\alpha,s,\theta)ds\right]\mu_c(d\theta) \tag{5.16}$$

is zero. The rest of the details are like those for Theorem 7.1.3 and are omitted. ■

8.6 Singular Controls

Only a few comments will be made. Some additional comments are in Chapter 9. First consider the model (3.6.1):

$$dx(t) = b(\bar{x}(t))dt + q_1(\bar{x}(t-))d\lambda(t) + \sigma(\bar{x}(t))dw(t) + dz(t). \tag{6.1}$$

In (6.1) there is no delay in the singular control $\lambda(\cdot)$. The methods of Section 6.6 for the no-delay problem can be carried over, but where the path memory segment is approximated by any of the methods of this chapter or of Chapter 7, and nothing more will be said about it.

Now suppose that there is a delay in the singular control, say with the model

$$\begin{aligned} dx(t) &= c(x(t))dt + dt\int_{-\bar\theta}^{0} b(x(t+\theta),\theta)d\mu_a(\theta) + \sigma(x(t))dw(t) \\ &\quad + q_0(x(t-))d\lambda(t) + dt\int_{\theta=-\bar\theta}^{0} q_2(x((t+\theta)-),\theta)d_\theta\lambda(t+\theta) + dz(t). \end{aligned} \tag{6.2}$$

Now the memory issue for the delayed control enters. The methods that have been used to approximate the memory segment for the control can all be adapted to the singular control problem, but there are some significant differences.

To introduce these differences, consider an analog of the periodic approximations (1.1) for the original system (4.4.3). Let $l\delta \le t_n^h < l\delta+\delta$. Then define an approximation to the control memory segment as follows.

$$\tilde\lambda^h(t,\theta) = \begin{cases} \lambda^h([l\delta,t)), & \theta\in[-\delta/2,0), \\ \lambda^h([l\delta-\delta,l\delta)), & \theta\in[-3\delta/2,-\delta/2), \\ \vdots & \\ \lambda^h([l\delta-Q_\delta^+\delta, l\delta-Q_\delta^+\delta+\delta)), & \theta\in[-\bar\theta,-\bar\theta+\delta). \end{cases} \tag{6.3}$$

The first issue is that the differences $\lambda^h([l\delta-\delta,l\delta))$, and so forth, are not necessarily bounded, so they will have to be truncated. A second issue is that their values are not necessarily confined to a finite set. Thirdly, one needs to track the passage of time in $[l\delta,t)$, as for the ordinary control case. The truncation issue can be handled by estimating the likely maximum values of the differences, and would in any case be done adaptively in that the level would be adjusted until further increases had a negligible effect on the results.

To deal with the problem of the range of values of the increments, one could restrict the jumps in the components of $\lambda(\cdot)$ to be integral multiples of some small $c_0>0$. The passage of time could be dealt with by an adaptation of the periodic-Erlang procedure of Section 4. Convergence theorems can be proved for such methods.

8.7 Neutral Equations

The approximation methods of Chapter 7 and of this chapter can be carried over to neutral equations, and we will comment only briefly on the forms. The results are only a beginning, and much further work is required. Consider, in particular, the model (3.2.17), which we rewrite as

$$x(t) - F(\bar{x}(t)) = x(0) - F(\bar{x}(0)) + \int_0^t b(\bar{x}(s), u(s))ds + \sigma(\bar{x}(s))dw(s) + z(t). \tag{7.1}$$

Assume (A3.2.7), the continuity, boundary, and uniqueness of solutions conditions used in Theorem 2.1, and that $F(\cdot)$ is a continuous $\mathbb{R}^r$-valued function on $C(G; [-\bar{\theta}, 0])$. One can add delays in the control, if desired. The approximation results of Chapter 4 carry over for the various approximations used there.

We will only describe the numerical algorithm for the simplest path memory segment approximation, that defined in Section 7.1, and illustrate the modifications that are necessary for this case. The forms for other cases of interest are modified similarly. Let ξ_n^h and $\Delta t^h(\cdot)$ denote the approximating chain and interpolation interval with $\bar{\xi}_n^h$ denoting the path memory segment, as in Section 7.1. If ξ_n^h is a reflection state, then it is sent to a state in G_h, and the mean of $\xi_{n+1}^h - \xi_n^h$, conditioned on the data to time n, is a reflection direction at the point ξ_n^h. In particular, (6.2.2) holds.

If $\xi_n^h \in G_h$, then we take a slightly indirect approach. The basic transition probabilities are first defined for the variables

$$\Xi_n^h = \xi_n^h - F(\bar{\xi}_n^h),$$

and the transition probabilities for the chain will then be obtained from these. First, let us rewrite the form (7.1.12). Define $\tilde{x} = x + \delta x$, and write the right side of (7.1.12) as

$$\frac{N^h(b(\bar{\xi}_n^h, u_n^h), a(\bar{\xi}_n^h), \xi_n^h + \delta x)}{D^h(b(\bar{\xi}_n^h, u_n^h), a(\bar{\xi}_n^h))}. \tag{7.2}$$

Now, given $\bar{\xi}_n^h$, compute Ξ_n^h. Then in lieu of using the transition probabilities (7.1.12) to compute the conditional distribution of ξ_{n+1}^h, we use them to compute the conditional distribution of Ξ_{n+1}^h, as follows:

$$P\left\{\Xi_{n+1}^h - \Xi_n^h = \delta\Xi \middle| \bar{\xi}_n^h, \xi_n^h, \alpha\right\} = \frac{N^h(b(\bar{\xi}_n^h, \alpha), a(\bar{\xi}_n^h), \xi_n^h + \delta\Xi)}{D^h(b(\bar{\xi}_n^h, \alpha), a(\bar{\xi}_n^h))}. \tag{7.3}$$

The range of values of $\delta\Xi$ in (7.3) is taken to be the same as that of the $\delta\tilde{x}$ in (7.2). Because ξ_n^h is determined by $\bar{\xi}_n^h$, it is redundant on the left side of (7.3).

Analogously to the situation in Section 7.1, we have local consistency in G in that the conditional mean of $\Xi_{n+1}^h - \Xi_n^h$ (with $u_n^h = \alpha$) is

$$b_h(\bar{\xi}_n^h, \alpha)\Delta t_n^h = b(\bar{\xi}_n^h, \alpha)\Delta t_n^h + o(\Delta t_n^h),$$

and its conditional covariance is

$$a_h(\bar{\xi}_n^h)\Delta t_n^h = a(\bar{\xi}_n^h)\Delta t_n^h + o(\Delta t_n^h).$$

By definition

$$\xi_{n+1}^h - F(\bar{\xi}_{n+1}^h) = \Xi_{n+1}^h, \tag{7.4}$$

where the conditional distribution of Ξ^h_{n+1} is obtained from (7.3). By the "gap" condition (A3.2.7), for small h, $F(\bar{\xi}^h_{n+1})$ is a function of $\bar{\xi}^h_n$. Hence from (7.3) and (7.4) we can compute the conditional probabilities $P\{\xi^h_{n+1} = \tilde{x}|\bar{\xi}^h_n, \alpha\} = p^h(\bar{\xi}^h_n, \tilde{x}|\alpha)$.

In the above development, it was supposed that the values of ξ^h_{n+1} that are computed from (7.4) are on the grid. If that is not the case, then we get the actual ξ^h_{n+1} by randomizing between nearest neighbors to attain the desired conditional means, although this procedure will add "numerical noise."

The dynamical system for the chain can be written as (neglecting the numerical noise due to any randomization that might be required)

$$\begin{aligned} \xi^h_n - F(\bar{\xi}^h_n) &= \xi^h_0 - F(\bar{\xi}^h_0) \\ &\quad + \sum_{i=0}^{n-1} b_h(\bar{\xi}^h_i, u^h_i)\Delta t^h_i + \sum_{i=0}^{n-1} \beta^h_i + \sum_{i=0}^{n-1} \Delta z^h_i, \end{aligned} \tag{7.5}$$

where the conditional covariance of the martingale difference term β^h_n is $a_h(\bar{\xi}^h_n)\Delta t^h_n$. The analog of the continuous time interpolation (7.1.14) is (again, neglecting the numerical noise)

$$\begin{aligned} \psi^h(t) - F(\bar{\xi}^h(q^h_\tau(t))) &= \xi^h_0 - F(\bar{\xi}^h_0) \\ &\quad + \int_0^t \int_{U^h} b_h(\bar{\xi}^h(q^h_\tau(s)), \alpha) r^h_\tau(d\alpha\, ds) + B^h_\tau(t) + z^h_\tau(t), \end{aligned} \tag{7.6}$$

where $B^h_\tau(\cdot)$ is a martingale with quadratic variation process

$$\int_0^t a_h(\bar{\xi}^h(q^h_\tau(s)))ds.$$

Working with (7.6) and using (A3.2.7) where needed, the various approximations to the path memory segments of this and of Chapter 7 can all be adapted.

8.8 The Ergodic Cost Problem

Assume the conditions on the model and cost function that were used in Chapter 5. In particular, the cost function is (5.1.2). For the numerical approximation, we use the periodic-Erlang form of Section 2 for the path memory segment, and the control is not delayed in either the dynamics or the cost function. At step n of the chain, the Markov state is $\Lambda^{h,\delta_0,\delta}_n \equiv (\bar{\xi}^{h,\delta_0,\delta}_{e,n}, L^{h,\delta_0,\delta}_n)$. This can also be written as $\Lambda^{h,\delta_0,\delta}_n \equiv (\tilde{X}^{h,\delta_0,\delta}_{e,n}, L^{h,\delta_0,\delta}_n)$. where $\tilde{X}^{h,\delta_0,\delta}_{e,n}$ is the vector of values of $\bar{\xi}^{h,\delta_0,\delta}_{e,n}$, as defined in (2.1), and we will use both forms. Because the controls are feedback, suppose that the control at step n has the form $u_n = u(\bar{\xi}^{h,\delta_0,\delta}_{e,n}, L^{h,\delta_0,\delta}_n)$ for some feedback control $u(\cdot)$.

The transition probability can be partitioned as (with $u_n^{h,\delta_0,\delta} = \alpha$)

$$P\left\{\xi_{n+1}^{h,\delta_0,\delta} = \xi_1, L_{n+1}^{h,\delta_0,\delta} = L_1 \middle| \bar{\xi}_{e,n}^{h,\delta_0,\delta}, L_n^{h,\delta_0,\delta}, \alpha\right\}$$
$$= P\left\{\xi_{n+1}^{h,\delta_0,\delta} | = \xi_1 \middle| \bar{\xi}_{e,n}^{h,\delta_0,\delta}, L_n^{h,\delta_0,\delta}, \alpha\right\} P\left\{L_{n+1}^{h,\delta_0,\delta} = L_1 \middle| \bar{\xi}_{e,n}^{h,\delta_0,\delta}, L_n^{h,\delta_0,\delta}, \alpha\right\}.$$

The transition probabilities for the current state $\xi_n^{h,\delta_0\delta}$ is (7.1.12), where, for the no-delay problem, the transition probabilities defined by the ratios

$$\frac{N^h(b(x,\alpha), a(x), \tilde{x})}{D^h(b(x,\alpha), a(x))}$$

are assumed to be locally consistent. The transition probabilities for the Erlang and memory segment states are given by (2.2) and (2.3).

Under the assumed conditions on the model and state space G for the no-delay problem (in particular, the nondegeneracy condition), the usual methods (say those discussed in [58]) of getting the transition probabilities ensure that the approximating chain is ergodic for each feedback control. Owing to this and to the random way that the Erlang state changes, w.l.o.g. we can suppose that for each feedback control the state space of the process $\{\Lambda_n^{h,\delta_0,\delta}\}$ consists of a transient set and a single ergodic set.

The cost function for the chain is the analog of (6.7.7) and (6.7.8); namely, for an arbitrary admissible control sequence $u = \{u_n^{h,\delta_0\delta}\}$,

$$\gamma^{h,\delta_0,\delta}(\hat{\xi}, u) = \limsup_n \frac{E_{\hat{\xi}}^{h,\delta_0,\delta,u} \sum_{i=0}^{n} \left[k(\bar{\xi}_{e,i}^{h,\delta_0\delta}, u_i^{h,\delta_0\delta})\Delta t_i^{h,\delta_0\delta} + k_0^h(\xi_i^{h,\delta_0\delta})h\right]}{E_{\hat{\xi}}^{h,\delta_0,\delta,u} \sum_{i=0}^{n} \Delta t_i^{h,\delta_0\delta}}, \tag{8.1}$$

where $k_0^h(x)$ is the cost at a reflection state $x \notin G_h$, and it is zero for $x \in G_h$. For a feedback control $u(\cdot)$, the chain $\Lambda_n^{h,\delta_0\delta}$ has a unique invariant measure that we denote by $\pi^{h,\delta_0\delta}(u)$, with values $\pi^{h,\delta_0\delta}(\hat{\xi}, L, u)$. Then (8.1) has a limit that is

$$\gamma^{h,\delta_0\delta}(u) = \frac{\sum_{\hat{\xi},L} \left[k(\hat{\xi}, u(\hat{\xi}, L))\Delta t^h(\hat{\xi}, u(\hat{\xi}, L)) + k_0^h(\hat{\xi}(0))h\right] \pi^{h,\delta_0\delta}(\hat{\xi}, L, u)}{\sum_{\hat{\xi}.L} \Delta t^h(\hat{\xi}, u(\hat{\xi}, L))\pi^{h,\delta_0\delta}(\xi, L, u)}. \tag{8.2}$$

The Bellman equation for the optimal cost and control for the chain is analogous to that in Section 6.7, and the comments made there concerning the necessity of the centering procedure to get a suitable contraction operator in this equation all hold. See [58, Chapter 7]. Let $\bar{\gamma}^{h,\delta_0\delta}$ denote the optimal cost over all feedback controls.

Theorem 8.1. *Assume the conditions of this section and that $h = o(\delta_0)$. As $h \to 0$ and $\delta_0 \to 0$, with δ fixed, we have $\bar{\gamma}^{h,\delta_0\delta} \to \bar{\gamma}^{\delta}$, the minimal cost for the periodic approximation of Section 5.4.*

Proof. The proof follows that for the no-delay case, which is outlined at the end of Section 6.7.[5] Let $u^{h,\delta_0\delta}(\cdot)$ be the optimal control. Write the interpolation that is analogous to (7.1.14):

$$\psi^{h,\delta_0,\delta}(t) = \xi_0^{h,\delta_0,\delta} + \int_0^t \int_{U^h} b_h(\bar{\xi}_e^{h,\delta_0,\delta}(q_\tau^{h,\delta_0,\delta}(s)), \alpha) r_\tau^{h,\delta_0,\delta}(d\alpha\, ds) + B_\tau^{h,\delta_0,\delta}(t) + z_\tau^{h,\delta_0,\delta}(t), \tag{8.3}$$

where $r_\tau^{h,\delta_0,\delta}(d\alpha\, ds)$ is the relaxed control representation of the interpolation (intervals $\{\Delta\tau_n^{h,\delta_0\delta}\}$) of the control sequence $u_n^{h,\delta_0\delta} = u^{h,\delta_0\delta}(\bar{\xi}_{e,n}^{h,\delta_0\delta}, L_n^{h,\delta_0\delta})$. As noted below (6.3.8) and (7.1.14), there is a martingale $w^{h,\delta_0\delta}(\cdot)$ with quadratic variation It and that converges weakly to a Wiener process such that, modulo an asymptotically negligible error,

$$B_\tau^{h,\delta_0\delta}(t) = \int_0^t \sigma(\psi^{h,\delta_0\delta}(s)) dw^{h,\delta_0\delta}(s). \tag{8.4}$$

Now consider the stationary process $\psi^{h,\delta_0\delta}(\cdot)$; i.e., where the distribution of the initial value of the Markov state is the invariant distribution. Then the cost can be written as

$$\bar{\gamma}^{h,\delta_0\delta} = E \int_0^1 \left[\int_{U^h} k(\bar{\xi}_e^{h,\delta_0,\delta}(q_\tau^{h,\delta_0,\delta}(s)), \alpha) r_\tau^{h,\delta_0,\delta}(d\alpha\, ds) + q' dy_\tau^{h,\delta_0\delta}(s) \right]. \tag{8.5}$$

For each fixed $\delta > 0$, the set of processes $(\psi^{h,\delta_0,\delta}(\cdot), r_\tau^{h,\delta_0,\delta}(\cdot), z_\tau^{h,\delta_0,\delta}(\cdot), w^{h,\delta_0,\delta}(\cdot))$ associated with the stationary system is tight. Let $(x(\cdot), r(\cdot), z(\cdot), w(\cdot))$ denote the limit of a weakly convergent subsequence as $h \to 0$ and $\delta_0 \to 0$. Then the limit is stationary in that the distribution of $(x(t+\cdot), r(t+\cdot) - r(t), z(t+\cdot) - z(t), w(t+\cdot) - w(t))$ does not depend on t. The $w(\cdot)$ is a Wiener process, and the other processes are nonanticipative with respect to it. The timescale equivalence Theorem 7.3.1 does not cover the current case the way that it is written. But a very similar proof, where $(h, \delta_0) \to 0$ and δ is fixed, shows that the process $\bar{\xi}_e^{h,\delta_0,\delta}(q_\tau^{h,\delta_0,\delta}(\cdot))$ converges weakly to $\bar{x}_p^\delta(\cdot)$, the periodic memory segment that is obtained from the stationary limit process $x(\cdot)$.

The limit set satisfies

$$dx(t) = \int_U b(\bar{x}_p^\delta(t), \alpha) r'(d\alpha, t) dt + \sigma(x(t)) dw(t) + dz(t), \tag{8.6}$$

where as in the proof in Section 5 of Theorems 7.1.1 and 7.1.2 for the reflecting boundary case, $z(\cdot)$ is the reflection process. The limit of (8.5) is

$$E \int_0^1 \left[\int_U k(\bar{x}_p^\delta(s), \alpha) r'(d\alpha, s) ds + q' dy(s) \right],$$

[5] See [58, Chapter 11] for fuller details for the no-delay problem.

which is the cost $\gamma^\delta(r)$ associated with the stationary limit system. By (A5.4.3), the infimum $\bar\gamma^\delta$ of the costs over periodic relaxed feedback controls cannot be improved by using general relaxed controls. Thus $\gamma^\delta(r) \geq \bar\gamma^\delta$. It follows that

$$\liminf_{h,\delta_0\to 0} \bar\gamma^{h,\delta_0\delta} \geq \bar\gamma^\delta. \tag{8.7}$$

To prove that $\lim_{h,\delta_0\to 0} \bar\gamma^{h,\delta_0\delta} \to \bar\gamma^\delta$, we need to show that

$$\limsup_{h,\delta_0\to 0} \bar\gamma^{h,\delta_0\delta} \leq \bar\gamma^\delta. \tag{8.8}$$

Consider the periodic model in Section 5.4. Theorem 5.4.6 says that the cost for the original system with a periodic memory segment can be well approximated by using a subset of U that contains only finitely many points. Recall the definitions $\tau^\delta(t) = t(\text{mod } \delta)$ and that of $\tilde X^\delta$ from the beginning of Section 5.4. If U contains only finitely many points, then Theorem 5.4.7 says that for each $\epsilon > 0$ there is a periodic relaxed feedback control with values $m^\epsilon(\tilde X^\delta, \tau^\delta, \{\alpha\})$[6] that is $\epsilon/2$-optimal and is continuous in the variables $\tilde X^\delta$ and τ^δ for each value of α. By first discretizing U, and then using Theorem 5.4.7, we can suppose that we have an ϵ-optimal control $m^\epsilon(\tilde X^\delta, \tau^\delta, d\alpha)$ that is continuous in $(\tilde X^\delta, \tau^\delta)$ for each value of α. Also, without loss of generality, we can suppose that $U^h = U$.

Let us adapt this control for use on the chain. First, note that the interpolation intervals for the reflecting states is zero, and no control is applied there. Hence we can work with the revised chain that omits these states. Their effects will still be contained in the processes $z^{h,\delta_0\delta}(\cdot)$ and $z_\tau^{h,\delta_0\delta}(\cdot)$. For use on the chain, the control will take the values $m^\epsilon(\tilde X_n^{h,\delta_0\delta}, \delta_0 L_n^{h,\delta_0\delta}, d\alpha)$, which might also be written as $m^\epsilon(\bar\xi_{e,n}^{h,\delta_0\delta}, \delta_0 L_n^{h,\delta_0\delta}, d\alpha)$, without ambiguity. Let us work with the stationary system associated with this control. The system can be written as (8.3), except that the control with values $m^\epsilon(\bar\xi_e^{h,\delta_0,\delta}(q_\tau^{h,\delta_0,\delta}(s)), L_\tau^{h,\delta_0\delta}(s), d\alpha)$ replaces $r_\tau^{h,\delta_0\delta,\prime}(d\alpha, s)$. The initial condition has the stationary distribution associated with the control $m^\epsilon(\cdot)$ used on the chain. Then the cost can be written as

$$\begin{aligned}
&\gamma^{h,\delta_0\delta}(m^\epsilon) = \\
&E\int_0^1 \Big[\int_{U^h} k(\bar\xi_e^{h,\delta_0,\delta}(q_\tau^{h,\delta_0,\delta}(s)),\alpha) m^\epsilon(\bar\xi_e^{h,\delta_0,\delta}(q_\tau^{h,\delta_0,\delta}(s)), \delta_0 L_\tau^{h,\delta_0\delta}(s), d\alpha) ds \\
&\qquad + q' dy_\tau^{h,\delta_0\delta}(s)\Big].
\end{aligned} \tag{8.9}$$

The sequence of stationary processes $\delta_0 L_\tau^{h,\delta_0\delta}(\cdot)$ converges to the stationary process $\tau^\delta(\cdot)$, where the initial value $\tau^\delta(0)$ is the limit of the sequence $\delta_0 L_\tau^{h,\delta_0\delta}(0)$. Now take a weakly convergent subsequence of the new processes

[6] Recall the convention concerning notation that was discussed below Theorem 5.4.6, where $m(\tilde X^\delta(t), \tau^\delta(t), d\alpha)$ and $m(\bar x_p^\delta(t), \tau^\delta(t), d\alpha)$ are used interchangeably.

$(\psi^{h,\delta_0\delta}(\cdot),\ w^{h,\delta_0\delta}(\cdot),\ z_\tau^{h,\delta_0\delta}(\cdot), \delta_0 L_\tau^{h,\delta_0\delta}(\cdot))$, with limit $(x(\cdot), w(\cdot), z(\cdot), \tau^\delta(\cdot))$. Due to the continuity of $m^\epsilon(\cdot)$, the limit is the unique stationary process under the control $m^\epsilon(\cdot)$. The limit system can be written as

$$dx(t) = \int_U b(\bar{x}_p^\delta(t), \alpha) m^\epsilon(\bar{x}_p^\delta(t), \tau^\delta(t), d\alpha)dt + \sigma(x(s))dw(s) + dz(t), \quad (8.10)$$

where $z(\cdot)$ is the reflection term. The cost (8.9) converges to the stationary cost

$$\gamma^\delta(m^\epsilon) = E \int_0^1 \left[\int_U k(\bar{x}_p^\delta(s), \alpha) m^\epsilon(\bar{x}_p^\delta(s), \tau^\delta(s), d\alpha)ds + q'dy(s) \right].$$

Because $\gamma^{h,\delta_0\delta}(m^\epsilon) \geq \bar{\gamma}^{h,\delta_0\delta}$ and $\epsilon > 0$ is arbitrary, (8.8) follows. ■

9

A Wave Equation Approach

9.0 Outline of the Chapter

In Chapters 7 and 8, the state of the model, as needed for the numerical procedure, consists of the segment of the path over the delay interval and of the control path as well (if the control is also delayed). Delayed reflection terms were not dealt with. Convergence theorems were proved, and "numerically efficient" representations of the state data were developed that reduced the memory requirements to manageable size for low-dimensional problems, if the path only were delayed. If the control and/or reflection terms are also delayed, and the control can take more than two or three values, then the memory requirements with those methods can be prohibitive at this time. In particular, one would have to keep track of the values of the control (relaxed control representation) or the reflection terms, whichever appear in delayed form, over the delay intervals and approximate them by finite-valued discrete-time processes that do not lose too much information. These approximations become part of the state space of the approximating chain and can lead to very large state spaces.

In this chapter we will take an alternative approach that reduces the memory requirements for general nonlinear stochastic problems where the control and reflection terms, as well as the path variables, might be delayed. The approach was suggested by the work in [94],[1] which dealt only with the linear deterministic system with a quadratic cost function, and the development depended heavily on the linear structure. The idea can be extended to the problems of concern here. With this method, the delay equation is replaced by a type of stochastic wave equation with no delays, and its numerical solution yields the optimal costs and controls for the original model. It appears that, with the use of appropriate numerical algorithms, the memory requirements are much reduced over more direct methods, although there is little practical numerical experience with such methods at present.

[1] The author would like to thank Kasi Itô for bringing [94] to his attention.

H.J. Kushner, *Numerical Methods for Controlled Stochastic Delay Systems*,
doi: 10.1007/978-0-8176-4621-9_9,

As in Chapters 7 and 8, we work mainly with the reflecting boundary case. The modifications that are required when the boundary is absorbing are minor. One drops the conditions on the reflection directions and adds the conditions on the time of first contact with the boundary. The optimal stopping problem is treated similarly.

The model and assumptions are in Section 1. Section 2 is concerned with a representation of the solution in terms of a type of stochastic wave equation without delay terms. This is an extension to the general nonlinear stochastic system of the idea in [94] for the deterministic linear problem, and the reference contains a history of the idea for that problem. With this representation, the delays are eliminated, but one must solve a PDE. It is shown that the representation is equivalent to the original problem in that any solution to one yields a solution to the other. To prepare ourselves for what will be required for the numerical approximations, a discrete-time approximation is developed in Section 3. This will suggest the correct scaling and illustrate the type of algebraic manipulations that are required. The adaptation of the Markov chain approximation method and the local consistency conditions are discussed in Section 4. Motivated by the ideas in Section 3, it is shown how to construct the approximating chains for the representation introduced in Section 2.

In Section 5, the size of the state space that is needed for the solution of the Bellman equation is discussed, and it is seen that the approach does moderate the requirements considerably. Although the form of the algorithm is motivated by those that are used for the no-delay problem, it is more complicated. But then the delay problem is substantially more complicated and the proposed algorithm appears to be quite promising in the sense of considerably reducing the memory requirements, if the control and/or reflection terms are delayed. The proof of convergence of the numerical procedure is started in Section 6, where various representations of the interpolated chains are derived. These are used in Section 7 where the proof of convergence is completed.

The periodic and periodic-Erlang approximations of Chapters 4 and 8 can be profitably adapted to the format of this chapter, and this is discussed in Section 8. Little detail concerning these particular adaptations is given as the approach of this chapter is in its infancy, and a great deal of detail had been provided in the earlier chapters for the problems of interest there. Some comments on the singular control problem are in Section 9.

9.1 The Model and Assumptions

The model. Recall the definition of the measure $\mu_a(\cdot)$ given at the beginning of Subsection 3.2.2, and that $d\mu_a(\theta) = \mu_a(\theta+d\theta)-\mu_a(\theta)$. Until further notice, we will concentrate on the model (3.2.11), which we rewrite for convenience:

$$
\begin{aligned}
dx(t) &= c(x(t), u(t))dt \\
&\quad + dt \int_{-\bar{\theta}}^{0} b(x(t+\theta), u(t+\theta), \theta) d\mu_a(\theta) + \sigma(x(t))dw(t) + dz(t) \\
&\quad + dt \int_{\theta=-\bar{\theta}}^{0} p(x(t+\theta), \theta) d_\theta y(t+\theta).
\end{aligned}
\tag{1.1}
$$

As noted in Chapter 3, the last integral in (1.1) is with respect to $d\theta$ in the sense that the interpretation

$$
p(x(t+\theta), \theta) d_\theta y(t+\theta) = p(x(t+\theta), \theta) \left[y(t+\theta+d\theta) - y(t+\theta) \right]
$$

is used. Conditions (A3.2.1) and (A3.2.2) on the boundary and reflection directions and (A3.2.4)–(A3.2.6) on the dynamical functions are assumed to hold. The model with drift term (3.2.13) can be treated with only notational changes.

The relaxed control form of (1.1) is

$$
\begin{aligned}
dx(t) &= dt \int_U c(x(t), \alpha) r'(d\alpha, t) \\
&\quad + dt \int_{-\bar{\theta}}^{0} \int_U b(x(t+\theta), \alpha, \theta) r'(d\alpha, t+\theta) d\mu_a(\theta) + \sigma(x(t))dw(t) \\
&\quad + dz(t) + dt \int_{\theta=-\bar{\theta}}^{0} p(x(t+\theta), \theta) d_\theta y(t+\theta).
\end{aligned}
\tag{1.2}
$$

If the boundary is absorbing rather than reflecting, then drop the reflection term and (A3.2.1) and (A3.2.2), and add (A3.4.1) and (A3.4.2).

A discounted cost function. Recall that $\hat{x}$ and $\hat{u}$ denote the canonical value of the path and control memory segments, resp., on $[-\bar{\theta}, 0]$. Let $\hat{r}$ denote the relaxed control representation of the initial control segment $\hat{u}$. For simplicity, the initial control segment will always be an ordinary control. Let $\hat{z}$ denote the initial segment $\{z(s), s \le 0\}$. Let $E^u_{\hat{x},\hat{u},\hat{z}}$ denote the expectation given the initial condition $(\hat{x}, \hat{u}, \hat{z})$ and that control $u(\cdot)$ is used on $[0, \infty)$. We concentrate the following discounted cost function. For $\beta > 0$, some vector q with nonnegative components and satisfying (A3.4.3), and control process $u(\cdot)$ on $[0, \infty)$, the cost function is to be the special case of (3.4.5):

$$
W(\hat{x}, \hat{u}, \hat{z}, u) = E^u_{\hat{x},\hat{u},\hat{z}} \int_0^\infty e^{-\beta t} \left[k(x(t), u(t))dt + q'dy \right]. \tag{1.3}
$$

If a relaxed $r(\cdot)$ control is used on $[0, \infty)$, then we write $W(\hat{x}, \hat{u}, \hat{z}, r)$ and $E^r_{\hat{x},\hat{u},\hat{z}}$.

One could proceed with the general form (3.4.2), but (1.3) is usually adequate in applications. Lemma 3.2.1 implies that the reflection term component of the cost is well defined. Let $V(\hat{x}, \hat{u}, \hat{z})$ denote the infimum of the costs over all controls.

Existence of an optimal control. Under (A3.2.1), (A3.2.2), (A3.2.4)–(A3.2.6), and (A3.4.3), there is an optimal relaxed control. The proof is similar to that of Theorem 3.5.1. One takes a minimizing sequence of controls and then a weakly convergent subsequence of the (path, relaxed control, Wiener process, reflection term) and shows that the limit satisfies (1.2) and that the minimizing sequence of costs converges to the cost for the limit processes. Furthermore ([31, Theorem 2.3]), the infimum of the cost over the relaxed controls is equal to the infimum of the costs over ordinary controls. If we replace the reflecting boundary and its conditions by an absorbing boundary with cost function (3.4.2) and conditions (A3.4.1) and (A3.4.2), then there is an optimal control as well.

9.2 A Key Representation of $x(\cdot)$

Delay equations and the wave equation. There is a close connection between delay equations and certain forms of the wave equation, as seen in Chapter 1. Recall the example in Chapter 1 of a system of temperature regulation. Hot water from a source flows into a pipe of length $\bar{\theta}$. At the entrance to the pipe the hot water is mixed with a fixed flow of cold water. The flow in the pipe has unit velocity. There are no thermal losses, and the only mixing occurs at the beginning of the pipe. The temperature is measured at the end of the pipe. Based on the measurement, a signal is sent (instantaneously) to a valve that controls the flow of hot water into the pipe. The goal is to reach and maintain a given temperature at the end of the pipe. Let $T(t,\theta)$ denote the temperature in the pipe at time t at a distance θ from the entry point. Then, for small $\delta > 0$, $T(t+\delta,\theta) = T(t,\theta-\delta)$ or $T(t+\delta,\theta) - T(t,\theta) = T(t,\theta-\delta) - T(t,\theta)$. Dividing by δ and letting $\delta \to 0$, we formally obtain the partial differential equation $\partial T(t,\theta)/\partial t = -\partial T(t,\theta)/\partial\theta$. This PDE represents the effects of the time delay. An extension of this example that effectively allows for additional inputs along the length of the pipe will play an important role in the development of this chapter.

9.2.1 A Representation of the Solution

Next we formally define a pair of processes $(\chi^0(\cdot), \chi^1(\cdot))$ that will be the basis of the approximation procedure. The process $\chi^0(\cdot)$ is parameterized by time. The process $\chi^1(\cdot)$ is parameterized by both time and a variable $\theta \in [-\bar{\theta}, 0]$. For $\theta \in [-\bar{\theta}, 0]$, formally define $\chi^0(\cdot)$ and $\chi^1(\cdot)$ by

$$d\chi^0(t) = \chi^1(t,0)dt + c(\chi^0(t), u(t))dt + \sigma(\chi^0(t))dw(t) + dz^0(t), \tag{2.1}$$

$$\begin{aligned} d_t\chi^1(t,\theta) = {} & -d_\theta\chi^1(t,\theta) + b(\chi^0(t), u(t), \theta)\left[\mu_a(\theta+dt) - \mu_a(\theta)\right] \\ & + p(\chi^0(t),\theta)dy^0(t). \end{aligned} \tag{2.2}$$

The cost function is (1.3) with $\chi^0(t)$ replacing $x(\cdot)$.

The stochastic partial differential equation (2.2) is in "symbolic" form. Its precise interpretation will be given by (2.7), (2.8), which redefines it in terms of the "shift" semigroup of the wave equation component $d_t\chi^1(t,\theta) = -d_\theta\chi^1(t,\theta)$. The process $\chi^0(\cdot)$ takes values in the constraint set G, which is subject to the boundary conditions (A3.2.1) and (A3.2.2), and $z^0(\cdot)$ is the reflection term for $\chi^0(\cdot)$. With the interpretation (2.7), (2.8), Theorem 2.1 shows that $\chi^0(\cdot) = x(\cdot)$. If there is no delayed reflection term, then the values of $\chi^1(t,\theta)$ will be seen to be bounded. If this process is not bounded, then for numerical purposes, "numerical" bounds will have to be added, and this issue is discussed in Section 5. With the use of relaxed controls, (2.1) and (2.2) are rewritten as

$$d\chi^0(t) = \chi^1(t,0)dt + \int_U c(\chi^0(t),\alpha)r'(d\alpha,t)dt + \sigma(\chi^0(t))dw(t) + dz^0(t), \quad (2.3)$$

$$\begin{aligned} d_t\chi^1(t,\theta) = -d_\theta\chi^1(t,\theta) &+ \int_U b(\chi^0(t),\alpha,\theta)r'(d\alpha,t)\left[\mu_a(\theta+dt) - \mu_a(\theta)\right] \\ &+ p(\chi^0(t),\theta)dy^0(t). \end{aligned} \quad (2.4)$$

These processes will be the basis of the numerical method that is to be developed. Instead of delays in the state, control, and reflection terms, we have the additional variable θ and the process $\chi^1(\cdot)$. The numerical approximations will not be simple, but we avoid the difficulties encountered in Chapter 8 in representing the memory segments of the control (and what would be the more difficult problem of representing a delayed reflection term, if any).

The initial conditions for (2.1) and (2.2) are $\chi^0(0) = x(0)$ and, for arbitrary $z^0(s), s \le 0$,

$$\chi^1(0,\theta) = \int_{-\bar\theta}^{\theta} b(x(\gamma-\theta),u(\gamma-\theta),\gamma)d\mu_a(\gamma) + \int_{-\bar\theta}^{\theta} p(x(\gamma-\theta),\gamma)d_\gamma y^0(\gamma-\theta). \quad (2.5)$$

Additionally, we have the boundary condition $\chi^1(t,-\bar\theta) = 0$.

The linear and deterministic forms of (2.1), (2.2) were introduced in [94] for the treatment of the linear deterministic problem with a quadratic cost function and no constraint set G. Numerical approximations were not a concern. See the references in [94] for a history of the idea for linear systems.

9.2.2 Comments on the Dimension and the System State

The dimension of the component $\chi^0(\cdot)$ is that of $x(\cdot)$, but there are no delayed terms in (2.1). The dimension of $\chi^1(\cdot)$ is equal to the number of components of $x(\cdot)$ whose dynamical terms contain delays. This dimension does not otherwise depend on the number of controls or on the number of individual terms that appear in a delayed form. Thus, if the expression for only one component of

$x(\cdot)$ in (1.1) or (1.2) contains (any number of) dynamical terms with delays, the process $\chi^1(\cdot)$ has only one component. For the components $x_i(\cdot)$ whose dynamical terms do not have delays, simply define $\chi_i^1(\cdot) = 0$, and define $\chi_i^0(\cdot)$ by the ith component of (2.1) with $\chi_i^1(t, 0) = 0$.

Suppose that delayed values of components $x_i(\cdot), i = 1, \ldots, r_1$, $u_i(\cdot), i = 1, \ldots, r_2$, and $y_i(\cdot), i = 1, \ldots, r_3$, are required. For the original problem, the full system state consists of the initial condition $x(0)$ and the memory segments of the $x_i(\cdot), i = 1, \ldots, r_1$, $u_i(\cdot), i = 1, \ldots, r_2$, and $y_i(\cdot), i = 1, \ldots, r_3$. In general, we know little *a priori* of the regularity properties of the controls, and a numerical approximation of the control memory segment might require a lot of memory, as seen in Chapter 8. The same thing can be said for the memory segments of the delayed reflection terms, if any. On the other hand, the full system or memory state for (2.1), (2.2) at time t is just $\chi^0(t)$ and the current values $\chi^1(t, \theta)$, $-\bar{\theta} \le \theta \le 0$. These considerations illustrate the promise of the representation (2.1), (2.2). But keep in mind that there is very little practical experience with the type of approximations that will be developed in the later sections.

To develop the idea a little further, suppose that the dynamics of only one component of $x(\cdot)$, say $x_1(\cdot)$, contains delayed path, control, and/or reflection terms. The delayed path or reflection terms that appear in that component of $x(\cdot)$ might arise from other components of $x(\cdot)$, and the control might be vector-valued. The state space for the representation (1.1) or (1.2) would consist of the memory segments of all of these processes. Working directly with the system (1.1) requires that we discretize these memory segments. The implicit, periodic, and periodic-Erlang approximation procedures that were discussed in Chapters 7 and 8 greatly alleviated the problem when only the path is delayed and in some cases where the control is delayed, but where the control takes only a few values. On the other hand, in this example $\chi^1(t, \theta)$ is real-valued. The procedure to be discussed will not eliminate the memory problem, as the variable θ will have to be discretized. But it does provide an alternative that has the potential of greatly reducing the needed memory. The main question is how to approximate $\chi^1(\cdot)$. This will be dealt with in Section 4, and the relative memory requirements are discussed in Section 5.

9.2.3 Proof of the Representation

The precise definition of (2.1), (2.2). The part $d_t\chi^1(t, \theta) = -d_\theta\chi^1(t, \theta)$ of the expression (2.2) is a type of wave equation, and its semigroup will play a major role in defining the solutions. Let $f(\cdot)$ be either a real or vector-valued function of $\theta \in [-\bar{\theta}, 0]$. Following the idea in [94], define the semigroup $\Phi(\cdot)$ by the constrained shift

$$\Phi(t)f(\theta) = \begin{cases} f(\theta - t), & \text{for } -\bar{\theta} \le \theta - t \le 0, \\ 0, & \text{otherwise.} \end{cases} \tag{2.6}$$

In expressions of the form $\Phi(t)f(t,\theta)$, $\Phi(\cdot)$ will act only on the argument θ.

The system (2.1), (2.2) is always defined by the "variation of constants solution" form

$$d\chi^0(t) = \chi^1(t,0)dt + c(\chi^0(t),u(t))dt + \sigma(\chi^0(t))dw(t) + dz^0(t), \tag{2.7}$$

$$\begin{aligned}\chi^1(t,\theta) = \Phi(t)\chi^1(0,\theta) &+ \int_0^t \Phi(t-s)b(\chi^0(s),u(s),\theta)\left[\mu_a(\theta+ds)-\mu_a(\theta)\right] \\ &+ \int_0^t \Phi(t-s)p(\chi^0(s),\theta)dy^0(s).\end{aligned} \tag{2.8}$$

The construction of the numerical approximations will use the formal dynamical representation (2.1), (2.2) as a heuristic guide, but the precise interpretations will always have a form such as (2.7), (2.8).

The integral involving $\mu_a(\cdot)$ in (2.8). The integration with respect to the measure $\mu_a(\cdot)$ is to be done after the operation by $\Phi(t-s)$. The integral is well defined by the interpretation

$$\begin{aligned}&\int_0^t \Phi(t-s)b(\chi^0(s),u(s),\theta)\left[\mu_a(\theta+ds)-\mu_a(\theta)\right] = \\ &\int_0^t b(\chi^0(s),u(s),\theta-t+s)I_{\{-\bar\theta\le\theta-t+s\le 0\}}\left[\mu_a(\theta-t+s+ds)-\mu_a(\theta-t+s)\right].\end{aligned} \tag{2.9}$$

On the right-hand side of (2.9) the integration is with respect to the variable s, with the quantity $\theta-t$ fixed. A change of variable in (2.9) leads to

$$\int_{\max\{\theta-t,-\bar\theta\}}^{\theta} b(\chi^0(\gamma+t-\theta),u(\gamma+t-\theta),\gamma)d\mu_a(\gamma). \tag{2.10}$$

For the relaxed control form of the control term in (2.8), use the expression

$$\int_0^t\int_U \Phi(t-s)b(\chi^0(s),\alpha,\theta)r'(d\alpha,s)\left[\mu_a(\theta+ds)-\mu_a(\theta)\right],$$

and in (2.7) replace $c(\chi^0(t),u(t))dt$ with $dt\int_U c(\chi^0(t),\alpha)r'(d\alpha,t)$.

On the θ-dependence of $\chi^1(t,\theta)$. The smoothness of $\chi^1(t,\theta)$ in θ and t depends heavily on the measure $\mu_a(\cdot)$. Consider the expression (2.10) and, for simplicity in the notation, suppose that t is large enough so that the lower limit of integration is $-\bar\theta$. First, let $\mu_a(\cdot)$ be Lebesgue measure on $[-\bar\theta,0]$. Then (2.10) can be written as

$$\int_{-\bar\theta}^{\theta} b(\chi^0(\gamma+t-\theta),u(\gamma+t-\theta),\gamma)d\gamma = \int_{-\bar\theta+t-\theta}^{t} b(\chi^0(\tau),u(\tau),\tau-t+\theta)d\tau, \tag{2.11}$$

which is continuous in θ. If $b(x,u,\cdot)$ is Lipschitz continuous, uniformly in x, α, then (2.11) is Lipschitz continuous in θ, uniformly in the values of $\chi^0(\cdot)$ and $u(\cdot)$.

If $\mu_a(\cdot)$ is concentrated on a finite number of points, the situation is quite different, and (2.10) has no continuity properties in general. To see this, consider the special case where $\mu_a(\cdot)$ is concentrated on $-\bar{\theta}$. Then (2.10) can be written as

$$b(\chi^0(-\bar{\theta}+t-\theta), u(-\bar{\theta}+t-\theta), -\bar{\theta}), \tag{2.12}$$

and unless the control is continuous, it will not be continuous.

The next theorem shows that ((2.7), (2.8)) is indeed a representation of (1.1) (or, in relaxed control notation, of (1.2)).

Theorem 2.1. *Assume* (A3.2.1), (A3.2.2), *and* (A3.2.4)–(A3.2.6). *Then, for initial condition* (2.5) *with arbitrary* $z^0(s), s \le 0$, *and* $\chi^0(0) = x(0)$, (2.7) *and* (2.8) *have the weak-sense unique solution*

$$\chi^0(\cdot) = x(\cdot), \tag{2.13}$$

$$\begin{aligned} \chi^1(t,\theta) = &\int_{-\bar{\theta}}^{\theta} b(\chi^0(\gamma+t-\theta), u(\gamma+t-\theta), \gamma) d\mu_a(\gamma) \\ &+ \int_{-\bar{\theta}}^{\theta} p(\chi^0(\gamma+t-\theta), \gamma) d_\gamma y^0(\gamma+t-\theta). \end{aligned} \tag{2.14}$$

The analogous result holds for the relaxed control form, where we use

$$\int_{-\bar{\theta}}^{\theta}\int_U b(\chi^0(\gamma+t-\theta), \alpha, \gamma) r'(d\alpha, \gamma+t-\theta) d\mu_a(\gamma)$$

in place of the first term on the right side of (2.14). *The result continues to hold if the reflecting boundary and the conditions on it are replaced by an absorbing boundary and* (A3.4.1), (A3.4.2).

Comment on the uniqueness of the solution to ((2.7), (2.8)). Note that if $t = 0$, then (2.14) reduces to the initial condition (2.5). By the proof of the theorem, any solution to ((2.7), (2.8)) must have the form ((2.13), (2.14)).. We assumed (in (A3.2.6)) that (1.1) and (1.2) have unique weak-sense solutions for each admissible ordinary or relaxed control and initial condition. Hence, the solution to ((2.7), (2.8)) is also weak-sense unique for each admissible control and initial condition of the form (2.5).

Proof. To keep the notation relatively simple, we will work with ordinary rather than relaxed controls. The development for relaxed controls involves only minor notational changes. From the remarks above regarding uniqueness we need only show that if $(\chi^0(\cdot), \chi^1(\cdot))$ are measurable and nonanticipative processes satisfying ((2.7), (2.8)), then (2.13) and (2.14) hold. Consider the representation (2.8) and define $\chi^0(t) = x(t)$ for $-\bar{\theta} \le t \le 0$. The component of (2.8) that is due to the initial condition in (2.5) is

$$\begin{aligned}
\Phi(t)\chi^1(0,\theta) &= \int_{-\bar{\theta}}^{\theta-t} b(\chi^0(\gamma+t-\theta), u(\gamma+t-\theta),\gamma) I_{\{-\bar{\theta}\le\theta-t\le 0\}} d\mu_a(\gamma) \\
&\quad + \int_{-\bar{\theta}}^{\theta-t} p(\chi^0(\gamma+t-\theta),\gamma) I_{\{-\bar{\theta}\le\theta-t\le 0\}} d_\gamma y^0(\gamma+t-\theta) \\
&= \int_{-\bar{\theta}}^{\max\{\theta-t,-\bar{\theta}\}} b(\chi^0(\gamma+t-\theta), u(\gamma+t-\theta),\gamma) d\mu_a(\gamma) \\
&\quad + \int_{-\bar{\theta}}^{\max\{\theta-t,-\bar{\theta}\}} p(\chi^0(\gamma+t-\theta),\gamma) d_\gamma y^0(\gamma+t-\theta).
\end{aligned} \tag{2.15}$$

At $\theta = 0$, (2.15) is

$$\begin{aligned}
&\int_{-\bar{\theta}}^{\max\{-t,-\bar{\theta}\}} b(\chi^0(\gamma+t), u(\gamma+t),\gamma) d\mu_a(\gamma) \\
&\quad + \int_{-\bar{\theta}}^{\max\{-t,-\bar{\theta}\}} p(\chi^0(\gamma+t),\gamma) d_\gamma y^0(\gamma+t).
\end{aligned} \tag{2.16}$$

Now consider the delayed reflection term in (2.8), namely,

$$\begin{aligned}
&\int_0^t \Phi(t-s) p(\chi^0(s),\theta) dy^0(s) = \int_0^t p(\chi^0(s),\theta-t+s) I_{\{-\bar{\theta}\le\theta-t+s\le 0\}} dy^0(s) \\
&= \int_{\theta-t}^{\theta} p(\chi^0(\gamma+t-\theta),\gamma) I_{\{-\bar{\theta}\le\gamma\le 0\}} d_\gamma y^0(\gamma+t-\theta) \\
&= \int_{\max\{\theta-t,-\bar{\theta}\}}^{\theta} p(\chi^0(\gamma+t-\theta),\gamma) d_\gamma y^0(\gamma+t-\theta).
\end{aligned} \tag{2.17}$$

The (noninitial condition part of the) term in (2.8) involving the measure $\mu_a(\cdot)$ was dealt with in (2.9), (2.10). Adding (2.15), (2.17), and (2.10) yields (2.14). Setting $\theta = 0$ in (2.14) and substituting it into (2.7) yields

$$\begin{aligned}
d\chi^0(t) &= c(\chi^0(t), u(t))dt + dt \int_{-\bar{\theta}}^0 b(\chi^0(t+\gamma), u(t+\gamma),\gamma) d\mu_a(\gamma) \\
&\quad + dt \int_{-\bar{\theta}}^0 p(\chi^0(t+\gamma),\gamma) d_\gamma y^0(t+\gamma) + \sigma(\chi^0(t))dw(t) + dz^0(t),
\end{aligned}$$

which is the equation for $x(\cdot)$. The same computations yield (2.14). ■

9.2.4 Extensions

Suppose that the form (3.2.13) replaces the $b(\cdot)$ term in (1.1) and (1.2). Then, with the obvious modifications in the notation, the result and proof are the same. The model (1.1) does not cover drift dynamical terms such as $x(t+\theta_1)x(t+\theta_2)$ or $x(t+\theta_1)u(t+\theta_2)$, where $\theta_1 \neq \theta_2$. Terms such as $x(t+\theta_1)x(t+\theta_2)$ could be accommodated by adding a term $c_0(\bar{x}(t), u(t))$. This entails adding $c_0(\bar{\chi}^0(t), u(t))$ to (2.1), in which case we will need to track the memory segment $\bar{\chi}^0(t)$ as well as $\chi^1(t,\theta), \theta \in [-\bar{\theta}, 0]$, and this would require much more memory in the numerical approximations. One can deal with delayed Wiener process values by adding the term

$$dt \int_{-\bar{\theta}}^{0} p_w(x(t+\theta), \theta) d_\theta w(t+\theta)$$

to (1.1).

9.3 A Discrete-Time Approximation

We will use the dynamical equations (2.1) and (2.2) as the basis of the numerical algorithms by approximating them by suitable Markov chains, analogously to the way that the model (6.1.1) was approximated in Chapter 6. But to verify the convergence to the correct values, we will need to show that the processes associated with the numerical approximations converge to the representations defined by (2.7) and (2.8), with $\chi^0(\cdot) = x(\cdot)$. Then, by Theorem 2.1, the resulting limit solves solves (1.1) or (1.2), depending on whether ordinary or relaxed controls are used.

The basic forms of the numerical algorithms for getting the costs or optimal costs will be discussed in Section 4. To prepare ourselves for that discussion and the types of algebraic manipulations that will be needed, it is useful to start with a formal finite-difference approximation to (2.1) and (2.2), and that will be done in this section. This approximation is not suitable for numerical computation of the cost functions, but it will provide helpful insights and guides and might be useful for simulations.

We now formally approximate (2.1) and (2.2) by a simple form, by discretizing time and θ. Because the purpose is motivational, for simplicity we will drop the $x(t)$-dependence of $p(\cdot)$. As (2.2) is in a "symbolic" form, the appropriate discrete-time approximation is not *a priori* clear, but we will use a form that will lead to the correct results, when adapted to the numerical procedure. Let δ be the discretization interval for θ and Δ the discretization interval for the time variable, with $\bar{\theta}$ being an integral multiple of δ. We are partly guided by the need to keep $\chi^{1,\delta,\Delta}(t, -\bar{\theta}) = 0$. The interpolations in time will be piecewise-constant and right continuous. Because we are interested in only the general idea of the approximation, let us suppose that the control $u(\cdot)$ is continuous and let $u^\Delta(\cdot)$ denote the piecewise-constant approximation of $u(\cdot)$. In the subsequent sections, the controls are arbitrary. Then, for $-\bar{\theta} < \theta \le 0$, $t = n\Delta, n = 0, 1, \ldots$, and letting $\chi^{0,\delta,\Delta}(\cdot)$ and $\chi^{1,\delta,\Delta}(\cdot)$ denote the approximations to $\chi^0(\cdot)$ and $\chi^1(\cdot)$, resp., the approximating processes are given by the recursions

$$
\begin{aligned}
\chi^{0,\delta,\Delta}(t+\Delta) - \chi^{0,\delta,\Delta}(t) &= \Delta\chi^{1,\delta,\Delta}(t,0) + \Delta c(\chi^{0,\delta,\Delta}(t), u^{\Delta}(t)) + \\
&\quad \sigma(\chi^{0,\delta,\Delta}(t))[w(t+\Delta) - w(t)] + [z^{0,\delta,\Delta}(t+\Delta) - z^{0,\delta,\Delta}(t)], \\
\chi^{1,\delta,\Delta}(t+\Delta,\theta) - \chi^{1,\delta,\Delta}(t,\theta) &= -\left[\chi^{1,\delta,\Delta}(t,\theta) - \chi^{1,\delta,\Delta}(t,\theta-\delta)\right]\frac{\Delta}{\delta} \\
&\quad + b(\chi^{0,\delta,\Delta}(t), u^{\Delta}(t), \theta-\delta)\left[\mu_a(\theta) - \mu_a(\theta-\delta)\right] \\
&\quad + p(\theta-\delta)\left[y^{0,\delta,\Delta}(t+\Delta) - y^{0,\delta,\Delta}(t)\right].
\end{aligned}
\tag{3.1}
$$

The initial condition is an appropriate discrete-time and discrete-θ approximation of (2.5), and the boundary condition is $\chi^{1,\delta,\Delta}(t,-\bar{\theta}) = 0$. The boundary conditions (A3.2.1) and (A3.2.2) are imposed on $\chi^{0,\delta,\Delta}(\cdot)$, and $z^{0,\delta,\Delta}(\cdot)$ is the associated reflection process. That is, if $\chi^{0,\delta,\Delta}(\cdot)$ ever leaves the set G, then it is immediately reflected back in accordance with the local reflection direction as defined in (A3.2.1), (A3.2.2). We use the representation $z^{0,\delta,\Delta}(\cdot) = \sum_i d_i y_i^{0,\delta,\Delta}(\cdot)$, where $d_i y_i^{0,\delta,\Delta}(\cdot)$ is the component of $z_i^{0,\delta,\Delta}(\cdot)$ due to reflection on the ith face of G. Note that the backward difference in θ is used for the θ-discretization, which is consistent with the discussion at the beginning of Section 2.

Solving (3.1) by iteration in time and letting $\delta, \Delta \to 0$ shows that we must have $\Delta = \delta$ if there is to be convergence to the correct limit. So, henceforth, let $\Delta = \delta$ and rewrite (3.1), for $\theta \geq -\bar{\theta} + \delta$, as (now, $t = n\delta, n = 0, 1, \ldots$)

$$
\begin{aligned}
\chi^{0,\delta}(t+\delta) &= \chi^{0,\delta}(t) + \delta\chi^{1,\delta}(t,0) + \delta c(\chi^{0,\delta}(t), u^{\delta}(t)) \\
&\quad + \sigma(\chi^{0,\delta}(t))\left[w(t+\delta) - w(t)\right] + \left[z^{0,\delta}(t+\delta) - z^{0,\delta}(t)\right],
\end{aligned}
\tag{3.2}
$$

$$
\begin{aligned}
\chi^{1,\delta}(t+\delta,\theta) &= \chi^{1,\delta}(t,\theta-\delta) + b(\chi^{0,\delta}(t), u^{\delta}(t), \theta-\delta)\left[\mu_a(\theta) - \mu_a(\theta-\delta)\right]) \\
&\quad + p(\theta-\delta)\left[y^{0,\delta}(t+\delta) - y^{0,\delta}(t)\right],
\end{aligned}
\tag{3.3}
$$

with the boundary condition $\chi^{1,\delta}(t,-\bar{\theta}) = 0$.

Comment on a special case. Suppose that there is no delayed reflection term and $\mu_a(\cdot)$ is the distribution function of a point mass concentrated at $\theta = -\bar{\theta}$. Thus, for this example we can suppose that $p(\cdot) = 0$ and $\mu_a(-\bar{\theta} + \delta) - \mu_a(-\bar{\theta}) = 1$ and $d\mu_a(\theta) = 0$ for all $-\bar{\theta} < \theta \leq 0$. Then, for $\theta - \delta > -\bar{\theta}$, $\chi^{1,\delta}(t+\delta,\theta) = \chi^{1,\delta}(t,\theta-\delta)$. The drift term enters only when $\theta = -\bar{\theta} + \delta$. An analogous result holds if there is a delayed reflection term and $p(\theta) = 0$ outside of a small neighborhood of $-\bar{\theta}$.

A representation in terms of a semigroup. A very convenient representation of the solution to (3.2) and (3.3) is in terms of an analog of the semigroup $\Phi(t)$ that was defined in (2.6). For functions $f(t,\theta)$, where $-\bar{\theta} \leq \theta \leq 0$, define the operator Φ^{δ} by

$$
\Phi^{\delta} f(t,\theta) = \begin{cases} f(t,\theta-\delta), & \text{for } -\bar{\theta} \leq \theta - \delta \leq 0, \\ 0, & \text{otherwise.} \end{cases}
\tag{3.4}
$$

Φ^δ acts only on the θ argument. Iterating (3.4) k times yields

$$[\Phi^\delta]^k f(t,\theta) = f(t,\theta - k\delta) I_{\{-\bar{\theta} \le \theta - k\delta \le 0\}}. \tag{3.5}$$

Using Φ^δ, rewrite (3.3) as

$$\begin{aligned}\chi^{1,\delta}(t+\delta,\theta) = \Phi^\delta \chi^{1,\delta}(t,\theta) + b(\chi^{0,\delta}(t), u^\delta(t), \theta - \delta)\left[\mu_a(\theta) - \mu_a(\theta-\delta)\right] \\ + p(\theta-\delta)\left[y^{0,\delta}(t+\delta) - y^{0,\delta}(t)\right].\end{aligned} \tag{3.6}$$

Iterating (3.6), starting at time zero, yields (where $[\Phi^\delta]^0$ is the identity)

$$\begin{aligned}\chi^{1,\delta}(n\delta,\theta) &= [\Phi^\delta]^n \chi^{1,\delta}(0,\theta) \\ &+ \sum_{i=0}^{n-1} [\Phi^\delta]^{n-i-1} \left[b(\chi^{0,\delta}(i\delta), u^\delta(i\delta), \theta-\delta)\right]\left[\mu_a(\theta) - \mu_a(\theta-\delta)\right] \\ &+ \sum_{i=0}^{n-1} [\Phi^\delta]^{n-i-1} p(\theta-\delta)\left[y^{0,\delta}(i\delta+\delta) - y^{0,\delta}(i\delta)\right],\end{aligned} \tag{3.7}$$

and

$$\begin{aligned}\chi^{0,\delta}(n\delta) = \chi^{0,\delta}(0) + \delta\sum_{i=0}^{n-1} \chi^{1,\delta}(i\delta,0) + \delta \sum_{i=0}^{n-1} c(\chi^{0,\delta}(i\delta), u^\delta(i\delta)) \\ + \sum_{i=0}^{n-1} \sigma(\chi^{0,\delta}(i\delta))\left[w(i\delta+\delta) - w(i\delta)\right] + z^{0,\delta}(n\delta).\end{aligned} \tag{3.8}$$

Approximating the cost function. Let us use the following analog of the approximation (7.1.13):

$$\begin{aligned}W^\delta(\hat{x}^\delta, \hat{u}^\delta, \hat{z}^\delta, u^\delta) = E^{\delta,u^\delta}_{\hat{x}^\delta,\hat{u}^\delta,\hat{z}^\delta} \sum_{n=0}^{\infty} e^{-\beta n\delta}\Big[k(\chi^{0,\delta}(n\delta), u^\delta(n\delta))\delta \\ + q'\left(y^{0,\delta}(n\delta+\delta) - y^{0,\delta}(n\delta)\right)\Big],\end{aligned} \tag{3.9}$$

where $\hat{x}^\delta$, $\hat{u}^\delta$, and $\hat{z}^\delta$, are piecewise-constant and right-continuous approximations to the (initial data) path, control, and reflection processes on $[-\bar{\theta}, 0]$, with intervals δ. Let $\chi^{0,\delta}(\cdot)$ and $u^\delta(\cdot)$ denote piecewise-constant and right-continuous interpolations of $\{\chi^\delta(n\delta)\}$ and $\{u^\delta(n\delta)\}$, resp., with intervals δ. The following theorem states that the costs converge to the correct value as $\delta \to 0$.

Theorem 3.1. *Assume* (A3.2.1), (A3.2.2), *and* (A3.2.4)–(A3.2.6). *Let the admissible controls $u^\delta(\cdot)$ converge to the continuous admissible control $u(\cdot)$, uniformly on each bounded time interval, and let the initial conditions converge uniformly to continuous limits. Then, as $\delta \to 0$, $\chi^{0,\delta}(\cdot)$ converges to $x(\cdot)$ and* *the costs converge to that for the process* (1.1) *with cost rate* $k(x(t), u(t))$ *and boundary reflection cost rate* $q'dy(t)$.

Proof. Let $\epsilon(t, \delta)$ denote a function that goes to zero as $\delta \to 0$, uniformly in (t, ω) on any bounded t-interval. Its value might change each time that it is used. For $t = n\delta$, the first sum in (3.7) is

$$\sum_{i=0}^{n-1} b(\chi^{0,\delta}(i\delta), u^\delta(i\delta), \theta - t + i\delta) I_{\{-\bar{\theta} \le \theta - t + i\delta + \delta \le 0\}} \\ \times \left[\mu_a(\theta - t + i\delta + \delta) - \mu_a(\theta - t + i\delta)\right]. \tag{3.10}$$

Using the continuity of $b(\cdot)$ to approximate the θ-dependence in (3.10), we can write (3.10) as

$$\begin{aligned} &\int_0^t b(\chi^{0,\delta}(s), u^\delta(s), \theta - t + s) I_{\{-\bar{\theta} \le \theta - t + s \le 0\}} \\ &\quad \times \left[\mu_a(\theta - t + s) - \mu_a(\theta - t + s - ds)\right] + \epsilon(t, \delta) \\ &= \int_{\max\{\theta - t, -\bar{\theta}\}}^{\theta} b(\chi^{0,\delta}(\gamma + t - \theta), u^\delta(\gamma + t - \theta), \gamma) d\mu_a(\gamma) + \epsilon(t, \delta). \end{aligned} \tag{3.11}$$

Because the $\chi^{0,\delta}(\cdot)$ and $u^\delta(\cdot)$ are piecewise-constant on the intervals $[k\delta, k\delta + \delta)$ and θ is a (negative) integral multiple of δ, the error term $\epsilon(t, \delta)$ arises from the θ dependence of $b(x, u, \theta)$.

The second sum in (3.7) is

$$\sum_{i=0}^{n-1} p(\theta - t + i\delta) \left[y^{0,\delta}(i\delta + \delta) - y^{0,\delta}(i\delta)\right] I_{\{-\bar{\theta} \le \theta - t + i\delta + \delta \le 0\}},$$

which equals

$$\int_{\max\{\theta - t, -\bar{\theta}\}}^{\theta} p(\gamma) d_\gamma y^{0,\delta}(\gamma + t - \theta), \tag{3.12}$$

modulo an error that is due to the approximation of $p(\cdot)$ by a piecewise-constant function. The error is bounded by $\epsilon(t, \delta)$ times the variation of $y^{0,\delta}(\cdot)$ on the interval $[t - \bar{\theta}, t - \bar{\theta}_0]$, where $\bar{\theta} > \bar{\theta}_0 > 0$ is defined in (A3.2.4). The initial condition is treated similarly. Analogously to the development in Theorem 2.1, the effect of the initial condition is to add terms of the form $\int_{-\bar{\theta}}^{\max\{\theta - t, -\bar{\theta}\}}$ to the expressions computed above.

Using (3.11), (3.12), and adding the contribution of the initial condition yields

$$\begin{aligned} \chi^{1,\delta}(t, \theta) &= \int_{-\bar{\theta}}^{\theta} b(\chi^{0,\delta}(t + \gamma - \theta), u^\delta(t + \gamma - \theta), \gamma) d\mu_a(\gamma) \\ &\quad + \int_{-\bar{\theta}}^{\theta} p(\gamma) d_\gamma y^{0,\delta}(t + \gamma - \theta) \\ &\quad + \epsilon(t, \delta) \left[1 + |z^{0,\delta}|(t - \bar{\theta}_0) - |z^{0,\delta}|(t - \bar{\theta})\right]. \end{aligned}$$

The value at $\theta = 0$ is

$$\begin{aligned}\chi^{1,\delta}(t,0) = & \int_{-\bar{\theta}}^{0} b(\chi^{0,\delta}(t+\gamma), u^\delta(t+\gamma), \gamma) d\mu_a(\gamma) \\ & + \int_{-\bar{\theta}}^{0} p(\gamma) d_\gamma y^{0,\delta}(t+\gamma) \\ & + \epsilon(t,\delta)\left[1 + |z^{0,\delta}|(t-\bar{\theta}_0) - |z^{0,\delta}|(t-\bar{\theta})\right].\end{aligned} \tag{3.13}$$

By (3.8), for $t = n\delta$,

$$\begin{aligned}\chi^{0,\delta}(t) = & x(0) + \int_0^t \chi^{1,\delta}(s,0) ds \\ & + \int_0^t c(\chi^{0,\delta}(s), u^\delta(s)) ds + \int_0^t \sigma(\chi^{0,\delta}(s)) dw(t) + z^{0,\delta}(t).\end{aligned} \tag{3.14}$$

Substituting (3.13) into (3.14), and letting t be arbitrary, yields that $(\chi^{0,\delta}(\cdot), u^\delta(\cdot), w(\cdot), z^{0,\delta}(\cdot))$ satisfies (1.1), modulo the error terms. It follows from Lemma 3.2.1 that the error terms go to zero as $\delta \to 0$ in that for each $T < \infty$

$$\lim_{\delta \to 0} E \sup_{t \le T} \epsilon(t,\delta)\left[1 + |z^{0,\delta}|(t-\bar{\theta}_0) - |z^{0,\delta}|(t-\bar{\theta})\right] = 0.$$

A weak convergence argument can be used to complete the proof. It follows from Lemma 3.2.2 that the sequence $\{z^{0,\delta}(\cdot)\}$ is tight in the Skorokhod topology and that all limit processes are continuous. Then the set of processes $\{(\chi^{0,\delta}(\cdot), u^\delta(\cdot), w(\cdot), z^{0,\delta}(\cdot))\}$ is tight in the Skorohod topology and all limits are continuous. One needs to show that the weak-sense limit $(\tilde{x}(\cdot), \tilde{u}(\cdot), \tilde{w}(\cdot), \tilde{z}(\cdot))$ of any weakly convergent subsequence satisfies (1.1), where $\tilde{z}(\cdot)$ is the reflection term, and all processes are nonanticipative with respect to the standard Wiener process $\tilde{w}(\cdot)$. The nonanticipativity is shown by the method used in the proof of Theorem 7.1.3 in Section 8.5. See also the proof of Theorem 3.5.1 in Chapter 3. That $\tilde{z}(\cdot)$ is the reflection process follows from Lemma 3.2.2. Clearly, the weak-sense limit $(\tilde{u}(\cdot), \tilde{w}(\cdot))$ has the same probability law as $(u(\cdot), w(\cdot))$. This and the weak-sense uniqueness of solutions implies that the original sequence $(\chi^{0,\delta}(\cdot), u^\delta(\cdot), w(\cdot), z^{0,\delta}(\cdot))$ converges to the weak-sense unique solution to(1.1).

The nonreflection part of the costs in (3.9) converges due to this weak convergence and the continuity and boundedness of the function $k(\cdot)$. The reflection component converges due to the weak convergence of $z^{0,\delta}(\cdot)$ and the fact that, for any $T < \infty$, Lemma 3.2.1 implies that the sequence of differences $\{z^{0,\delta}(lT+T) - z^{0,\delta}(lT); l, \delta\}$ of the reflection terms is uniformly (for small δ and $l \ge 0$) integrable. ■

9.4 The Markov Chain Approximation

The forms of the numerical approximations are to be motivated by the representations (2.3) and (2.4). With this representation, there are no issues with delays, but the variable θ needs to be discretized. The approximation of this system by a Markov chain follows the general procedure laid out in Chapter 6. As in Chapter 6, one first gets an approximating chain and cost function, then solves the associated control or optimal control problem, and finally shows that the cost functionals converge as the discretization parameter goes to zero. For simplicity in the notation, we will suppose that $z(s) = 0, s \leq 0$, although the general case can be treated by the same methods as well.

The main issue involves the treatment of the variable θ. It is seen from the discussion in Section 3 that the discretization of the time variable and of θ are closely connected. This is manifest particularly in the "shift" term $\Phi^{\delta}\chi^{1,\delta}(t,\theta)$ in (3.6) where the increase in time is associated with a shift in the value of θ, by the same amount. Thus the discretization level of the time variable and of θ had to be the same if there was to be convergence to the correct limit. Such a requirement would usually be onerous for the classical Markov chain approximation of, say, Section 6.4.[2] Nevertheless, we will use (3.2) and (3.6) as guides. The use of the implicit method of approximation in Section 7.3 alleviated the small time-interval problem for the approximation of the path memory values, by use of the interpolation based on the timescale implied by the interpolated "time" process $\phi^{h,\delta}(\cdot)$. The use of the implicit method of approximation in the current case will allow us to coordinate the discretization of time and of θ. The periodic and periodic-Erlang approximation procedures in Section 8.1 and 8.2 appear to have advantages over the pure implicit approximation method, and analogs will be developed in Section 8 for the model of this chapter.

The basic condition required for convergence is still local consistency, but for the pair $\chi^0(\cdot), \chi^1(\cdot)$. The discretization of $\chi^0(\cdot)$ is similar to that of $x(\cdot)$ in Chapter 6, as it does not contain any delays. Let h denote the discretization level for the $\chi^0(\cdot)$ and, until further notice, for the $\chi^1(\cdot,\theta)$ for the values of θ that will be used.[3]

For numerical purposes, the approximating chains must be bounded. This is discussed in the next subsection. Subsection 4.2 gives the local consistency condition for the implicit approximation procedure. It might look complicated but is actually a straightforward adaptation of the method of Chapter 6. The subsections that follow after these give the dynamical representations of the evolution of the chains, an approximation to the cost function, and the associated Bellman equation.

---

[2] This would require a constant interpolation interval. If h is the discretization level for the state variables and $\Delta t_n^h = O(h^2)$, then this would entail a level of order h^2 for the discretization of θ, which yields a state space that might be too large for practical use.

[3] The discretization level can depend on the coordinate and is quite flexible provided only that the local consistency condition holds. See [58] for a discussion of the possibilities in the no-delay problem. The same considerations apply here.

9.4.1 Preliminaries and Boundaries

Notation. Let $\bar{\theta}$ be an integral multiple of $\delta > 0$ and let the discretization of the variable θ take values in the set $T^\delta = \{-\bar{\theta} + \delta, \ldots, -\delta, 0\}$. Owing to the boundary condition $\chi^1(t, -\bar{\theta}) = 0$, there is no need for the value $\theta = -\bar{\theta}$. The Markov chain approximating $(\chi^0(t), \chi^1(t, \theta), \theta \in T^\delta)$ will be denoted by $(\xi_n^{0,h,\delta}, \xi_n^{1,h,\delta}(\theta), \theta \in T^\delta)$. The component $\xi_n^{0,h,\delta}$ will take values in $G_h \cup \partial G_h^+$ with instantaneous reflection back if it leaves G_h, in accordance with (A3.2.1) and (A3.2.2). For simplicity in the development, we suppose that for each $\theta \in T^\delta$, $\xi_n^{1,h,\delta}(\theta)$ takes values in a regular grid with spacing $h_1 = O(h)$.

Boundaries and bounds for $\boldsymbol{\xi_n^{1,h,\delta}(\theta)}$. The $\xi_n^{0,h,\delta}$ take values in G_h^+, hence they are bounded. If $p(\cdot) = 0$, then there is no delayed reflection term, and (2.14) shows that $|\chi^1(t,\theta)|$ is bounded by $B(\theta) = |b|\mu_a(\theta)$ where $|b| = \sup_{x,\alpha} |b(x,\alpha)|$, and $|\xi_n^{1,h,\delta}(\theta)|$ can be taken to be bounded by a slightly higher value, say $B^\epsilon(\theta) = B(\theta) + \epsilon$, for some small $\epsilon > 0$. In the limit, as $h, \delta \to 0$, the values of $\xi_n^{1,h,\delta}(\theta)$ would not exceed $B^\epsilon(\theta)$. If the process $|\xi_n^{1,h,\delta}(\theta)|$ ever exceeded $B^\epsilon(\theta)$, then one could either stop it or inject it back.

However, if there is a delayed reflection term, then the value of $\xi_n^{1,h,\delta}(\theta)$ is influenced by the last term in (2.14). This term is bounded by a constant times the variation of $z(\cdot)$ on $[t - \bar{\theta}, t - \bar{\theta}_0]$, which satisfies Lemma 6.3.1 and, in the limit, Lemma 3.2.1. For numerical purposes, we need to bound $|\xi_n^{1,h,\delta}(\theta)|$, and this entails an approximation and tradeoff: the larger the bound the more the computation, but the better the approximation. For specificity and to allow us to proceed, we will use the following simple procedure. Set a level $\bar{B}(\theta) > B^\epsilon(\theta)$ such that the probability is small that $\bar{B}(\theta)$ is exceeded by $|\chi^1(t,\theta)|$ on some time interval $[0, T]$. If the value of $|\xi_n^{1,h,\delta}(\theta)|$ exceeds $\bar{B}(\theta)$, then immediately reflect it back to the boundary. Until Theorem 7.1, ignore these boundaries.

9.4.2 Transition Probabilities and Local Consistency: An Implicit Approximation Procedure

We will develop a slightly different version of the implicit approximation method of Section 7.3, one that will be suitable for the problem of this chapter. As in Section 7.3, let δ denote the discretization level for the time variable, and let $\phi_n^{h,\delta}$ denote the value of the time component of the approximating chain at step n. As there, the steps can be divided into two classes, corresponding to whether the time variable advances or not. First we give the transition probabilities of $\xi_n^{1,h,\delta}(\theta)$ when the time variable advances, then give the general form of the transition probabilities for $\xi_n^{0,h,\delta}$, and finally we define the local consistency property and the transition probabilities for $\xi_n^{1,h,\delta}(\theta)$ when the time variable does not advance. The reader should be familiar with Section 7.3 before proceeding further.

The transitions in the Markov state. Introductory comments. Recall the forms (3.3) or (3.6), which exhibited the coordination between the advance of time and the shift in the value of θ. The approximating Markov chain will attempt to duplicate this behavior, although it will be done rather indirectly. The transitions in $(\xi_n^{0,h,\delta}, \xi_n^{1,h,\delta}(\theta), \theta \in T^\delta)$ depend on whether the time variable advances at step n or not and can be outlined as follows. If the time variable does advance at step n, then the transitions aim to duplicate the effects of the shift term $\chi^{1,\delta}(t+\delta,\theta) = \chi^{1,\delta}(t,\theta-\delta)$ of (3.3), with the drift and diffusion terms ignored. If the time variable does not advance, then the transitions are guided by the drift and diffusion terms in (3.2), (3.3), with the shift term in (3.3) ignored, in the sense that the $\chi^{1,\delta}(t,\theta-\delta)$ on the right side is dropped and the $\chi^{1,\delta}(t+\delta,\theta)$ on the left side is replaced by $\chi^{1,\delta}(t+\delta,\theta) - \chi^{1,\delta}(t,\theta)$. The approximation must be done so that it is locally consistent with (2.7), (2.8). The following description fills in the details.

The transitions when the time variable advances. If the time variable advances at step n, then the transitions are, for $\theta \geq -\bar{\theta} + \delta$,

$$\xi_{n+1}^{1,h,\delta}(\theta) = \xi_n^{1,h,\delta}(\theta-\delta), \quad \xi_{n+1}^{0,h,\delta} = \xi_n^{0,h,\delta}, \quad \phi_{n+1}^{h,\delta} = \phi_n^{h,\delta} + \delta. \tag{4.1a}$$

Thus the transition can be represented as

$$\xi_n^{1,h,\delta} \equiv \{\xi^1(-\bar{\theta}+\delta), \ldots, \xi^1(0)\} \to \xi_{n+1}^{1,h,\delta} = \{0, \xi^1(-\bar{\theta}+\delta), \ldots, \xi^1(-\delta)\}. \tag{4.1b}$$

The transition in $\xi_n^{0,h,\delta}$ if the time variable does not advance at step n will be defined next.

The transition probabilities for $\xi_n^{0,h,\delta}$ when the time variable does not advance. In the no-delay case of Chapter 6, the transition probabilities and interpolation interval for the explicit approximation procedure were given first. Then, in Subsection 6.5.1, Equation (6.5.6) defined the transition probabilities and interpolation interval for the implicit approximation procedure simply in terms of those for the explicit approximation procedure. This process was repeated for the model in Chapter 7. It was noted that, for the explicit approximation procedure for the no-delay case, the transition probabilities and interpolation interval can be expressed as ratios as in (7.1.10). This form was adapted to the explicit approximation procedure for the delay problem in (7.1.11). Then the transition probabilities and interpolation interval for the implicit approximation procedure were computed in terms of these in (7.3.1) and (7.3.2).

The identical (and unrestrictive) approach will be used here to get the transition probabilities for the approximating chain component $\xi_n^{0,h,\delta}$. The only difference between the drift term of the model for $x(\cdot)$ in Chapter 6 and that of $\chi^0(\cdot)$ in this chapter is the presence of the $\chi^1(t,0)$ component in the equation for $\chi^0(\cdot)$. Suppose that we have a transition probability and an interpolation interval for the model without delays that satisfy (7.1.11) and are locally consistent for the no-delay problem, in the sense of (6.2.1). To adapt this formula to the current case, define $\xi_n^{0,h,\delta} = x^0 \in G_h, \xi_n^{1,h,\delta}(0) = x^1$ and let control value α be used. Then the probability that $\xi_{n+1}^{0,h,\delta}$ takes the value $\tilde{x}^0$, *conditioned on the event that the time variable does not advance at step n*, is

$$p^h\left(x^0, \tilde{x}^0 \middle| \alpha, x^1\right) = \frac{N^h\left(x^1 + c(x^0,\alpha), a(x^0), \tilde{x}^0\right)}{D^h\left(x^1 + c(x^0,\alpha), a(x^0)\right)}, \qquad (4.2)$$
$$\left|\tilde{x}^0 - x^0\right| = O(h).$$

Thus, if the time variable does not advance at a step, then the transition probability for $\xi_n^{0,h,\delta}$ has the same dependence on the drift vector and covariance matrix as in the nondelay case; the quantity $x^1 + c(x^0,\alpha)$ simply replaces $b(x,\alpha)$ or $b(\hat{x},\alpha)$. Any of the algorithms in [58, Chapter 5] can be used to get the functions $N^h(\cdot)$ in (4.2). (The function $D^h(\cdot)$ is just a normalization.)

The probability that the time variable advances. In analogy to the definition in (7.1.10), define the interval

$$\Delta t^h(x^0, x^1, \alpha) = \frac{h^2}{D^h(x^1 + c(x^0,\alpha), a(x^0))}.$$

Then, in analogy to (7.3.1), the probability that the time variable advances at step n is

$$p^{h,\delta}(x^0, n\delta; x^0, n\delta + \delta | \alpha, x^1) = \frac{\Delta t^h(x^0, x^1, \alpha)}{\Delta t^h(x^0, x^1, \alpha) + \delta}. \qquad (4.3)$$

Also, analogously to (7.3.2), define the interpolation interval

$$\Delta t^{h,\delta}(x^0, x^1, \alpha) = \frac{\delta \Delta t^h(x^0, x^1, \alpha)}{\Delta t^h(x^0, x^1, \alpha) + \delta}. \qquad (4.4)$$

Let $u_n^{h,\delta}$ denote the control applied at step n, and redefine the interpolation intervals

$$\Delta t_n^h = \Delta t^h(\xi_n^{0,h,\delta}, \xi_n^{1,h,\delta}(0), u_n^{h,\delta}),$$
$$\Delta t_n^{h,\delta} = \Delta t^{h,\delta}(\xi_n^{0,h,\delta}, \xi_n^{1,h,\delta}(0), u_n^{h,\delta}).$$

An assumption on the delayed reflection term. Let $\mathcal{F}_n^{h,\delta}$ denote the minimal σ-algebra that measures the system data to step n, with the associated conditional expectation denoted by $E_n^{h,\delta}$. Define $\Delta z_n^{0,h,\delta} = [\xi_{n+1}^{0,h,\delta} - \xi_n^{0,h,\delta}] I_{\{\xi_n^{0,h,\delta} \notin G_h\}}$, and define $\Delta y_n^{0,h,\delta}$ by $\Delta z_n^{0,h,\delta} = \sum_i d_i \Delta y_n^{0,h,\delta}$. To simplify the discussion, we will make the following assumption on the delayed reflection term $p(\chi^0(t),\theta) dy^0(t)$. In applications in communication theory, the part of the reflection term that is delayed is usually that due to buffer overflows. In this case the boundary of concern represents a (scaled) buffer level and the reflection is directly back to the value that corresponds to a full buffer. Then, the corresponding components of $\Delta z_n^{0,h,\delta}$ and $\Delta y_n^{0,h,\delta}$ are known when $\xi_n^{0,h,\delta}$ is known. With this motivation, we suppose that, if $\xi_n^{0,h,\delta} \notin G_h$, then $p(\xi_n^{0,h,\delta}, \theta)\Delta y_n^{0,h,\delta}$ is known at step n and included in $\mathcal{F}_n^{h,\delta}$. Otherwise, the conditional (on $\mathcal{F}_n^{h,\delta}$) expectation of $\Delta y_n^{0,h,\delta}$ would be used in (4.8) below.

The local consistency condition. If $\xi_n^{0,h,\delta} = x^0 \notin G_h$, then the reflection back to G_h is instantaneous in that $\Delta t^h(x^0, x^1, \alpha) = 0$, and it is in accord with (6.2.2).

Let $I_n^{h,\delta}$ denote the indicator function of the event that the time variable advances at step n. By the local consistency of the transition probabilities (4.2) when the time variable does not advance, for $\xi_n^{0,h,\delta} \in G_h$ it follows that

$$\begin{aligned} E\left[\xi_{n+1}^{0,h,\delta} - \xi_n^{0,h,\delta} \big| I_n^{h,\delta} = 0, \mathcal{F}_n^{h,\delta}\right] &= \Delta t_n^h \left[\xi_n^{1,h,\delta}(0) + c(\xi_n^{0,h,\delta}, u_n^{h,\delta})\right] + o(\Delta t_n^h), \\ \text{covar}\left[\xi_{n+1}^{0,h,\delta} - \xi_n^{0,h,\delta} \big| I_n^{h,\delta} = 0, \mathcal{F}_n^{h,\delta}\right] &= a(\xi_n^{0,h,\delta})\Delta t_n^h + o(\Delta t_n^h). \end{aligned} \tag{4.5}$$

By averaging over $I_n^{h,\delta}$, given $\mathcal{F}_n^{h,\delta}$, the Δt_n^h changes to $\Delta t_n^{h,\delta}$ and we have

$$\begin{aligned} E\left[\xi_{n+1}^{0,h,\delta} - \xi_n^{0,h,\delta} \big| \mathcal{F}_n^{h,\delta}\right] &= \Delta t_n^{h,\delta} \left[\xi_n^{1,h,\delta}(0) + c(\xi_n^{0,h,\delta}, u_n^{h,\delta})\right] + o(\Delta t_n^{h,\delta}), \\ \text{covar}\left[\xi_{n+1}^{0,h,\delta} - \xi_n^{0,h,\delta} \big| \mathcal{F}_n^{h,\delta}\right] &= a(\xi_n^{0,h,\delta})\Delta t_n^{h,\delta} + o(\Delta t_n^{h,\delta}). \end{aligned} \tag{4.6}$$

The transitions for $\xi_n^{1,h,\delta}$ when the time variable does not advance. The rule for the transition probabilities of the component $\xi_n^{0,h,\delta}$ was easy to establish as $\chi^0(t)$ evolves as a diffusion, so the methods of Chapter 6 and any of the algorithms in [58] could be readily adapted. The rules for the transitions of the $\xi_n^{1,h,\delta}(\theta)$ are only a little more complicated. If the time variable advances then, as stated by (4.1), we simply shift as $\xi_{n+1}^{1,h,\delta}(\theta) = \xi_n^{1,h,\delta}(\theta - \delta)$. The local consistency condition for the transition in $\xi_n^{1,h,\delta}(\theta)$ when the time variable does not advance is obtained similarly to what was done for $\xi_n^{0,h,\delta}$: In particular, the conditional mean of the change in $\xi_n^{1,h,\delta}(\theta)$ must be the nonshift part of (3.3) or (3.6), rescaled to a single step of the chain, namely, for $\theta \geq -\bar{\theta} + \delta$,

$$E\left[\xi_{n+1}^{1,h,\delta}(\theta) - \xi_n^{1,h,\delta}(\theta) \big| I_n^{h,\delta} = 0, \mathcal{F}_n^{h,\delta}\right] = q_n^{h,\delta}(\theta), \tag{4.7}$$

where

$$q_n^{h,\delta}(\theta) = b(\xi_n^{0,h,\delta}, u_n^{h,\delta}, \theta - \delta)\frac{[\mu_a(\theta) - \mu_a(\theta - \delta)]}{\delta}\Delta t_n^h + p(\xi_n^{0,h,\delta}, \theta - \delta)\Delta y_n^{0,h,\delta}. \tag{4.8}$$

This need hold only modulo $o(\Delta t_n^h) + o(\Delta y_n^{0,h,\delta})$. Hence, in view of the θ-continuity of $b(\xi^0, \alpha, \cdot)$ and $p(\xi^0, \cdot)$, the $\theta - \delta$ arguments in $b(\cdot)$ and $p(\cdot)$ will be replaced by θ, for notational simplicity. The motivation for the form (4.8) will be apparent in the proof of convergence in Sections 6 and 7.

Only one of the terms on the right side of (4.8) can be nonzero at a time. The boundary condition is $\xi_n^{1,h,\delta}(-\bar{\theta}) = 0$. The complexity of the computation of the transition probabilities is the same whether the path, control, reflection term, or any combination of them are delayed. Both the cases where δ is fixed and is of order $O(h)$ are of interest.

On averaging out the conditioning event $I_n^{h,\delta} = 0$ in (4.7), the Δt_n^h in (4.8) is replaced by the interval $\Delta t_n^{h,\delta}$. The expression (4.7) defines local consistency for $\xi_n^{1,h,\delta}(\theta)$ when the time variable does not advance at step n. Any transition probabilities that satisfy (4.5) and (4.7) (modulo the allowed small errors) when the time variable does not advance can be used.

Example. Suppose that there is no delayed reflection term and that there is only a point delay of $\bar{\theta}$ for the path and/or control. Then $\mu_a(\cdot)$ is the distribution function of a point mass at $-\bar{\theta}$, and for the approximation we can suppose that $\mu_a(\theta+\delta) - \mu_a(\theta) = 1$ for $\theta = -\bar{\theta}+\delta$ and it is zero for $\theta > -\bar{\theta}+\delta$. Then

$$q_n^{h,\delta}(-\bar{\theta}+\delta) = b(\xi_n^{0,h,\delta}, u_n^{h,\delta}, -\bar{\theta}+\delta)\frac{\Delta t_n^h}{\delta}$$

and it is zero otherwise. In this case the computation of the transition probabilities and the Bellman equation is particularly simple as we need to update $\xi_n^{1,h,\delta}(\theta)$ only for $\theta = -\bar{\theta}+\delta$ between updates of the time variable.

Realizing (4.7). The transition probability that is used to attain (4.7) can be viewed as being a randomization among neighboring grid points. The simplest approach is to randomize independently in θ at each step n. One could also consider any set of grid points in whose convex hull $\{q_n^{h,\delta}(\theta), \theta \in T^\delta\}$ lies, and randomize among those points. A useful method for coordinating the updates for all θ is discussed at the end of the subsection.

A note on the rule for updating $\xi_n^{1,h,\delta}(\theta)$. The method for updating the $\xi_n^{1,h,\delta}(\theta)$ when the time variable does not advance, based on (4.7), (4.8), was chosen because at this time it seemed to be about as simple as possible from a numerical point of view. One way or another, between the steps at which the time variable advances, one would have to keep track of the approximations to the integrals $\int b(\chi^0(s), u(s), \theta)\,[\mu_a(\theta+ds) - \mu_a(\theta)]$ and $\int p(\chi^0(s), \theta) dy^0(s)$ from (2.2) between those steps. This needs to be done with a reasonable requirement on the system memory. If we sought to add the effects of the approximations to these terms only at the steps when the time variable is advanced, we would need to keep track of the running sums of the $q_n^{h,\delta}(\theta)$ between such time variable advances, which would amount to an additional state component. Clearly, much more work is needed to find the best forms.

Summary: The full method for the transition probabilities for the implicit approximation procedure. At the current step n we first decide, according to the probability (4.3), whether the time variable advances or not. If the time variable advances, then the transitions are given by (4.1). Now suppose that the time variable does not advance at the current step. Then (4.5) is all that $\xi_{n+1}^{0,h,\delta} - \xi_n^{0,h,\delta}$ must satisfy. This is assured if the transition probabilities are any ones that are locally consistent for the no-delay problem, with drift term $x^1 + c(x,\alpha)$ replacing $b(x,\alpha)$. The transitions of the component $\xi_n^{1,h,\delta}(\theta)$ need only satisfy (4.7), (4.8), modulo an error whose conditional mean and covariance are $o(\Delta t_n^h) + o(\Delta y_n^{0,h,\delta})$. The transitions for the components $\xi_n^{0,h,\delta}$ and $\xi_n^{1,h.\delta} = \{\xi_n^{1,h.\delta}(\theta), \theta \in T^\delta\}$ are determined simultaneously and independently.

Randomization errors. Using (4.6)–(4.8), define the martingale difference term $\rho_n^{h,\delta}(\theta)$ by

$$\begin{aligned}\rho_n^{h,\delta}(\theta) &= \left[\left(\xi_{n+1}^{1,h,\delta}(\theta) - \xi_n^{1,h,\delta}(\theta)\right) - E_n^{h,\delta}\left(\xi_{n+1}^{1,h,\delta}(\theta) - \xi_n^{1,h,\delta}(\theta)\right)\right]\left(1 - I_n^{h,\delta}\right) \\ &= \left[\left(\xi_{n+1}^{1,h,\delta}(\theta) - \xi_n^{1,h,\delta}(\theta)\right) - q_n^{h,\delta}(\theta)\right]\left(1 - I_n^{h,\delta}\right).\end{aligned} \tag{4.9}$$

The variance of $\rho_n^{h,\delta}(\theta)$ for each θ is minimized by realizing the conditional mean (4.7) by randomizing between the grid points that are closest to $\xi_n^{1,h,\delta}(\theta) + q_n^{h,\delta}(\theta)$. When analyzing the processes ξ_n^h or $\xi_n^{h,\delta}$ in Chapters 6 and 7, the errors due to any randomization that was done to attain the conditional mean vanished asymptotically, whereas those due to the attainment of the correct local covariance led to the diffusion term in the limit. Due to the θ-shift term when the time variable advances, the analysis of the asymptotic effects of $\rho_n^{h,\delta}(\theta)$ needs to be treated differently, and this will be done in Section 6. Since there is no term in (2.2) that is due to a Wiener process, it will be seen that the effects of the randomization will vanish asymptotically.

A method for coordinating the updates of $\xi_n^{1,h,\delta}(\theta)$ for all θ simultaneously. Let $\{\gamma_n\}$ be a sequence of i.i.d. random variables, each uniformly distributed on $[0,1]$, and with $\gamma_i, i \geq n$, being independent of the systems data up to time n. We can work with one component of $\xi_n^{1,h,\delta}(\theta)$ at a time, so assume that it is real-valued in this paragraph. To attain the conditional mean $q_n^{h,\delta}(\theta)$, we first center so that $q_n^{h,\delta}(\theta)$ can be taken to lie in $[0,h_1]$. Then, with this centering, the probability that $\xi_{n+1}^{1,h,\delta}(\theta) = h_1$ is $q_n^{h,\delta}(\theta)/h_1$, and it takes the value zero otherwise. Choose $\xi_{n+1}^{1,h,\delta}(\theta) = h_1$ if $q_n^{h,\delta}(\theta)/h_1 \leq \gamma_n$. The same γ_n is used for all $\theta \in T^\delta$. This method is not needed for the convergence proofs, but it will be useful in Section 6 and is one way of controlling the randomization errors. If this method is used, the γ_n themselves do not appear in the algorithms; only their distribution appears.

9.4.3 Dynamical Representations, the Cost Function and Bellman Equation

In preparation for the convergence proof in Section 7, let us write a dynamical representation of the evolution of the approximating chain.

Note that, as in (6.5.7),

$$E_n^{h,\delta}\left[\phi_{n+1}^{h,\delta} - \phi_n^{h,\delta}\right] = \Delta t_n^{h,\delta}.$$

Analogously to the definitions above (6.5.8), define the martingale differences

$$\tilde{\beta}_n^{0,h,\delta} = \left(\xi_{n+1}^{0,h,\delta} - \xi_n^{0,h,\delta} - E\left[\xi_{n+1}^{0,h,\delta} - \xi_n^{0,h,\delta} \middle| \mathcal{F}_n^{h,\delta}, I_n^{h,\delta} = 0\right]\right) I_{\{\xi_n^{0,h,\delta} \in G_h\}},$$

$$\beta_n^{0,h,\delta} = \left(\xi_{n+1}^{0,h,\delta} - \xi_n^{0,h,\delta} - E\left[\xi_{n+1}^{0,h,\delta} - \xi_n^{0,h,\delta} \middle| \mathcal{F}_n^{h,\delta}\right]\right) I_{\{\xi_n^{0,h,\delta} \in G_h\}}.$$

We have

$$\begin{aligned} \text{covar}\left[\tilde{\beta}_n^{0,h,\delta} \middle| \mathcal{F}_n^{h,\delta}, I_n^{h,\delta} = 0\right] &= a(\xi_n^{0,h,\delta})\Delta t_n^h + o(\Delta t_n^h), \\ \text{covar}\left[\beta_n^{0,h,\delta} \middle| \mathcal{F}_n^{h,\delta}\right] &= a(\xi_n^{0,h,\delta})\Delta t_n^{h,\delta} + o(\Delta t_n^{h,\delta}). \end{aligned} \tag{4.10}$$

With the definition of $\beta_{0,n}^{h,\delta}$ as above (6.5.8), we can write

$$\phi_{n+1}^{h,\delta} = \phi_n^{h,\delta} + \Delta t_n^{h,\delta} + \beta_{0,n}^{h,\delta}. \tag{4.11}$$

The conditional covariance of the martingale difference $\beta_{0,n}^{h,\delta}$ is $o(\Delta t^{h,\delta})$. Using the above definitions, we can decompose $\xi_{n+1}^{0,h,\delta} - \xi_n^{0,h,\delta}$ as

$$\begin{aligned} \xi_{n+1}^{0,h,\delta} &= \xi_n^{0,h,\delta} \\ &+ \left[\Delta t_n^h \xi_n^{1,h,\delta}(0) + \Delta t_n^h c(\xi_n^{0,h,\delta}, u_n^{h,\delta}) + \tilde{\beta}_n^{0,h,\delta} + \Delta z_n^{0,h,\delta}\right](1 - I_n^{h,\delta}) + o(\Delta t_n^h). \end{aligned}$$

Alternatively, by centering $I_n^{h,\delta}$ about its conditional expectation, the above expression can be written as

$$\xi_{n+1}^{0,h,\delta} = \xi_n^{0,h,\delta} + \Delta t_n^{h,\delta}\xi_n^{1,h,\delta}(0) + \Delta t_n^{h,\delta} c(\xi_n^{0,h,\delta}, u_n^{h,\delta}) + \beta_n^{0,h,\delta} + \Delta z_n^{0,h,\delta} + o(\Delta t_n^h). \tag{4.12}$$

The cost function. Let $\hat{x} = \{x(s), -\bar{\theta} \le s \le 0\}$ be the initial path segment, and $\hat{u} = \{u(s), -\bar{\theta} \le s < 0\}$ the initial control segment. If the path is not delayed, then write $x = \hat{x}(0)$ for $\hat{x}$. If the control is not delayed, then drop $\hat{u}$. The cost function for the approximating chain is a discretization of (1.3), and an analog of (3.9), but with $z^0(s) = 0$ for $s \le 0$. The initial data $\chi^0(0), \chi^1(0, \cdot)$ for (2.1), (2.2), is the function of the initial path and control segments $\hat{x}, \hat{u}$, resp., as given by (2.5). Keep in mind that it is only (2.5) that needs to be discretized and not $\hat{x}$ or $\hat{u}$ otherwise. Let $\xi^0 = \xi_0^{0,h,\delta}$ and $\xi^1 = \{\xi_0^{1,h,\delta}(\theta), \theta \in$

$T^\delta\}$ denote the discretization of the initial condition that is used for the chain. As usual, we suppose that the reflection term is zero for $t \leq 0$. If control $u^{h,\delta} = \{u_n^{h,\delta}, 0 \leq n < \infty\}$ is used, then the cost function for the approximating chain can be written in a form analogous to (7.3.12), namely,

$$W^{h,\delta}(\xi^0, \xi^1, u^{h,\delta}) \\ = E_{\xi^0,\xi^1}^{h,\delta,u^{h,\delta}} \sum_{n=0}^{\infty} e^{-\beta t_n^{h,\delta}} \left[k(\xi_n^{0,h,\delta}, u_n^{h,\delta})\Delta t_n^{h,\delta} + q'\Delta y_n^{0,h,\delta}\right], \tag{4.13}$$

where $E_{\xi^0,\xi^1}^{h,\delta,u^{h,\delta}}$ denotes the expectation given the approximations ξ^0, ξ^1 to the initial data, and the use of control sequence $u^{h,\delta}$ on $[0,\infty)$. An alternative form is an analog of (7.3.11), namely,

$$\hat{W}^{h,\delta}(\xi^0, \xi^1, u^{h,\delta}) \\ = E_{\xi^0,\xi^1}^{h,\delta,u^{h,\delta}} \sum_{i=0}^{\infty} e^{-\beta \phi_n^{h,\delta}} \left[k(\xi_n^{0,h,\delta}, u_n^{h,\delta})\delta I_{\{\phi_{n+1}^{h,\delta} \neq \phi_n^{h,\delta}\}} + q'\Delta y_n^{0,h,\delta}\right]. \tag{4.14}$$

Without loss of generality, we can suppose that the values of ξ^0 are on the h-grid, and the values of $\xi^1(\theta)$ are on the h_1-grid for each θ of interest. Let $V^{h,\delta}(\xi^0, \xi^1)$ denote the infimum of the costs over all controls for initial condition ξ^0, ξ^1.

A Bellman equation. Recall the definition (4.3) of the probability that the time variable advances at the current step. Let (ξ^0, ξ^1) denote the initial condition and (ξ_1^0, ξ_1^1) the canonical value at the next step. Let the expression $p^{h,\delta}\left(\xi^0, \phi; \xi^0, \phi + \delta | \alpha, \xi^1(0)\right)$ denote the probability that the time variable advances and $p^{h,\delta}\left(\xi^0, \xi^1; \xi_1^0, \xi_1^1 | \alpha\right)$ the probability that time does not advance and the next state is (ξ_1^0, ξ_1^1), all given the initial data and control value α.

The Bellman equation that is based on the representation (4.14) has the form, for $\xi^0 \in G_h$,

$$V^{h,\delta}(\xi^0, \xi^1) = \inf_{\alpha \in U^h} \Big[\sum_{\xi_1^0, \xi_1^1} p^{h,\delta}\left(\xi^0, \xi^1; \xi_1^0, \xi_1^1 | \alpha\right) V^{h,\delta}\left(\xi_1^0, \xi_1^1\right) \\ + e^{-\beta\delta} p^{h,\delta}\left(\xi^0, \phi; \xi^0, \phi + \delta | \alpha, \xi^1(0)\right) V^{h,\delta}\left(\xi^0, \tilde{\xi}_1^1\right) \\ + k(\xi^0, \alpha)\Delta t^{h,\delta}(\xi^0, \xi^1(0), \alpha)\Big], \tag{4.15a}$$

where $\tilde{\xi}_1^1$ is obtained from ξ^1 by a shift, analogously to how $\xi_{n+1}^{1,h,\delta}$ was obtained from $\xi_n^{1,h,\delta}$ in (4.1b). Let $\xi^0 \in \partial G_h^+$, the set of reflecting states. Then as above (6.2.6), we can define $Y^{h,\delta}(\xi^0, \xi_1^0) = q'\Delta y_0^{h,\delta}$. Analogously to the second line of (6.2.6), for $\xi^0 \notin G_h$ the Bellman equation is

$$V^{h,\delta}(\xi^0, \xi^1) = \sum_{\xi_1^0} p^h(\xi^0, \xi_1^0)\left[V^{h,\delta}(\xi_1^0, \xi^1) + Y^{h,\delta}(\xi^0, \xi_1^0)\right]. \tag{4.15b}$$

We use the the transition probability notation $p^h(\xi^0, \xi_1^0)$ for the reflecting states, as for those states the values of α, δ, and ξ^1 play no role.

Note that a Bellman equation based on (4.13) can also be used, analogously to (7.3.14), as by the timescale equivalences in Theorems 6.5.1 and 7.3.2, the two forms are asymptotically equal. Due to the discounting, there is a unique solution to the Bellman equation.

9.5 Size of the State Space for the Approximating Chain

The comments concerning dimension and memory in Subsection 2.2 all apply to the numerical procedures. The complexity of the computation of $\xi_n^{1,h,\delta}(\theta)$ is not heavily dependent on the dimension of the control variable, or on components of $x(\cdot)$ that do not have delay components. The dimension of $\xi^{1,h,\delta}(\cdot)$ is just the number of components of $x(\cdot)$ whose dynamical terms contain delays. This is one of its key advantages. The size of the state space is the product of what is needed for $\xi_n^{0,h,\delta}$ and $\xi_n^{1,h,\delta}(\theta)$, where θ takes $\bar{\theta}/\delta$ values.

Consider a one-dimensional problem where $p(\cdot) = 0$ and let the discretization level for the $\xi_n^{1,h,\delta}(\theta)$ be $h_1 = O(h)$ for each θ. Then there is a $C_0 < \infty$ such that $|\chi^1(t, \theta)| \leq C_0(\bar{\theta} + \theta), -\bar{\theta} \leq \theta \leq 0$. Without loss of generality, suppose that $x(t)$ has been centered so that it lies in an interval $[0, B_0]$ for some $B_0 < \infty$. The state space for $\xi_n^{0,h,\delta}$ has $[B_0/h+3]$ points. There are $\bar{\theta}/\delta$ values for θ. Thus, if we bound $\xi_n^{1,h,\delta}(\theta)$ by $C_0(\bar{\theta}+\theta)$, the maximum number of points (which includes the reflecting states) is

$$[B_0/h + 3]\frac{C_0\bar{\theta}}{h_1}\frac{C_0(\bar{\theta} - \delta)}{h_1}\cdots\frac{C_0\delta}{h_1}. \tag{5.1}$$

Though large, this is better (in terms of memory, for given h, δ) than the procedure in Chapter 8 when there are delays in the control, and the control takes more than two or three values.

The approximation for fixed δ is discussed in Section 8, and, under its assumptions, that approach requires a smaller memory than what we get from (5.1) as the level of discretization of θ is fixed.

Using differences for the states, and a truncation procedure. The size of the state space can be reduced by the use of differences and truncations. The best approach is not clear, but the possibilities are suggested by the following computations that make use of estimates of the randomization errors in the realization of the $\xi_n^{1,h,\delta}(\theta)$.

Let $\delta = O(h), \Delta t^h(\cdot) = O(h^2)$, and suppose that $b(\cdot)$ is Lipschitz continuous in its arguments. Because $\xi_n^{1,h,\delta}(-\bar{\theta}) = 0$, the state space can consist of the differences $\xi_n^{1,h,\delta}(\theta) - \xi_n^{1,h,\delta}(\theta - \delta)$, $-\bar{\theta} + \delta \leq \theta \leq 0$. Consider the special case represented by (2.11). The expression (2.11) corresponds to $\mu_a(\cdot)$ being Lebesgue measure and the form is typical of the case where $\mu_a(\cdot)$ is absolutely continuous with respect to Lebesgue measure. Then $q_n^{h,\delta}(\theta) = \Delta t_n^h b(\xi_n^{0,h,\delta}, u_n^{h,\delta}, \theta)$, which we can suppose (for small h) is bounded by h_1^2. We have

$$|\chi^1(t,\theta) - \chi^1(t,\theta-\delta)| = O(\delta). \tag{5.2}$$

An analogous relation will be seen to hold for the approximating chain. Recall the definition (6.5.13), where $v_0^{h,\delta} = 0$ and, for $n \geq 0$, $v_{n+1}^{h,\delta} = \min\{i > v_n^{h,\delta} : \phi_{i+1}^{h,\delta} - \phi_i^{h,\delta} = \delta\}$. When $v_l^{h,\delta}$ is used as a subscript, we write it simply as v_l.

Neglecting the effects of the initial condition, w.l.o.g., we have the difference in the mean values at the time of (but not including) the nth shift:

$$\begin{aligned} & E\left[\xi_{v_n}^{1,h,\delta}(\theta) - \xi_{v_n}^{1,h,\delta}(\theta-\delta)\right] \\ & \quad = E\sum_{i=0}^{n-1} \left[\Phi^\delta\right]^{n-i-1} \sum_{l=v_i+1}^{v_{i+1}-1} \left[q_l^{h,\delta}(\theta) - q_l^{h,\delta}(\theta-\delta)\right]. \end{aligned} \tag{5.3}$$

Only the most recent $\bar{\theta}/\delta$ shifts yield nonzero results in (5.3). Using this fact, the fact that $E[v_{i+1}^{h,\delta} - v_i^{h,\delta}] = O(1/h)$, and the Lipschitz condition on $b(\cdot)$, we have the upper bound $O(\delta)$ (equivalently, $O(h)$).

The sum of the differences of the conditional means, given by the expression

$$\sum_{i=0}^{n-1} \left[\Phi^\delta\right]^{n-i-1} \sum_{l=v_i+1}^{v_{i+1}-1} \left[q_l^{h,\delta}(\theta) - q_l^{h,\delta}(\theta-\delta)\right], \tag{5.4}$$

is also $O(\delta)$. The next step is the estimation of the randomization errors in the realization of the difference $\xi_{v_n}^{1,h,\delta}(\theta) - \xi_{v_n}^{1,h,\delta}(\theta-\delta)$, and it is given by

$$\begin{aligned} \sum_{i=0}^{n-1} \left[\Phi^\delta\right]^{n-i-1} \sum_{l=v_i+1}^{v_{i+1}-1} & \left[\left(\xi_{l+1}^{1,h,\delta}(\theta) - \xi_l^{1,h,\delta}(\theta) - q_l^{h,\delta}(\theta)\right)\right. \\ & \left. - \left(\xi_{l+1}^{1,h,\delta}(\theta-\delta) - \xi_l^{1,h,\delta}(\theta-\delta) - q_l^{h,\delta}(\theta-\delta)\right)\right]. \end{aligned} \tag{5.5}$$

This is the value of $\xi_{v_n}^{1,h,\delta}(\theta) - \xi_{v_n}^{1,h,\delta}(\theta-\delta)$, but where each of the summands is centered about its conditional mean.

The summands in (5.5) are martingale differences. The mean square value of (5.5) is the expectation of the squares of the summands as in

$$\sum_{i=0}^{n-1} \left[\Phi^\delta\right]^{n-i-1} E \sum_{l=v_i+1}^{v_{i+1}-1} E_l^{h,\delta}\left[l\text{th term}\right]^2.$$

This can be bounded by twice the sum of the expectations

$$\sum_{i=0}^{n-1} \left[\Phi^\delta\right]^{n-i-1} E \sum_{l=v_i+1}^{v_{i+1}-1} E_l^{h,\delta} \left|q_l^{h,\delta}(\theta) - q_l^{h,\delta}(\theta-\delta)\right|^2 \tag{5.6}$$

and

$$\sum_{i=0}^{n-1} \left[\Phi^\delta\right]^{n-i-1} E \sum_{l=v_i+1}^{v_{i+1}-1} \tag{5.7}$$
$$E_l^{h,\delta} \left| \left(\xi_{l+1}^{1,h,\delta}(\theta) - \xi_l^{1,h,\delta}(\theta)\right) - \left(\xi_{l+1}^{1,h,\delta}(\theta-\delta) - \xi_l^{1,h,\delta}(\theta-\delta)\right)\right|^2 .$$

The sum (5.6) is $O(h^4)$. To evaluate (5.7), we will use the method of choosing the updates for $\xi_n^{1,h,\delta}(\theta)$ that was discussed at the end of Subsection 4.2, and, w.l.o.g., we can suppose that $\xi_n^{1,h,\delta}(\theta)$ is real-valued. The probability that different values are chosen for the two terms $\xi_{n+1}^{1,h,\delta}(\theta) - \xi_n^{1,h,\delta}(\theta)$ and $\xi_{n+1}^{1,h,\delta}(\theta-\delta) - \xi_n^{1,h,\delta}(\theta-\delta)$ is the difference of the probabilities

$$P_n^{h,\delta}(\theta) = \left|q_n^{h,\delta}(\theta)/h_1 - q_n^{h,\delta}(\theta-\delta)/h_1\right| = O(h^2). \tag{5.8}$$

If the two terms do differ, then the difference is $O(h)$. Thus the expression (5.7) is $O(h^2)$. From the evaluations of (5.4), (5.6), and (5.7), we can conclude that the mean of $\xi_n^{1,h,\delta}(\theta) - \xi_n^{1,h,\delta}(\theta-\delta)$ is bounded by $O(h)$ and the variance is bounded by $O(h^2)$. This yields the estimate

$$P\left\{\left|\xi_n^{1,h,\delta}(\theta) - \xi_n^{1,h,\delta}(\theta-\delta)\right| \geq N h_1\right\} \leq \frac{O(h^2)}{N^2 h_1^2} = \frac{O(1)}{N^2}. \tag{5.9}$$

Thus we can get a good approximation to the difference by letting N be large.

The estimates are crude but suggest ways of reducing the size of the state space. One simple possibility is to save $\xi_n^{1,h,\delta}(0) - \xi_n^{1,h,\delta}(-\delta)$ in lieu of $\xi_n^{1,h,\delta}(-\delta)$, and truncate the difference.

9.6 Proof of Convergence: Preliminaries

9.6.1 The Randomization Errors

For the convergence proof in the next section, we will need to know that the effects of the martingale difference terms $\rho_n^{h,\delta}$ defined in (4.9) and due to the realization of the conditional expectation in (4.7) by randomization between adjacent grid points are asymptotically negligible. This is implied by (6.1) in the next theorem. The expression (6.2) shows that $\xi_n^{1,h,\delta}(0)$ changes little *between* the steps that the time variable changes. Recall that, for the implicit procedure, $\xi_n^{0,h,\delta}$ does not change at the times $v_i^{h,\delta}$ that the time variable advances, that the components of $\xi_n^{1,h,\delta}$ shift at these times, and that for each $\theta \in T^\delta$, $\xi_n^{1,h,\delta}(\theta)$ takes values in a regular grid with spacing $h_1 = O(h)$. We continue to use $z(s) = 0, s \leq 0$, for notational simplicity only.

Theorem 6.1. *Assume* (A3.2.1), (A3.2.2), *and* (A3.2.4)–(A3.2.6), *and that* $\Delta t^h(x^0, x^1, \alpha) = O(h^2)$. *Then*

$$\lim_{h,\delta\to 0} \sup_{u^{h,\delta},\hat{x},\hat{u},\theta} \sup_n E\left|\sum_{i=0}^{n-1}[\Phi^\delta]^{n-i-1}R_i^{h,\delta}(\theta)\right|^2 = 0, \tag{6.1}$$

where

$$R_i^{h,\delta}(\theta) = \sum_{l=v_i^{h,\delta}+1}^{v_{i+1}^{h,\delta}-1} \rho_l^{h,\delta}(\theta),$$

where $\rho_l^{h,\delta}(\theta)$ *is defined in* (4.9). *Also, for each* $t<\infty$,

$$\lim_{h,\delta\to 0} \sup_{u^{h,\delta},\hat{x},\hat{u}} \sup_{n:\phi^{h,\delta}_{v_n^{h,\delta}}\le t} E \sup_{v_n^{h,\delta}+1\le l\le v_{n+1}^{h,\delta}} \left|\xi_l^{1,h,\delta}(0)-\xi^{1,h,\delta}_{v_n^{h,\delta}+1}(0)\right|^2 = 0. \tag{6.2}$$

Proof. Write $v_n^{h,\delta}$ as v_n, for notational simplicity. We will get an upper bound to the mean square error in (6.1) by considering the contributions of the reflection term and the drift term separately. First, let $\xi_n^{0,h,\delta} \notin G_h$, so that we are at a reflection step, and consider the randomization noise associated with realizing the conditional mean $q_n^{h,\delta}(\theta) = p(\xi_n^{0,h,\delta},\theta)\Delta y_n^{0,h,\delta}$. Without loss of generality, suppose that $p(\xi_n^{0,h,\delta},\theta)\Delta y_n^{0,h,\delta}$ is real-valued and lies in $[l_n(\theta)h_1, l_n(\theta)h_1+h_1]$ where $l_n(\theta)$ is an integer, either positive or negative. Then the transition probability gets $\xi_{n+1}^{1,h,\delta}(\theta)$ by randomizing between the end points of the interval so that the desired conditional mean value $p(\xi_n^{0,h,\delta},\theta)\Delta y_n^{0,h,\delta}$ is achieved. To evaluate the conditional (on $\mathcal{F}_n^{h,\delta}$) variance, we can suppose that we have shifted the means so that $l_n(\theta)=0$. Then the probability of selecting h_1 is $p(\xi_n^{0,h,\delta},\theta)\Delta y_n^{0,h,\delta}/h_1$, and, as $\Delta y_n^{0,h,\delta}=O(h)$, the conditional variance of $\rho_n^{h,\delta}(\theta)$ is

$$\begin{aligned}&\left[h_1 - p(\xi_n^{0,h,\delta},\theta)\Delta y_n^{0,h,\delta}\right]^2 \frac{p(\xi_n^{0,h,\delta},\theta)\Delta y_n^{0,h,\delta}}{h_1}\\ &\quad+\left[p(\xi_n^{0,h,\delta},\theta)\Delta y_n^{0,h,\delta}\right]^2\left[1-\frac{p(\xi_n^{0,h,\delta},\theta)\Delta y_n^{0,h,\delta}}{h_1}\right] = O(h)|p(\xi_n^{0,h,\delta},\theta)\Delta y_n^{0,h,\delta}|.\end{aligned} \tag{6.3}$$

The contribution of the randomization error due to the reflection steps between the nth and $(n+1)$st update of the time variable is

$$Q_n^{h,\delta}(\theta) = \sum_{l=v_n+1}^{v_{n+1}-1}\left[\left(\xi_{l+1}^{1,h,\delta}(\theta)-\xi_l^{1,h,\delta}(\theta)\right)-p(\xi_l^{0,h,\delta},\theta)\Delta y_l^{0,h,\delta}\right] I_{\{\xi_l^{0,h,\delta}\notin G_h\}}.$$

Then, because

$$\xi_{v_i+1}^{1,h,\delta}(\theta) = \Phi^\delta \xi_{v_i}^{1,h,\delta}(\theta) = \Phi^\delta\left[\xi^{1,h,\delta}_{v_{i-1}+1}(\theta) + \sum_{v_{i-1}+1}^{v_i-1}\left(\xi_{l+1}^{1,h,\delta}(\theta)-\xi_l^{1,h,\delta}(\theta)\right)\right],$$

the sum

$$\sum_{i=0}^{n-1} [\Phi^\delta]^{n-i-1} Q_i^{h,\delta}(\theta) \tag{6.4}$$

is the total contribution of the randomization errors due to the reflection steps to the value of $\xi_{v_n}^{1,h,\delta}(\theta)$. For each n, the summands in (6.4) and the one for $Q_n^{h,\delta}(\theta)$ above are martingale differences as

$$E_{v_i+l}^{h,\delta}\left[\left(\xi_{v_i+l+1}^{1,h,\delta}(\theta) - \xi_{v_i+l}^{1,h,\delta}(\theta)\right) - p(\xi_{v_i+l}^{0,h,\delta},\theta)\Delta y_{v_i+l}^{0,h,\delta}\right]$$
$$\times I_{\{\xi_{v_i+l}^{0,h,\delta}\notin G_h\}} I_{\{v_i+l<v_{i+1}\}} = 0.$$

By this martingale property, the mean square value of (6.4) can be written as

$$E\sum_{i=0}^{n-1}[\Phi^\delta]^{n-i-1}$$
$$\left(E_{v_i}^{h,\delta}\sum_{l=v_i+1}^{v_{i+1}-1}\left|\left(\xi_{l+1}^{1,h,\delta}(\theta)-\xi_l^{1,h,\delta}(\theta)\right)-p(\xi_l^{0,h,\delta},\theta)\Delta y_l^{0,h,\delta}\right|^2 I_{\{\xi_l^{0,h,\delta}\notin G_h\}}\right).$$

Now, using the evaluation of the conditional variance in (6.3) and the fact that $[\Phi^\delta]^k f(\theta) = 0$ for $k > \bar{\theta}/\delta$, we have that the mean square value of (6.4) is bounded by

$$O(h)E\left[\left|z^{0,h,\delta}\right|(t_{v_n}^{h,\delta}) - \left|z^{0,h,\delta}\right|(t_{v_{n-\bar{\theta}/\delta}}^{h,\delta})\right],$$

where $z^{0,h,\delta}(\cdot)$ is the interpolation of $\{\Delta z_n^{0,h,\delta}\}$ with intervals $\{\Delta t_n^{h,\delta}\}$. Lemma 6.3.1 does not directly apply to this expression, but a slight modification does. Let $t = n\delta$. Then using the random nature of the time shifts and Lemma 6.3.1, one can show that, as $h \to 0$ and independently of δ, n, and the initial conditions, the difference between the above expression and

$$O(h)E\left[\left|z^{0,h,\delta}\right|(t) - \left|z^{0,h,\delta}\right|(t-\bar{\theta})\right]$$

goes to zero, and that (6.1) holds for the reflection component.

Now consider the component of $q_n^{h,\delta}(\theta)$ in (4.8) that is defined by

$$\hat{q}_n^{h,\delta}(\theta) = \frac{[\mu_a(\theta) - \mu_a(\theta-\delta)]}{\delta} b(\xi_n^{0,h,\delta}, u_n^{h,\delta}, \theta)\Delta t_n^h.$$

Again, suppose that we have centered so that its values are in $[0, h_1]$. Then to attain the desired conditional (on $\mathcal{F}_n^{h,\delta}$) mean of $\xi_{n+1}^{1,h,\delta}(\theta) - \xi_n^{1,h,\delta}(\theta)$, the transition probability effectively selects the value h_1 with probability $\hat{q}_n^{h,\delta}(\theta)/h_1$. The conditional variance of the difference between the true value of $\xi_{n+1}^{1,h,\delta}(\theta) - \xi_n^{1,h,\delta}(\theta)$ and its conditional expectation is $O(h)\hat{q}_n^{h,\delta}(\theta)$. Redefine

$$Q_n^{h,\delta}(\theta) = \sum_{l=v_n+1}^{v_{n+1}-1}\left[\left(\xi_{l+1}^{1,h,\delta}(\theta) - \xi_l^{1,h,\delta}(\theta)\right) - \hat{q}_l^{h,\delta}(\theta)\right] I_{\{\xi_l^{0,h,\delta}\in G_h\}}. \tag{6.5}$$

The next step is to evaluate (6.4) with this new definition of $Q_n^{h,\delta}(\theta)$. For each n, the summands are martingale differences in that

$$E_{v_i+l}^{h,\delta}\left[\left(\xi_{v_i+l+1}^{1,h,\delta}(\theta)-\xi_{v_i+l}^{1,h,\delta}(\theta)\right)-\hat{q}_{v_i+l}^{h,\delta}(\theta)\right]I_{\{\xi_{v_i+l}^{0,h,\delta}\in G_h\}}I_{\{v_i+l<v_{i+1}\}}=0.$$

With the new definition of $Q_n^{h,\delta}(\theta)$, the mean square value of (6.4) is

$$E\sum_{i=0}^{n-1}[\Phi^\delta]^{n-i-1}E_{v_i}^{h,\delta}\sum_{l=v_i+1}^{v_{i+1}-1}\left|\left(\xi_{l+1}^{1,h,\delta}(\theta)-\xi_l^{1,h,\delta}(\theta)\right)-\hat{q}_l^{h,\delta}(\theta)\right|^2 I_{\{\xi_l^{0,h,\delta}\in G_h\}}. \tag{6.6}$$

Now, using the above evaluation of the conditional variance, the orders of δ and Δt_l^h in h, and the fact that $E_{v_i}^{h,\delta}(v_{i+1}-v_i)=O(1/h)$, yields that the (conditioned on $\mathcal{F}_{v_n}^{h,\delta}$) expectation of the inner sum in (6.6) is bounded by

$$O(h^2)\left[\mu_a(\theta)-\mu_a(\theta-\delta)\right]E_{v_i}^{h,\delta}[v_{i+1}-v_i]=O(h)\left[\mu_a(\theta)-\mu_a(\theta-\delta)\right].$$

Now taking the shift $[\Phi^\delta]^{n-i-1}$ in (6.6) into account and noting that $\mu_a(\theta)-\mu_a(\theta-\delta)=0$ for $\theta\le-\bar{\theta}$, we see that (6.6) is $O(h)$, uniformly in n. Thus (6.1) holds. The verification of (6.2) is straightforward and the details are omitted.
■

9.6.2 Continuous Time Interpolations

Definitions. Recall the continuous time interpolation $\psi^{h,\delta}(\cdot)$ in (6.5.12) and the definition of the intervals $\Delta\tau_n^{h,\delta}$. As there, define $\tau_n^{h,\delta}=\sum_{i=0}^{n-1}\Delta\tau_i^{h,\delta}$. Define the interpolation $\psi^{0,h,\delta}(\cdot)$ by setting $\psi^{0,h,\delta}(t)=\xi_n^{0,h,\delta}$ for $t\in[\tau_n^{h,\delta},\tau_{n+1}^{h,\delta})$, and define the interpolations $\psi^{1,h,\delta}(t,\theta)$, $u_\tau^{h,\delta}(\cdot)$, $\phi_\tau^{h,\delta}(\cdot)$, $z_\tau^{0,h,\delta}(\cdot)$, and $y_\tau^{0,h,\delta}(\cdot)$, analogously. Let $r_\tau^{h,\delta}(\cdot)$ denote the relaxed control representation of $u_\tau^{h,\delta}(\cdot)$, with the derivative at t denoted by $r_\tau^{h,\delta,\prime}(\cdot,t)$.

Analogously to (6.5.12), we can write

$$\begin{aligned}\psi^{0,h,\delta}(t)&=\xi_0^{0,h,\delta}+\int_0^t\psi^{1,h,\delta}(s,0)ds\\&+\int_0^t\int_{U^h}c(\psi^{0,h,\delta}(s),\alpha)r_\tau^{h,\delta,\prime}(d\alpha,s)ds+B_\tau^{h,\delta}(t)+z_\tau^{0,h,\delta}(t)+\epsilon_1^{h,\delta}(t),\end{aligned} \tag{6.7}$$

where $B_\tau^{h,\delta}(\cdot)$ is a martingale with quadratic variation process

$$\int_0^t a(\psi^{0,h,\delta}(s))ds+\epsilon_2^{h,\delta}(t),$$

and the $\epsilon_i^{h,\delta}(\cdot)$ satisfy, for any $T<\infty$,

$$\lim_{h,\delta\to0}\ \sup_{u^{h,\delta},\hat{x},\hat{u}}\ \sup_{t\le T}E\sup_{t\le T}|\epsilon_i^{h,\delta}(t)|=0. \tag{6.8}$$

As noted below (6.3.8) in Section 6.3, there is a martingale $w^{h,\delta}(\cdot)$ with quadratic variation process It and that converges weakly to a Wiener process as $(h,\delta) \to 0$ such that

$$B_\tau^{h,\delta}(t) = \int_0^t \sigma(\psi^{0,h,\delta}(s))dw^{h,\delta}(s) + \epsilon_3^{h,\delta}(t), \tag{6.9}$$

where $\epsilon^{h,\delta}(\cdot)$, satisfies (6.8). The $\epsilon^{h,\delta}(\cdot)$ error terms are due to the $o(\Delta t^{h,\delta})$ terms in (4.6). Recall the discussion of the boundaries on $\xi_n^{1,h,\delta}(\theta)$ at the beginning of Subsection 4.1. They are ignored in the next theorem but reintroduced in Theorem 7.1. For $\theta = 0$, note the similarity of (6.10) to the integral over time of (2.14).

Theorem 6.2. *Assume* (A3.2.1), (A3.2.2), *and* (A3.2.4)–(A3.2.6). *Let* $\xi_n^{1,h,\delta}$ *be bounded, and suppose that* $\Delta t^h(x^0, x^1, \alpha) = O(h^2)$, $\delta = O(h)$, *and* $h_1 = O(h)$. *Then*

$$\begin{aligned}\int_0^t \psi^{1,h,\delta}(s,0)ds &= \int_0^t ds \int_{-\bar\theta}^0 d\mu_a(\gamma) \int_{U^h} b(\psi^{0,h,\delta}(\gamma+s),\alpha,\gamma) r_\tau^{h,\delta,\prime}(d\alpha,\gamma+s) \\ &+ \int_0^t ds \int_{-\bar\theta}^0 p(\psi^{0,h,\delta}(\gamma+s),\gamma) d_\gamma y_\tau^{0,h,\delta}(\gamma+s) + \rho_0^{h,\delta}(t),\end{aligned} \tag{6.10}$$

where $\rho_0^{h,\delta}(t)$ *satisfies, for each* $T < \infty$,

$$\lim_{h,\delta\to 0} \sup_{u^{h,\delta},\hat{x},\hat{u}} \sup_{t\le T} E|\rho_0^{h,\delta}(t)| = 0. \tag{6.11}$$

Proof. For notational simplicity, we work with ordinary, rather than relaxed, controls, and continue to use the notation $v_n = v_n^{h,\delta}$ in subscripts and superscripts. Recall that the nth shift occurs at index $v_n^{h,\delta}$, so it first affects the value of the next iterate, and that the value at index $v_{n+1}^{h,\delta}$ is the sum of the value at index $v_n^{h,\delta}+1$ and the contributions of the iterates taken in the interval $[v_n^{h,\delta}+1, v_{n+1}^{h,\delta})$. The value at iterate $v_n^{h,\delta}+1$ is $\Phi\xi_{v_n}^{1,h,\delta}(\theta) = \xi_{v_n}^{1,h,\delta}(\theta-\delta)$. Thus we can write

$$\xi_{v_{n+1}}^{1,h,\delta}(\theta) = \xi_{v_n}^{1,h,\delta}(\theta-\delta) + P_n^{h,\delta}(\theta) + B_n^{h,\delta}(\theta) + R_n^{h,\delta}(\theta), \tag{6.12}$$

where $R_n^{h,\delta}(\theta)$ was defined below (6.1), and we define

$$P_n^{h,\delta}(\theta) = \sum_{l=v_n+1}^{v_{n+1}-1} p(\xi_l^{0,h,\delta},\theta)\Delta y_l^{0,h,\delta},$$

$$B_n^{h,\delta}(\theta) = [\mu_a(\theta) - \mu_a(\theta-\delta)] \sum_{l=v_n+1}^{v_{n+1}-1} b(\xi_l^{0,h,\delta}, u_l^{h,\delta}, \theta)\frac{\Delta t_l^h}{\delta}.$$

Until further notice ignore the effects of the initial condition (2.5).

First, let us determine the contribution of the $p(\xi_i^{0,h,\delta},\theta)\Delta y_l^{0,h,\delta}$ terms to the value of $\psi^{1,h,\delta}(s,\theta)$. For $n \geq 1$, their total contribution to $\xi_{v_n}^{1,h,\delta}(\theta)$ is

$$\sum_{i=0}^{n-1}[\Phi^\delta]^{n-i-1}P_i^{h,\delta}(\theta) = \sum_{i=0}^{n-1}P_i^{h,\delta}(\theta - n\delta + i\delta + \delta). \tag{6.13}$$

Recall the definition $d_\tau^{h,\delta}(s) = \max\{n : \tau_n^{h,\delta} \leq s\}$ from (6.5.23) and define $N^{h,\delta}(t) = \phi_\tau^{h,\delta}(t)/\delta$, the number of time advances that have occurred when interpolated time (in the $\tau_n^{h,\delta}$-scale) t is reached.

The values of $\psi^{1,h,\delta}(s,\theta)$ and $\xi_{v_{N^{h,\delta}(s)}+1}^{1,h,\delta}(\theta)$ differ by the sum of the differences $\xi_{l+1}^{1,h,\delta}(\theta) - \xi_l^{1,h,\delta}(\theta)$ for indices l in the interval $[v_{N^{h,\delta}(s)}^{h,\delta}+1, d_\tau^{h,\delta}(s)-1]$, those that occur before interpolated time s is reached but at or after the last update of the time variable before interpolated time s is reached. By (6.2), the contributions of these terms for $\theta = 0$ is asymptotically negligible. Ignoring these "end terms" and using $n = N^{h,\delta}(s)$ in (6.13) yields that (6.13) is asymptotically equal to

$$\sum_{l=0}^{d_\tau^{h,\delta}(s)-1} p(\xi_l^{0,h,\delta}, \theta - \phi^{h,\delta}(s) + \phi_l^{h,\delta} + \delta)\Delta y_l^{0,h,\delta}, \tag{6.14}$$

in the sense that the difference of the processes goes to zero as $h,\delta \to 0$.

By Theorem 6.5.1, $\phi_l^{h,\delta}$ is asymptotically equal to $t_l^{h,\delta}$ and $\tau_l^{h,\delta}$ in mean square in that, for any T and $t_l^{h,\delta} \leq T$ or $\tau_l^{h,\delta} \leq T$, the mean square value of the sup of the differences on $[0,T]$ goes to zero as $h,\delta \to 0$. This fact and the continuity of $p(\cdot)$ imply that (6.14) equals

$$\sum_{l=0}^{d_\tau^{h,\delta}(s)-1} p(\xi_l^{0,h,\delta}, \theta - s + \tau_l^{h,\delta})\Delta y_l^{0,h,\delta} \tag{6.15}$$

modulo an error that is bounded by $\epsilon(h,\delta,s)\left[|z_\tau^{0,h,\delta}|(s) - |z_\tau^{0,h,\delta}|(s-\bar{\theta}-\delta)\right]$, where $\epsilon(h,\delta)(s) \to 0$ uniformly in (ω, s) as $h,\delta \to 0$. It follows from Lemma 6.3.1 that this last error term satisfies (6.11). Then by a change of variable and using the fact that $p(x^0,\theta) = 0$ for $\theta < -\bar{\theta}$, we can write (6.15), for $\theta = 0$, as

$$\int_0^s p(\psi^{0,h,\delta}(v), -s+v)dy_\tau^{0,h,\delta}(v) = \int_{\max\{-s,-\bar{\theta}\}}^0 p(\psi^{0,h,\delta}(s+\gamma),\gamma)d_\gamma y_\tau^{0,h,\delta}(s+\gamma). \tag{6.16}$$

It can be shown that adding the effects of the initial condition changes the right-hand integral to $\int_{-\bar{\theta}}^0$. Doing this and integrating the result over $s \in [0,t]$ yields the contribution of the reflection term to (6.10).

Now consider the contribution of the terms involving $b(\cdot)$ to the integral $\int_0^t \psi^{1,h,\delta}(s,0)ds$, ignoring the effects of the initial condition until later. Ignore the effects of the reflection term, which have already been accounted for. By (6.2), to evaluate this integral we can suppose that $\psi^{1,h,\delta}(\cdot, 0)$ is constant on the intervals $(\tau_{v_n}^{h,\delta}, \tau_{v_{n+1}}^{h,\delta}]$ between successive increases of the time variable. Then, making the piecewise-constant approximation, the integral is (modulo an error satisfying (6.11)),

$$\sum_{n=0}^{N^{h,\delta}(t)-1} \xi_{v_n}^{1,h,\delta}(0) \left[\tau_{v_{n+1}}^{h,\delta} - \tau_{v_n}^{h,\delta}\right].$$

This equals

$$\sum_{n=0}^{N^{h,\delta}(t)-1} \xi_{v_n}^{1,h,\delta}(0)\delta, \tag{6.17}$$

modulo an error that satisfies (6.11). For $n \geq 1$ and arbitrary θ,

$$\xi_{v_n}^{1,h,\delta}(\theta) = \sum_{i=0}^{n-1} B_i^{h,\delta}(\theta - n\delta + i\delta + \delta).$$

By the definition of $B_i^{h,\delta}(\theta)$ below (6.12), this expression equals

$$\begin{aligned}\xi_{v_n}^{1,h,\delta}(\theta) = \sum_{i=0}^{n-1} & \frac{[\mu_a(\theta - n\delta + i\delta + \delta) - \mu_a(\theta - n\delta + i\delta)]}{\delta} \\ & \times \sum_{l=v_i+1}^{v_{i+1}-1} b(\xi_l^{0,h,\delta}, u_l^{h,\delta}, \theta - n\delta + i\delta + \delta)\Delta t_l^h.\end{aligned} \tag{6.18}$$

Let $\theta = 0$. Then the sum (6.17) will only be changed by a quantity satisfying (6.11) if we redefine $\xi_{v_n}^{1,h,\delta}(0)$ by replacing the inner sum in (6.18) by

$$\hat{B}_i^{h,\delta}(-n\delta + i\delta) = \sum_{l=v_i}^{v_{i+1}-1} b(\xi_l^{0,h,\delta}, u_l^{h,\delta}, -n\delta + i\delta + \delta)\Delta\tau_l^{h,\delta}.$$

By a change of variable $n - i = q$, the use of $\hat{B}_i^{h,\delta}(\cdot)$, and a change in the order of summation, we can write (6.17) as

$$\sum_{q=1}^{N^{h,\delta}(t)-1} [\mu_a(-q\delta + \delta) - \mu_a(-q\delta)] \sum_{n=q}^{N^{h,\delta}(t)-1} \hat{B}_{n-q}^{h,\delta}(-q\delta). \tag{6.19}$$

We can write the inner sum of (6.19) as (modulo an error satisfying (6.11))

$$\int_0^{t-q\delta} b(\psi^{0,h,\delta}(s), u_\tau^{h,\delta}(s), -q\delta + \delta)ds.$$

Recall that $b(\cdot,\theta) = 0$ and $\mu_a(\theta) = 0$ for $\theta < -\bar{\theta}$. Next, using this last expression and the continuity of $b(\cdot)$, (6.19) can be approximated by (modulo an error satisfying (6.11))

$$\int_{\max\{-t,-\bar{\theta}\}}^{0} d\mu_a(\gamma) \int_0^{t+\gamma} b(\psi^{0,h,\delta}(s), u_\tau^{h,\delta}(s), \gamma) ds,$$

which is equal to

$$\int_{\max\{-t,-\bar{\theta}\}}^{0} d\mu_a(\gamma) \int_{-\gamma}^{t} b(\psi^{0,h,\delta}(s+\gamma), u_\tau^{h,\delta}(s+\gamma), \gamma) ds. \tag{6.20}$$

By a change in the order of integration, (6.20) can be written as

$$\int_0^t ds \int_{\max\{-s,-\bar{\theta}\}}^{0} b(\psi^{0,h,\delta}(s+\gamma), u_\tau^{h,\delta}(s+\gamma), \gamma) d\mu_a(\gamma). \tag{6.21}$$

Adding the effects of the initial condition changes the lower limit $\max\{-s,-\bar{\theta}\}$ to $-\bar{\theta}$. Then, reverting to relaxed control notation, we have the contribution of the drift term to (6.10). By Theorem 6.1, the contribution of the randomization errors $R_l^{h,\delta}(\theta)$ to (6.10) satisfies (6.11). ■

9.7 Convergence of the Numerical Algorithm

The next assumption is all that we require on the initial condition.

A7.1. *The initial condition $\xi_0^{1,h,\delta}(\theta)$ converges to $\chi^1(0,\theta)$ given by (2.5), uniformly in $\theta \in [-\bar{\theta}, 0]$, as $h,\delta \to 0$, and has values that that are on the discrete grid for each θ. Also $|\xi_0^{0,h,\delta} - x(0)| \to 0$.*

The cost function (4.13) can be written as (modulo an asymptotically negligible error)

$$W^{h,\delta}(\xi_0^{0,h,\delta}, \xi_0^{1,h,\delta}, u^{h,\delta}) = E_{\xi_0^{0,h,\delta},\xi_0^{1,h,\delta}}^{h,\delta,u^{h,\delta}} \int_0^\infty \int_{U^h} e^{-\beta s} \left[k(\psi^{0,h,\delta}(s), \alpha) r_\tau^{h,\delta}(d\alpha\, ds) + q' dy_\tau^{0,h,\delta}(s) \right]. \tag{7.1}$$

The next theorem shows that the optimal values computed by the numerical algorithm converge to the optimal value of the original problem as $h, \delta \to 0$.

Theorem 7.1. *Assume* (A7.1), (A3.2.1), (A3.2.2), (A3.2.4)–(A3.2.6), (A3.4.3), *and* $\Delta t^h(x^0, x^1, \alpha) = O(h^2)$, $\delta = O(h)$, $h_1 = O(h)$. *Suppose that there is no delayed reflection term and that the boundaries on $\xi_n^{1,h,\delta}(\theta)$ are large enough so that $\chi^1(t,\theta)$ would not reach them. For any sequence of controls for the* *chain, the set* $(\psi^{0,h,\delta}(\cdot), r_\tau^{h,\delta}(\cdot), w^{h,\delta}(\cdot), z_\tau^{0,h,\delta}(\cdot))$ *is tight, and any weakly convergent subsequence converges to a solution to* (1.2). *The optimal costs for the chain* $\{\xi_n^{0,h,\delta}, \xi_n^{1,h,\delta}(\theta), \theta \in T^\delta\}$ *and cost function* (7.1) (*or* (4.13) *or* (4.14)) *converge to the optimal cost for original process* (1.2) *and cost function* (1.3).

Now add the delayed reflection term and recall the discussion on boundaries at the end of Subsection 4.1. *The sequence* $(\xi^{0,h,\delta}(\cdot), r_\tau^{h,\delta}(\cdot), w^{h,\delta}(\cdot), z_\tau^{0,h,\delta}(\cdot))$ *is tight, and any weakly convergent subsequence converges to a solution to* (1.2). *The limits of the optimal costs for the chain are arbitrarily close to that for the original process if the boundaries are large enough.*

Now consider the model with an absorbing rather than a reflecting boundary. Drop the assumptions on the reflection directions and assume (A3.4.1), (A3.4.2). *Then the conclusions continue to hold.*

Proof. With the preparation in Theorem 6.2 in hand, the proof follows those in Section 8.5 closely. Fix a control sequence, with relaxed control representations (in the interpolation using intervals $\{\Delta\tau_n^{h,\delta}\}$) $r_\tau^{h,\delta}(\cdot)$. We comment only on the case with reflecting boundaries. The absorbing boundary case is analogous, subject to the treatment of the absorbing boundary in the proof of Theorem 7.1.3 in Section 8.5. The sequence of martingales $B_\tau^{h,\delta}(\cdot)$ in (6.7) is tight in the Skorokhod topology. Then, as the increments $\beta_n^{0,h,\delta}$ are $O(h)$, any weak-sense limit has continuous paths with probability one. The sequence $w^{h,\delta}(\cdot)$ in (6.9) is tight and the weak sense limit is a standard vector-valued Wiener process. The sequence of processes defined by the first two integrals on the right of (6.7) are also tight, and any weak-sense limit must be continuous with probability one. By (6.8) the sequence $\epsilon_1^{h,\delta}(\cdot)$ in (6.7) converges to the "zero" process. Any sequence $r_\tau^{h,\delta}(\cdot)$ of relaxed controls is tight. As in Theorem 3.5.5, the sequence of reflection processes $z_\tau^{0,h,\delta}(\cdot)$ is tight, and the limit of any weakly convergent subsequence is continuous.

These facts and the boundedness of $\{\xi_n^{1,h,\delta}(0)\}$ implies the tightness of $\psi^{0,h,\delta}(\cdot)$ and the asymptotic continuity of any weak-sense limit. Now extract a weakly convergent subsequence of $\{\psi^{0,h,\delta}(\cdot), r_\tau^{h,\delta}(\cdot), w^{h,\delta}(\cdot), z_\tau^{0,h,\delta}(\cdot)\}$. with limit denoted by $(x(\cdot), r(\cdot),\ w(\cdot), z(\cdot))$. Abusing notation, let h, δ also index this subsequence and use the Skorokhod representation so that we can suppose that the convergence is w.p.1 in the topology of the path spaces. The proof of nonanticipativity in the proof of Theorem 7.1.3 in Section 8.5 can be used to show that $(x(\cdot), w(\cdot), r(\cdot), z(\cdot))$ is nonanticipative with respect to $w(\cdot)$. Then the approximation argument in that proof yields the convergence $\int_0^t \sigma(\psi^{h,\delta}(s))dw^{h,\delta}(s) \to \int_0^t \sigma(x(s))dw(s)$. The argument of Theorem 3.5.5 yields that the limit process $z(\cdot)$ is a reflection process for $x(\cdot)$.

We have shown that the second line of (6.7) converges to

$$\int_0^t \int_U c(x(s), \alpha) r'(d\alpha, s) ds + \int_0^t \sigma(x(s)) dw(s) + z(t). \tag{7.2}$$

The limit of the sequence of processes defined by $\int_0^t \psi^{1,h,\delta}(s,0)ds$ is continuous as the integral terms on the right side of (6.10) converge to

$$\int_0^t ds \int_{-\bar{\theta}}^0 d\mu_a(\gamma) \int_U b(x(\gamma+s),\alpha,\gamma)r'(d\alpha,\gamma+s) \\ + \int_0^t ds \int_{-\bar{\theta}}^0 p(x(\gamma+s),\gamma)d_\gamma y(\gamma+s). \tag{7.3}$$

The last two sentences and (6.11) imply that the process $\rho_0^{h,\delta}(\cdot)$ in (6.10) converges to the zero process (which is not implied by (6.11) alone). Finally, the convergences (7.2) and (7.3) imply that $x(\cdot)$ solves (1.2) with relaxed control $r(\cdot)$. The weak convergences and the integrability properties of $y_\tau^{0,h,\delta}(\cdot)$ implied by Lemma 6.3.1 imply that $W^{h,\delta}(\xi_0^{0,h,\delta},\xi_0^{1,h,\delta},u^{h,\delta}) \to W(\hat{x},\hat{u},r)$.

If $u^{h,\delta}(\cdot)$ is an optimal control for the chain, then this last convergence implies that

$$\liminf_{h,\delta\to 0} V^{h,\delta}(\xi_0^{0,h,\delta},\xi_0^{1,h,\delta}) \geq V(\hat{x},\hat{u}).$$

The reverse inequality

$$\limsup_{h,\delta\to 0} V^{h,\delta}(\xi_0^{0,h,\delta},\xi_0^{1,h,\delta}) \leq V(\hat{x},\hat{u})$$

is proved just as it was in the proof of Theorem 7.1.3 in Section 8.5, which used the representation (8.5.10) of a particular ϵ-optimal control for (1.2) that is a continuous function of its arguments. That result was based on one for the no-delay problem in [58, Chapters 11 and 12]. The presence of delays does not materially change the structure of the proof. ■

9.8 Alternatives: Periodic and Periodic-Erlang Approximations

9.8.1 A Periodic Approximation

There are analogs of the periodic and periodic-Erlang approximations of Sections 4.2, 8.1, and 8.2. We will drop the delay in the reflection term. There is a similar result if a delayed reflection term is included, but the development is more complicated. Let $\delta > 0$ be fixed and consider a modification of (2.3), (2.4) or, more formally, of (2.7), (2.8), where the shift occurs only at integral multiples of δ, and $\bar{\theta} = Q_\delta\delta$, where Q_δ is an integer, as in Chapters 4 and 7.

The idea is to approximate the original processes $\chi^0(\cdot)$ and $\chi^1(\cdot)$ as follows. Run the approximating processes with no shift on the time intervals $[l\delta, l\delta+\delta), l = 0,1,\ldots,$ and then shift at time $l\delta, l = 1,2\ldots.$ In particular, on $[l\delta, l\delta+\delta)$ and for $\theta = 0,-\delta,\ldots,-\bar{\theta}+\delta$, define the processes $\chi^{0,\delta}(\cdot)$ and $\chi^{1,\delta}(\cdot,\theta)$ by

$$d\chi^{0,\delta}(t) = \chi^{1,\delta}(t,0)dt + c(\chi^{0,\delta}(t),u(t))dt + \sigma(\chi^{0,\delta}(t))dw(t) + dz^{0,\delta}(t), \tag{8.1}$$

$$d_t\chi^{1,\delta}(t,\theta) = b(\chi^{0,\delta}(t), u(t), \theta)\frac{\mu_a(\theta) - \mu_a(\theta-\delta)}{\delta}dt, \tag{8.2}$$

or their relaxed control counterparts. At $t = l\delta$ we have the shift

$$\chi^{1,\delta}(l\delta,\theta) = \Phi^\delta\chi^{1,\delta}(l\delta-,\theta) = \chi^{1,\delta}(l\delta-,\theta-\delta)I_{\{-\bar{\theta}\le\theta-\delta\le 0\}}. \tag{8.3}$$

The first shift is at time δ.

The initial condition is (2.5), namely: $\chi^{0,\delta}(0) = x(0)$, $z^{0,\delta}(s) = 0$ for $s \le 0$, and

$$\chi^{1,\delta}(0,\theta) = \int_{-\bar{\theta}}^{\theta} b(x(\gamma-\theta), u(\gamma-\theta), \gamma)d\mu_a(\gamma). \tag{8.4}$$

Additionally, as in Section 2, we have the boundary condition $\chi^{1,\delta}(t,-\bar{\theta}) = 0$. It will be seen in the discussion after the proof that this format yields an analog of the periodic Approximation 3 of Chapter 4, in which the maximum delay varied periodically between $\bar{\theta} - \delta/2$ and $\bar{\theta} + \delta/2$.

The following assumption will be used. Under (A8.1), the main advantage of the periodic approximation is that the dimension of the approximating chain is fixed as $h \to 0$, as δ is fixed. The value of δ is determined by the tolerable errors in the approximation of the delays. See the comments in the next subsection.

A8.1. *$\mu_a(\cdot)$ is concentrated on the points $-\bar{\theta}, -\bar{\theta}+\delta, \ldots, -2\delta$, and $p(\cdot) = 0$. For the initial condition, $x(\cdot)$ and $u(\cdot)$ are constant on the intervals $[-i\delta, -i\delta+\delta)$, $-\bar{\theta} \le -i\delta \le -\delta$.*

Theorem 8.1. *Fix $\delta > 0$. Assume* (A8.1), (A3.2.1), (A3.2.2), (A3.2.4)–(A3.2.6) *and* (A3.4.3). *If the boundary is not reflecting, then drop the boundary conditions and the process $z^{0,\delta}(\cdot)$ and add* (A3.4.1) *and* (A3.4.2). *For any integer $n \ge 0$, we have*

$$\chi^{0,\delta}(n\delta) = x(0) + \int_0^{n\delta} c(\chi^{0,\delta}(s), u(s))ds +$$

$$\int_0^{n\delta} ds \int_{-\bar{\theta}}^{0} b(\chi^{0,\delta}(s+\theta), u(s+\theta), \theta)d\mu_a(\theta) + \int_0^{n\delta} \sigma(\chi^{0,\delta}(s))dw(s) + z^{0,\delta}(n\delta), \tag{8.5a}$$

or its relaxed control counterpart. For $n\delta < t < n\delta + \delta$, the second integral is replaced by a linear interpolation of the values at $n\delta$ and $n\delta + \delta$. Thus $\chi^{0,\delta}(t)$ equals (8.5a) *plus the term*

$$\frac{t-n\delta}{\delta}\int_{n\delta}^{n\delta+\delta} ds \int_{-\bar{\theta}}^{0} b(\chi^{0,\delta}(s+\theta), u(s+\theta), \theta)d\mu_a(\theta). \tag{8.5b}$$

Proof. Define $\Delta\mu_a(\theta) = \mu_a(\theta) - \mu_a(\theta-\delta)$. We will work with ordinary rather than relaxed controls for notational simplicity. The form with relaxed controls should be obvious. For $t \in [n\delta, n\delta+\delta)$ (i.e., after n shifts from the start), (8.2) and (8.3) yield

$$\begin{aligned}\chi^{1,\delta}(t,\theta) &= [\Phi^\delta]^n \chi^{1,\delta}(0,\theta) + \sum_{i=0}^{n-1} [\Phi^\delta]^{n-i} B_i^\delta(\theta) \frac{\Delta\mu_a(\theta)}{\delta} \\ &\quad + \int_{n\delta}^{t} b(\chi^{0,\delta}(s), u(s), \theta) ds \frac{\Delta\mu_a(\theta)}{\delta},\end{aligned} \tag{8.6}$$

where

$$B_i^\delta(\theta) = \int_{i\delta}^{i\delta+\delta} b(\chi^{0,\delta}(s), u(s), \theta) ds.$$

Because it is $\chi^{1,\delta}(t,0)$ that appears in the equation for $\chi^{0,\delta}(\cdot)$, and $\Delta\mu_a(0) = \mu_a(0) - \mu_a(-\delta) = 0$ the term in the second line of (8.6) will equal zero when we set $\theta = 0$, and it can be ignored henceforth. Thus $\chi^{1,\delta}(\cdot,\theta)$ can be taken to be constant on the intervals $[n\delta, n\delta+\delta), n = 0, 1, \ldots$.

The rest of the development has the form of the manipulations in Theorem 6.2. Set $\theta = 0$ and, until further notice, ignore the effects of the initial condition. With these simplifications and $\theta = 0$, (8.6) can be written as

$$\sum_{i=0}^{n-1} B_i^\delta(-n\delta + i\delta) \frac{\Delta\mu_a(-n\delta + i\delta)}{\delta}.$$

Let $t \in [N\delta, N\delta + \delta)$. Then

$$\begin{aligned}\int_0^t \chi^{1,\delta}(s,0) ds &= \int_0^{N\delta} \chi^{1,\delta}(s,0) ds + \chi^{1,\delta}(N\delta,0)(t - N\delta) \\ &= \delta \sum_{n=0}^{N-1} \chi^{1,\delta}(n\delta,0) + \chi^{1,\delta}(N\delta,0)(t - N\delta),\end{aligned}$$

which equals

$$\begin{aligned}&\sum_{n=0}^{N-1} \sum_{i=0}^{n-1} B_i^\delta(-n\delta + i\delta) \Delta\mu_a(-n\delta + i\delta) \\ &\quad + \sum_{i=0}^{N-1} B_i^\delta(-N\delta + i\delta) \Delta\mu_a(-N\delta + i\delta) \frac{(t - N\delta)}{\delta}.\end{aligned}$$

With a change of variable and in the order of summation, this can be written as

$$\sum_{q=1}^{N-1} \Delta\mu_a(-q\delta) \sum_{n=q}^{N-1} B_{n-q}^\delta(-q\delta) + \frac{(t - N\delta)}{\delta} \sum_{q=1}^{N} B_{N-q}^\delta(-q\delta) \Delta\mu_a(-q\delta). \tag{8.7}$$

The first term of (8.7) equals

$$\sum_{q=1}^{N-1} \Delta\mu_a(-q\delta) \int_0^{N\delta-q\delta} b(\chi^{0,\delta}(s), u(s), -q\delta)ds. \tag{8.8}$$

Because $\Delta\mu_a(0) = 0$, the lower index in the sum in (8.8) can be set to zero. As $\mu_a(\cdot)$ is concentrated on integral multiples of δ, we can write (8.8) as

$$\int_{\max\{-N\delta+\delta,-\bar{\theta}\}}^{0} d\mu_a(\gamma) \int_0^{N\delta+\gamma} b(\chi^{0,\delta}(s), u(s), \gamma)ds,$$

which equals

$$\begin{aligned}\int_{\max\{-N\delta+\delta,-\bar{\theta}\}}^{0} & d\mu_a(\gamma) \int_{-\gamma}^{N\delta} b(\chi^{0,\delta}(s+\gamma), u(s+\gamma), \gamma)ds \\ &= \int_0^{N\delta} ds \int_{\max\{-s,-N\delta+\delta,-\bar{\theta}\}}^{0} b(\chi^{0,\delta}(s+\gamma), u(s+\gamma), \gamma)d\mu_a(\gamma).\end{aligned} \tag{8.9}$$

Because $\mu_a(0) - \mu_a(-\delta) = 0$, (8.9) involves the process $\chi^{0,\delta}(s)$ only for $s \le N\delta - \delta$.

Now consider the effects of the initial condition. At the value $\theta = 0$, we have

$$\delta \sum_{n=0}^{N-1} [\Phi^\delta]^n \chi^{1,\delta}(0,\theta)\Big|_{\theta=0} = \delta \sum_{n=0}^{N-1} \int_{-\bar{\theta}}^{\max\{-n\delta,-\bar{\theta}\}} b(x(\gamma+n\delta), u(\gamma+n\delta), \gamma)d\mu_a(\gamma).$$

Because $n \le N-1$, we can change $\max\{-n\delta, -\bar{\theta}\}$ to $\max\{-n\delta, -N\delta+\delta, -\bar{\theta}\}$. By the facts that $x(\cdot)$ and $u(\cdot)$ are piecewise-constant on $[-\bar{\theta}, 0]$ with intervals δ and that $\mu_a(\cdot)$ is concentrated on integral multiples of δ, this last expression can be written as

$$\int_0^{N\delta} ds \int_{-\bar{\theta}}^{\max\{-s,-N\delta+\delta,-\bar{\theta}\}} b(x(\gamma+s), u(\gamma+s), \gamma)d\mu_a(\gamma). \tag{8.10}$$

Using $\chi^{0,\delta}(s) = x(s)$ for $s \le 0$ and adding (8.10) to (8.9) yields the right side of (8.9) with $\max\{-s, -N\delta+\delta, -\bar{\theta}\}$ changed to $-\bar{\theta}$, which is

$$\int_0^{N\delta} ds \int_{-\bar{\theta}}^{0} b(\chi^{0,\delta}(s+\gamma), u(s+\gamma), \gamma)d\mu_a(\gamma), \tag{8.11}$$

the double integral in (8.5a) when $n = N$.

Now return to the term on the right of (8.7). The sum in this term can be shown to equal

$$\int_{N\delta}^{N\delta+\delta} ds \int_{-\bar{\theta}}^{0} b(\chi^{0,\delta}(s+\gamma), u(s+\gamma), \gamma)d\mu_a(\gamma). \tag{8.12}$$

Adding this term to (8.11) yields (8.11) with $N\delta$ changed to $N\delta+\delta$. Thus (8.7), with the effects of the initial condition added, is just a linear interpolation of the values at $t = N\delta$ and $t = N\delta + \delta$, and the proof is completed. ■

9.8.2 The Effective Delay and Numerical Procedures

The effective delay. For $t \in [n\delta, n\delta + \delta)$, the term in (8.5) that contains the delays is the double integral in (8.5a) plus (8.5b), and its time derivative[4] is

$$\frac{1}{\delta}\int_{n\delta}^{n\delta+\delta} ds \int_{-\bar{\theta}}^{0} b(\chi^{0,\delta}(s+\theta), u(s+\theta), \theta) d\mu_a(\theta). \tag{8.13}$$

Suppose, for example, that $\mu_a(\cdot)$ is concentrated (with unit mass) at $-\bar{\theta}$. Then (8.13) equals

$$\frac{1}{\delta}\int_{n\delta}^{n\delta+\delta} b(\chi^{0,\delta}(s-\bar{\theta}), u(s-\bar{\theta}), -\bar{\theta}) ds. \tag{8.14}$$

The values of $\chi^{0,\delta}(\cdot)$ that appear in (8.14) are those on the interval $[n\delta-\bar{\theta}, n\delta+\delta-\bar{\theta}]$, no matter what $t \in (n\delta, n\delta+\delta)$ is. Thus, when $t = (n\delta)^+$, the effective delay is uniformly distributed on the interval $[\bar{\theta}-\delta, \bar{\theta}]$. When $t = (n\delta+\delta)^-$, the effective delay is uniformly distributed on the interval $[\bar{\theta}, \bar{\theta}+\delta]$. Hence the delay is periodic, with values varying in the range $[\bar{\theta}-\delta, \bar{\theta}+\delta]$.

Periodic-Erlang approximations. The procedure that was described in this section required that the time since the last shift be monitored. This can be discretized by using an approximation such as the periodic-Erlang approximation of Chapter 4. Introduce the Erlang state process $L_n^{\delta_0,\delta}$, with δ/δ_0 being an integer. Then do the shift at the renewal times for the Erlang process. The resulting process will converge to that defined by (8.5) as $\delta_0 \to 0$.

The numerical approximation. The Markov chain approximation method of Section 4 is readily adapted to the problem at hand. Introduce the Erlang state process $L_n^{h,\delta_0,\delta}$ as in Section 8.2, and let $\xi_n^{0,h,\delta_0,\delta}, \xi_n^{1,h,\delta_0,\delta}(\theta)$, $\theta = -\bar{\theta}+\delta, , \ldots, -\delta, 0$, denote the approximating processes. One shifts $\xi_n^{1,h,\delta_0,\delta}(\cdot)$ at the renewal times of the process $L_n^{h,\delta_0,\delta}$ as in Section 8.2. Between shifts, the updates of $\xi_n^{1,h,\delta_0,\delta}(\cdot)$ are as in Section 4. The computed value and optimal value functions converge to those for the model (8.5) as $h \to 0$ and $\delta_0 \to 0$.

9.9 Singular and Impulsive Controls

No delay in the singular control. Recall the model (8.6.2). Suppose that the term $q_0(x(t-))d\lambda(t)$ is added to (1.1), there is no delayed singular control, and that $\int_0^\infty e^{-\beta t} q'_\lambda d\lambda(t)$ is added to the cost function. Then the only change that is required in Section 2 is the addition of $q_0(\chi^0(t-))d\lambda(t)$ to (2.1). The changes in the numerical approximations that are required in Section 4

[4] It is the derivative that yields the dynamical term.

are guided by the ideas for the singular control problem in Section 6.6. The impulsive control is dealt with similarly, and the details are left to the reader.

Delay in the singular control. Now suppose that the terms

$$q_0(x(t-))d\lambda(t) + dt \int_{\theta=-\bar{\theta}}^{0} q_2(x((t+\theta)-),\theta)d_\theta\lambda(t+\theta)$$

from the model (8.6.2) are included in (1.1). Then add $q_0(\chi^0(t-))d\lambda(t)$ to (2.1) and $q_2(\chi^0(t-),\theta)d\lambda(t)$ to (2.2). If $\lambda(s)=0, s\leq 0$, then the initial condition is (2.5). Otherwise one needs to modify the initial condition analogously to what was done for the delayed reflection term in (2.5). Theorem 2.1 remains valid with these additions. Obviously the numerical procedure will be more complicated, as one has to adapt the methods of Section 6.6 to those of Section 4. This can be done. But we will not proceed further here as the approach of this chapter, though intriguing and promising, is still in its infancy.

References

1. E. Altman and H. J. Kushner. Heavy traffic analysis of AIMD models for congestion control. In S. Raghavan and G. Anandalingham, editors, *Telecommunications Planning: Innovations in Pricing, Network Design and Management.* Kluwer, Amsterdam, 2005. Selected papers from the 2004 INFORMS telecommunications conference.
2. C. T. H. Baker and E. Buckwar. Numerical analysis of explicit one-step methods for stochastic delay differential equations. *London Mathematical Society Journal of Computation and Mathematics*, 3:315–335, 2000.
3. H. T. Banks and J. A. Burns. Hereditary control problems: numerical methods based on averaging approximations. *SIAM J. Control Optimiz.*, 16:296–332, 1975.
4. J. J. Batzel, F. Kappel, D. Schneditz, and H. T. Tran. *Cardiovascular and Respiratory Systems.* SIAM, Philadelphia, 2006.
5. H. Bauer and U. Reidel. Stochastic control problems with delay. *Math. Methods in Operations Research*, 62:411–427, 2005.
6. A. Bellen and M. Zennaro. *Numerical Methods for Delay Differential Equations.* Oxford Science Publications, Oxford, 2003.
7. P. Billingsley. *Convergence of Probability Measures.* Wiley, New York, 1968.
8. P. Billingsley. *Convergence of Probability Measures; Second Edition.* Wiley, New York, 1999.
9. V. S. Borkar. *Optimal Control of Diffusion Processes.* Longman Scientific and Technical, Harlow, Essex, UK, 1989.
10. L. Breiman. *Probability Theory*, volume 7 of *Classics in Applied Mathematics*, A reprint of the 1968 Addison-Wesley edition. SIAM, Philadelphia, 1992.
11. M.-S. Chang and R. K. Youree. The European option with hereditary price structure: Basic theory. *Applied Math. and Computation*, 102:279–296, 1999.
12. A. Chojnowska-Michalik and B. Goldys. Existence, uniqueness and invariant measures for stochastic semilinear equations on hilbert spaces. *Probability Theory and Related Fields*, 102:331–356, 1995.
13. P.-L. Chow, J. L. Menaldi, and M. Robin. Additive control of stochastic linear systems with finite horizons. *SIAM J. Control Optimiz.*, 23:858–899, 1985.
14. K.-L. Chung. *Markov Chains with Stationary Transition Probabilities.* Springer-Verlag, Berlin and New York, 1960.

15. R. F. Curtain and H. J. Zwart. *An Introduction to Infinite-Dimensional Linear Systems Theory.* Springer-Verlag, Berlin and New York, 1995.
16. J. L. Doob. *Stochastic Processes.* Wiley, New York, 1953.
17. L. Dugard and E. I. Verriest. *Stability and Control of Time-Delay Systems.* Springer-Verlag, Berlin and New York, 1998.
18. T. E. Duncan, B. Pasik-Duncan, and L. Stettner. On the ergodic and adaptive control of stochastic differential delay systems. *J. of Optimization Theory and Applications*, 81:509–531, 1994.
19. N. Dunford and J. T. Schwartz. *Linear Operators, Part 1: General Theory.* Wiley-Interscience, New York, 1966.
20. P. Dupuis and H. Ishii. On Lipschitz continuity of the solution mapping to the Skorokhod problem, with applications. *Stochastics Stochastics Rep.*, 35:31–62, 1991.
21. P. Dupuis and H. Ishii. SDE's with oblique reflection on nonsmooth domains. *Ann. Probability*, 21:554–580, 1993.
22. A. El-Safty, M. S. Salim, and M. El-Khatib. Existence, uniqueness and stability of the spline approximation for delay controlled dynamical systems. *International J. of Computer Mathematics*, 77:629–640, 2001.
23. S. N. Ethier and T. G. Kurtz. *Markov Processes: Characterization and Convergence.* Wiley, New York, 1986.
24. W. Feller. *An Introduction to Probability Theory and its Applications, Volume 2.* Wiley, New York, 1966.
25. S. Floyd and V. Jacobson. Random early detection gateways for congestion avoidance. *IEEE-ACM Transactions on Networking*, 1:397–413, 1993.
26. J. Gibson. Linear-quadratic optimal control of hereditary differential systems: Infinite dimensional Ricatti equations and numerical approximations. *SIAM J. Control Optimiz.*, 21:95–139, 1983.
27. J. K. Hale and S. M. Verduyn Lunel. *Introduction to Functional Differential Equations.* Springer-Verlag, Berlin and New York, 1993.
28. J. K. Hale and S. M. Verduyn Lunel. Stability and control of feedback systems with time delays. *Int. J. of Systems Science*, 24:497–504, 2003.
29. J. M. Harrison and R.J. Williams. Brownian models of open queueing networks with homogeneous customer populations. *Stochastics*, 22:77–115, 1987.
30. K. Helmes and R. H. Stockbridge. Linear programming approach to the optimal stopping of singular stochastic processes. *Stochastics Stochastics Rep.*, 35:309–335, 2007.
31. H. J.Kushner. Numerical approximations for nonlinear stochastic systems with delays. *Stochastics Stochastics Rep.*, 77:211–240, 2005.
32. Y. Hu, S-E. A. Mahammed, and F. Yan. Discrete-time approximations of stochastic delay equations: The Milstein scheme. *Ann. Appl. Prob.*, 32:265–314, 2004.
33. Elsanosi I, B. Øksendal, and A. Sulem. Some solvable stochastic control problems with delays. *Stochastics Stochastics Rep.*, 71:69–90, 2000.
34. N. Ikeda and S. Watanabe. *Stochastic Differential Equations and Diffusion Processes,* First Ed. North-Holland, Amsterdam, 1981.
35. K. Ito and F. Kappel. *Evolution Equations and Approximations.* World Scientific, River Edge, NJ, 2002.
36. K. Ito and F. Kappel. Approximation of infinite delay and Volterra type equations. *Numerical Mathematics*, 54:415–444, 1989.

37. K. Ito and R. Teglas. Legendre-tau approximations for functional differential equations. *SIAM J. on Control Optimiz.*, 24:737–759, 1986.
38. K. Ito and R. Teglas. Legendre-T au approximations for functional differential equations Part II: The linear quadratic optimal control problem. *SIAM J. on Control Optimiz.*, 25:1379–1408, 1987.
39. A. F. Ivanov, Y. I. Kazmerchuk, and A. V. Swishchuk. Theory, stochastic stability and applications of stochastic delay differential equations: a survey of results. *Differential Equations and Dynamical Systems*, 11:55–115, 2003.
40. F. Kappel. Galerkin type approximations schemes for delay systems. *Ann. Differential Equations*, 1:57–82, 1985.
41. I. Karatzas. A class of singular stochastic control problems. *Adv. in Appl. Probab.*, 15:225–254, 1983.
42. I. Karatzas and S. E. Shreve. *Brownian Motion and Stochastic Calculus.* Springer-Verlag, New York, 1988.
43. S. Karlin and H. M. Taylor. *A First Course in Stochastic Processes, Second Edition.* Academic Press, New York, 1975.
44. V. B. Kolmanovskii and A. Myshkis. *Applied Theory of Functional Differential Equations.* Kluwer Academic Publishers, Dordrecht, 1992.
45. V. B. Kolmanovskii and V. R. Nosov. *Stability of Functional Differential Equations.* Academic Press, New York, 1986.
46. V. B. Kolmanovskii and L. E. Skaikhet. *Control Systems With Aftereffect.* American Mathematical Society, Providence, RI, 1996. Translations of Mathematical Monographs, Vol. 157.
47. U. Küchler and E. Platen. Strong discrete time approximation of stochastic differential equations with time delay. *Mathematics and Computers in Simulation*, 54:189–205, 2000.
48. U. Küchler and E. Platen. Weak discrete time approximation of stochastic differential equations with time delay. *Mathematics and Computers in Simulation*, 59:497–507, 2002.
49. T. G. Kurtz. *Approximation of Population Processes*, volume 36 of *CBMS-NSF Regional Conf. Series in Appl. Math.* SIAM, Philadelphia, 1981.
50. H. J. Kushner. Numerical approximations for stochastic systems with delays in the state and control. *Stochastics Stochastics Rep.*, 78:343–376, 2006.
51. H. J. Kushner. On the stability of stochastic differential-difference equations. *J. Diff. Eqns.*, 4:424–443, 1968.
52. H. J. Kushner. Stability and existence of diffusions with discontinuous or rapidly growing terms. *J. Math. Anal. Appl.*, 11:156–168, 1972.
53. H. J. Kushner. *Probability Methods for Approximations in Stochastic Control and for Elliptic Equations.* Academic Press, New York, 1977.
54. H. J. Kushner. Optimality conditions for the average cost per unit time problem with a diffusion model. *SIAM J. Control Optimiz.*, 16:330–346, 1978.
55. H. J. Kushner. *Weak Convergence Methods and Singularly Perturbed Stochastic Control and Filtering Problems*, volume 3 of *Systems and Control.* Birkhäuser, Boston, 1990.
56. H. J. Kushner. *Heavy Traffic Analysis of Controlled Queueing and Communication Networks.* Springer-Verlag, Berlin and New York, 2001.
57. H. J. Kushner and D. Barnea. On the control of linear functional-differential equations with quadratic cost. *SIAM J. Control Optimiz.*, 8:257–272, 1970.

58. H. J. Kushner and P. Dupuis. *Numerical Methods for Stochastic Control Problems in Continuous Time.* Springer-Verlag, Berlin and New York, 1992. Second Edition, 2001.
59. H. J. Kushner, D. Jarvis, and J. Yang. Controlled and optimally controlled multiplexing systems: A numerical exploration. *Queueing Systems*, 20:255–291, 1995.
60. H. J. Kushner and L. F. Martins. Numerical methods for stochastic singular control problems. *SIAM J. Control Optimiz.*, 29:1443–1475, 1991.
61. H. J. Kushner and G. Yin. *Stochastic Approximation Algorithms and Applications.* Springer-Verlag, Berlin and New York, 1997. Second Edition, 2003.
62. B. Larssen and N. H. Risebro. When are HJB equations in stochastic control of delay systems finite dimensional. *Stochastic Anal. and Appl.*, 21:643–671, 2003.
63. I. Lasiecka and A. Manitius. Differentiability and convergence rates of approximating semigroups for retarded functional-differential equations. *SIAM. J. of Numerical Analysis*, 25:883–907, 1988.
64. J. P. Lehoczky and S. E. Shreve. Absolutely continuous and singular stochastic control. *Stochastics*, 17:91–110, 1986.
65. R. Liptser and A. N. Shiryaev. *Statistics of Random Processes.* Springer-Verlag, Berlin and New York, 1977.
66. A. Manitius and T. Tran. Numerical approximations for hereditary systems with input and output delays: Convergence results and convergence rates. *SIAM J. Control Optimiz.*, 32:1332–1363, 1994.
67. X. Mao. *Stability of Stochastic Differential Equations With Respect to Semimartingales.* Longman Scientific and Technical, Harlow, Essex, UK, 1991.
68. X. Mao. *Stochastic Differential Equations and Applications.* Harwood Publishing, Chichester, UK, 1997.
69. X. Mao and S. Sabinis. Numerical solution of stochastic differential delay equations under local lipschitz conditions. *J. of Computational and Applied Mathematics*, 151:215–227, 2003.
70. J.-L. Menaldi and M. Taksar. Optimal correction problem for a multidimensional stochastic system. *Automatica*, 25:223–232, 1989.
71. S. P. Meyn and R. I. Tweedie. *Markov Chains and Stochastic Stability.* Springer-Verlag, Berlin and New York, 1994.
72. L. Mirkin. On the approximation of distributed-delay control laws. *Systems and Control Letters*, 51:331–342, 2004.
73. S-E. A. Mohammed. *Stochastic Functional Differential Equations.* Pitman, Boston, 1984.
74. S-E. A. Mohammed. Stochastic differential systems with memory: Theory, examples and applications. In L. Decreusefond, J. Gjende, B. Øksendal, and A. Üstinel, editors, *Stochastic Analysis and Related Topics VI; The Geilo Workshop.* Birkhäuser, Boston, 1998.
75. B. S. Mordukhovich and R. Trubnik. Stability of discrete approximations and necessary conditions for delay-differential inclusions. *Annals of Oper. Research*, 101:149–170, 2001.
76. J. Neveu. *Mathematical Foundations of the Calculus of Probability.* Holden-Day, San Francisco, 1965.
77. S.-I. Niculescu. *Delay Effects on Stability: A Robust Control Approach.* Springer, New York and Berlin, 2001. Lecture Notes in Control and Information Sciences, Vol. 269.

78. S.-I. Niculescu and K. Gu. *Advances in Time-Delay Systems: Volume 38, Lecture Notes in Computational Science and Engineering.* Springer-Verlag, Berlin and New York, 2004.
79. J. Nilsson, B. Bernhardsson, and B. Wittenmark. Stochastic analysis and control of real-time systems with random time delays. *Automatica*, 34:57–64, 1998.
80. B. Øksendal. *Stochastic Differential Equations.* Springer-Verlag, Berlin and New York, 1995.
81. C. V. Pao. Finite difference solutions of reaction diffusion equations with continuous time delays. *Computers and Mathematics With Applications*, 42:399–412, 2001.
82. W. B. Powell. *Approximate Dynamic Programming.* Wiley, New York, 2007.
83. G. Da Prato and J. Zabczyk. *Ergodicity for Infinite Dimensional Systems.* Cambridge University Press, Cambridge, UK, 1996.
84. M. I. Reiman and R. J. Williams. A boundary property of semimartingale reflecting Brownian motions. *Prob. Theory Rel. Fields*, 77:87–97, 1988.
85. D. Revuz. *Markov Chains.* North Holland, Amsterdam, 1984.
86. M. Scheutzow. Qualitative behavior of stochastic delay equations with a bounded memory. *Stochastics*, 12:41–80, 1984.
87. J. Si, A. G. Barto, W. B. Powell, and D. Wunch. *Handbook of Learning and Approximate Dynamic Programming.* IEEE Press, New York, 2004.
88. H. M. Soner and S. E. Shreve. Regularity of the value function for a two-dimensional singular stochastic control problem. *SIAM J. Control Optimiz.*, 27:876–907, 1989.
89. R. Srikant. Control of communication networks. In T. Samed, editor, *Perspectives in Control Engineering: Technologies, Applications, New Directions.* IEEE Press, New York, 1999.
90. R. Srikant. *The Mathematics of Internet Congestion Control.* Birkhäuser, Boston, 2003.
91. J. C. Strikwerda. *Finite Difference Schemes and Partial Differential Equations.* Wadsworth and Brooks/Cole, Pacific Grove, CA, 1989.
92. S. Tarbouriech, C. T. Abdallah, and M. Ariola. Bounded control of multiple-delay systems with applications to ATM networks. In S.-I. Niculescu and K. Gu, editors, *Advances in Time-Delay Systems: Volume 38, Lecture Notes in Computational Science and Engineering.* Springer-Verlag, Berlin and New York, 2004.
93. E. I. Verriest. Asymptotic properties of stochastic delay systems. In S.-I. Niculescu and K. Gu, editors, *Advances in Time-Delay Systems: Volume 38, Lecture Notes in Computational Science and Engineering.* Springer-Verlag, Berlin and New York, 2004.
94. R. B. Vinter and R. H. Kwong. The infinite time quadratic control problem for linear systems with state and control delays: An evolution equation approach. *SIAM J. Control Optimiz.*, 19:139–153, 1981.
95. Z. Xiong. Superconvergence of the continuous Galerkin finite element method for delay differential equations with several terms. *J. of Computational and Applied Mathematics*, 198:160–166, 2007.

Index

Symbol Index

Printed in the United States of America